NATURAL-BASED POLYMERS FOR BIOMEDICAL APPLICATIONS

NATURAL-BASED POLYMERS FOR BIOMEDICAL APPLICATIONS

Tatiana G. Volova, DSc
Yuri S. Vinnik, DSc
Ekaterina I. Shishatskaya, DSc
Nadejda M. Markelova, DSc
Gennady E. Zaikov, DSc

APPLE ACADEMIC PRESS

Apple Academic Press Inc.
3333 Mistwell Crescent
Oakville, ON L6L 0A2 Canada

Apple Academic Press Inc.
9 Spinnaker Way
Waretown, NJ 08758 USA

© 2017 by Apple Academic Press, Inc.
First issued in paperback 2021
No claim to original U.S. Government works
ISBN-13: 978-1-77463-632-9 (pbk)
ISBN-13: 978-1-77188-435-8 (hbk)

Library and Archives Canada Cataloguing in Publication

Volova, T. G. (Tatiana Grigorevna), author
Natural-based polymers for biomedical applications / Tatiana G. Volova, DSc, Yuri S. Vinnik, DSc, Ekaterina I. Shishatskaya, DSc, Nadejda M. Markelova, DSc, Gennady E. Zaikov, DSc.

Includes bibliographical references and index.
Issued in print and electronic formats.

ISBN 978-1-77188-435-8 (hardcover).--ISBN 978-1-315-36603-6 (PDF)
1. Polymers in medicine. 2. Biomedical materials--Research. I. Vinnik, Yuri S., author II. Shishatskaya, Ekaterina I., author III. Markelova, Nadejda M., author IV. Zaikov, G. E. (Gennadiĭ Efremovich), 1935-, author V. Title.

| R857.P6V65 2017 | 610.28 | C2017-901101-4 | C2017-901102-2 |

Library and Archives Canada Cataloguing in Publication

Names: Volova, T. G. (Tatiana Grigorevna), author.
Title: Natural-based polymers for biomedical applications/editors, Tatiana G. Volova, DSc [and four others].
Description: Toronto; New Jersey: Apple Academic Press, 2017. | Includes bibliographical references and index.
Identifiers: LCCN 2017005303 (print) | LCCN 2017007073 (ebook) | ISBN 9781771884358 (hardcover : alk. paper) | ISBN 9781315366036 (CRC/Taylor & Francis eBook)
Subjects: LCSH: Polymers in medicine. | Biomedical materials--Research.
Classification: LCC R857.P6 N39 2017 (print) | LCC R857.P6 (ebook) | DDC 610.28/4--dc23
LC record available at https://lccn.loc.gov/2017005303

Apple Academic Press also publishes its books in a variety of electronic formats. Some content that appears in print may not be available in electronic format. For information about Apple Academic Press products, visit our website at **www.appleacademicpress.com** and the CRC Press website at **www.crcpress.com**

CONTENTS

LIST OF ABBREVIATIONS

AFM	atomic-force microscopy
AT-MSCs	adipose tissue mesenchymal stem cells
BM-MSCs	bone marrow mesenchymal stem cells
CFU	colony-forming unit
COGEGA	EU General Confederation of Agricultural Cooperatives
COPA	EU Committee of Professional Agricultural Organizations
C_x	degree of crystallinity
Đ	polydispersity
DDSs	drug delivery systems
DOX	doxorubicin
DSC	differential scanning calorimetry
DTA	differential thermal analysis
EAC	Ehrlich ascites carcinoma
EE	encapsulation efficiency
EPR	electron paramagnetic resonance
ESC	Ehrlich solid carcinoma
ESR	erythrocyte sedimentation rate
FAs	fatty acids
FBGCs	foreign body giant cells
FDA	U.S. Food and Drug Administration
GIA	Global Industry Analysts
HDPE	high-density polyethylene
HPLC	high-performance liquid chromatography
IR	infrared spectroscopy
LPS	lipopolysaccharides
M_n	number average molecular weight
M_w	weight average molecular weight
NMR	nuclear magnetic resonance
P3HB	poly-3-hydroxybutyrate
P3HB/3HHx	poly-3-hydroxybutyrate/3-hydroxyhexanoate
P3HB/3HO	poly-3-hydroxybutyrate/3-hydroxyoctanoate

P3HB/3HV	poly-3-hydroxybutyrate/3-hydroxyvalerate
P3HB/4HB	poly-3-hydroxybutyrate/4-hydroxybutyrate
PCM	polymer composite materials
PDO	polydioxanone
PE	polyethylene
PECH	polyepichlorohydrin
PEG	polyethylene glycol
PEO	polyethylene oxide
PHA$_{LCL}$	long-chain-length PHA
PHA$_{MCL}$	medium-chain-length PHA
PHAs	polyhydroxyalkanoates
PHA$_{SCL}$	short-chain-length PHA
PLA	polylactide, polylactic acid
PMMA	polymethyl methacrylate
PNL	polymorphonuclear lymphocytes
PNX	poly-n-xylene
PP	polypropylene
PTFE	poly(tetrafluoroethylene)
PVA	polyvinyl alcohol
PVAc	polyvinyl acetate
SEM	scanning electron microscopy
SPU	segmented polyurethanes
TGI	tumor growth inhibition
TMC	tetramethyl cellulose
UHMWPE	ultra high molecular weight polyethylene

PREFACE

This book is devoted to the biomedical study of natural degradable bio-polymers – polyhydroxyalkanoates (PHAs). It gives an idea of the demand for medical grade materials and modern trends, showing the present position and potentials of polyhydroxyalkanoates. The authors present new results of their biomedical studies that have been carried out at the Institute of Biophysics of the Siberian Branch of the Russian Academy of Sciences, Siberian Federal University, and V.F. Voino-Yasenetsky Krasnoyarsk Federal Medical University (Russia). Special sections are devoted to the use of PHAs for construction and employment of new-generation sustained-release drug delivery systems and bioconstructions for the purposes of cellular and tissue engineering. The authors describe some original results of clinical trials, proving the effectiveness of using PHAs in repair of skin and bone defects and in general and abdominal surgery.

The book is intended for biotechnologists, material scientists, medical practitioners, and teachers and students of biological and medical departments of universities.

ABOUT THE AUTHORS

Tatiana G. Volova, DSc
Professor and Head, Department of Biotechnology,
Siberian Federal University, Krasnoyarsk, Russia

Tatiana G. Volova, DSc, is a doctor of biological sciences in microbiology. She is a Professor and Head of the Department of Biotechnology at the Siberian Federal University, Krasnoyarsk, Russia. She is the creator and head of the Laboratory of Chemoautotrophic Biosynthesis in the Institute of Biophysics, Siberian Branch of Russian Academy of Sciences. Professor Volova is conducting research in the field of physico-chemical biology and biotechnology and is a well-known expert in the field of microbial physiology and biotechnology. Tatiana Volova has created and developed a new and original branch in chemoautotrophic biosynthesis, in which the two main directions of the XXI century technologies are conjugate, hydrogen energy and biotechnology. The obtained fundamental results provided significant outputs and were developed by the unique biotechnical producing systems, based on hydrogen biosynthesis for single-cell protein, amino acids, and enzymes. Under the guidance of Professor Volova, the pilot production facility of single cell protein, utilizing hydrogen, had been created and put into operation. The possibility of involvement of man-made sources of hydrogen into biotechnological processes as a substrate, including synthesis gas from brown coals and vegetable wastes, was demonstrated in the research of Professor Volova. She had initiated and deployed in Russia the comprehensive research on microbial degradable bioplastics; the results of this research cover various aspects of biosynthesis, metabolism, physiological role, structure, and properties of these biopolymers and polyhydroxyalkanoates (PHAs), and have made a scientific basis for their biomedical applications and allowed them to be used for biomedical research. Professor Tatiana Volova is the author of more than 300 scientific works, including 12 monographs, 16 inventions, and a series of textbooks for universities.

Yuri S. Vinnik, DSc
Professor, Head, Department of General Surgery,
Krasnoyarsk State Medical School, Krasnoyarsk, Russia

Yuri S. Vinnik, DSc, is a doctor of medicine and Professor and Head of the Department of General Surgery at Krasnoyarsk State Medical School (named after Professor V. F. Voyno-Yasenetsky), from 2006 to the present time. He is an Honored Scientist of Russia, Honored Doctor of the Russian Federation, member of the International Academy of Ecology and Life Safety, and corresponding member of the New York Academy of Sciences. His research interests include hepato-biliary-pancreatic surgery, including treatment of acute pancreatitis, biliary reconstructive surgery, trauma of pancreas; cholelithiasis and its complications; jaundice; surgery of stomach and duodenum; herniology with the use of mesh implants; surgery of the colon (endovascular surgery); purulent surgery; diabetic foot syndrome; frostbite (the study of the pathogenesis of cold injury and treatment of its complications); questions of immunodeficiency; cytokine therapy; metabolic immunotherapy of surgical diseases; peritonitis; systemic inflammatory response syndrome; the use in surgery of the new biopolymers (polyhydroxyalkanoates); and medical devices for these various issues. Professor Vinnick is the founder of the scientific school and abdominal purulent surgery in Russia. He is the author of over 800 publications, 32 patents copyright in Russia, and 36 monographs.

Ekaterina I. Shishatskaya, DSc
Head, Department of Medical Biology, Siberian Federal University,
Krasnoyarsk, Russia

Ekaterina I. Shishatskaya, DSc, is a doctor of medicine and doctor of sciences in biotechnology. She is Head of the Department of Medical Biology at Siberian Federal University, Krasnoyarsk, Russia, and a leading researcher at the Institute of Biophysics, Siberian Branch of Russian Academy of Sciences, in Krasnoyarsk, where she supervises the direction of biomedical research of new materials. Dr. Shishatskaya's work includes an investigation of interaction mechanisms between biomaterials and biological objects and development of high-tech biomedical devices. This actual science direction is oriented to the development of new reconstructive biomedical technologies, which includes cell biology and tissue engineering and biodegradable polymers of carbon acids—polyhydroxyalkanoates (PHAs), a new class of materials for medicine. Dr. Shishatskaya is Russian leader in comprehensive medical and biological studies of this class of polymers. Her professional activity is focused on the implementation of the obtained results in practice, and she maintains communication with clinical centers and top specialists in regenerative medicine. Dr. Shishatskaya is author of about 150 research works, including five monographs and eight patents. She is the winner of the President of Russian Federation for youth in the field of science and innovations and holds a State Prize of the Krasnoyarsk Region in the area of high education and science. She is also a laureate of L'Oréal-UNESCO for young women in science.

Nadejda M. Markelova, DSc
Associate Professor, General Surgery Department,
Krasnoyarsk State Medical School, Krasnoyarsk, Russia

Nadejda M. Markelova, MD, is Associate Professor in the General Surgery Department at the Krasnoyarsk State Medical School (named after Professor V. F. Voyno-Yasenetsky), Krasnoyarsk, Russia. Her major research interests are reconstructive surgery, hepatobiliary, purulent surgery, experimental surgery, and development, implementation, and clinical testing of new biomaterials and medical devices based on them. Dr. Markelova has published 202 publications, including four monographs and four patents in Russia for inventions in the area of surgical treatment. Dr. Markelova graduated from Krasnoyarsk State Medical School in 2001 and passed her clinical internship at the Department of General Surgery. She defended her thesis on the topic "Prevention and treatment of infected pancreatic necrosis" in 2005 and her doctoral dissertation in surgery on "Justification of the application of high-tech medical products of biodegradable polymers in reconstructive surgery (experimentally clinical research)" in 2013.

Gennady E. Zaikov, DSc

Head of the Polymer Division, N. M. Emanuel Institute of Biochemical Physics, Russian Academy of Sciences, Moscow, Russia; Professor, Moscow State Academy of Fine Chemical Technology, Russia; Professor, Kazan National Research Technological University, Kazan, Russia

Gennady E. Zaikov, DSc, is Head of the Polymer Division at the N. M. Emanuel Institute of Biochemical Physics, Russian Academy of Sciences, Moscow, Russia, and Professor at Moscow State Academy of Fine Chemical Technology, Russia, as well as Professor at Kazan National Research Technological University, Kazan, Russia. He is also a prolific author, researcher, and lecturer. He has received several awards for his work, including the Russian Federation Scholarship for Outstanding Scientists. He has been a member of many professional organizations and is on the editorial boards of many international science journals. Dr. Zaikov has recently been honored with tributes in several journals and books on the occasion of his 80th birthday for his long and distinguished career and for his mentorship to many scientists over the years.

INTRODUCTION

Creation of environmentally friendly materials with useful properties remains one of the main challenges of the modern world. Development of science and technology has led to a wider use of different products synthesized by living systems. In recent years, researchers have paid increasing attention to biopolymers (polymers of biological origin). The main goal of this research field is to find and study novel biopolymers and develop a basis for constructing biological systems synthesizing polymers with target properties. One of the main goals of biotechnology today is to develop biocompatible polymer materials for reconstructive biomedical technologies. It is impossible to enhance the effectiveness of treatment in reconstructive medicine and improve the quality of life without using revolutionary technologies and novel, highly functional and specific, materials and systems capable of reproducing the functions of biological tissues. In spite of substantial success achieved over recent years, materials fully compatible with the living organism have not been created yet. The main obstacles to the wide use of much-needed biodegradable polymers are a rather narrow variety of these materials and the yet unsolved problem of controlling their degradation and functioning in vivo.

Among the biomaterials that have already been developed or are being developed now are aliphatic polyesters, polyamides, segmented polyester urethanes, polymers of lactic and glycolic acids, silicon, polyethylene terephthalate, and polyesters synthesized by microorganisms, the so-called polyhydroxyalkanoates (PHAs). The discovery of polyhydroxyalkanoates was a notable event for biotechnology of novel materials. In their physicochemical properties, PHAs are comparable with synthetic polymers such as polypropylene, which are widely used and abundantly manufactured but are not degraded in the environment. In addition to being thermoplastic, PHAs are biodegradable and biocompatible; they also have optical activity, antioxidant properties, and piezoelectricity. Potential areas of application of these polymers are very wide: reconstructive surgery, cellular and tissue engineering, transplant surgery, pharmacology, light and food industries, municipal engineering, agriculture, etc. Since the late 1980s to early 1990s, these polymers have been studied by research teams in all developed countries. Potential medical uses of PHAs may include manufacture of medical instruments and aids (nonwovens, disposables, sutures, wound dressings), pharmaceutics (controlled drug delivery systems), reconstructive surgery, and transplant surgery. PHAs are particularly promising for reconstructing tissue defects in reconstructive surgery, as scaffolds for cellular and tissue engineering, and for constructing new-generation sustained-release drug delivery systems.

Polyhydroxyalkanoates have physicochemical properties similar to those of some synthetic polymers (polypropylene, polyethylene), which are used widely but cannot be degraded in the environment. In addition to being thermoplastic, polyhydroxyalkanoates also exhibit optical activity, antioxidant properties, piezoelectricity, and, most important, biodegradability and biocompatibility. Polyhydroxyalkanoates differ very much in their structure and properties (flexibility, crystallinity, melting temperature, etc.), depending on the taxonomic position and physiological and biochemical properties of microorganisms producing them, conditions of biosynthesis, and the type of carbon source used. Moreover, by producing PHA composites with various natural and synthetic materials, which can have different structure and composition, one can vary the basic properties of the material: plasticity, mechanical strength, temperature characteristics, and other properties. This makes PHAs even more attractive for use in different areas. At the present time, the application of PHAs is limited by their rather high cost (almost an order of magnitude higher than that of polyolefins). However, the growing environmental concerns on the one hand, and the possibility of reducing the cost of biopolymers by increasing the production efficiency, on the other, make polyhydroxyalkanoates promising materials of the 21st century.

The first of the discovered polyhydroxyalkanoates was polymer of hydroxybutyric acid (polyhydroxybutyrate, PHB). Now, there are more than 150 known polyesters of this type. In 1976, ICI in Great Britain started the first commercial investigation of microbiological production of this polymer using sugar-containing substrates extracted from plant biomass. PHB is, however, brittle and has a low extension to break. The lack of flexibility and thermal stability hampers its processing and, thus, limits its range of applications, and if P(3HB) were the only existing polyhydroxyalkanoate, it would not be a promising material. However, a polyester with properties that differed from those of polyhydroxybutyrate was isolated from activated sludge. Detailed chromatographic analysis revealed the presence of 3-hydroxybutyric acid as a major component and 3-hydroxyvaleric and 3-hydroxyhexanoic acids as minor components of the new compound. That was the first PHA heteropolymer. The discovery of the ability of microorganisms to synthesize PHA heteropolymers gave a strong impetus to extensive investigations of these polyesters. Variation in monomer fractions of a PHA leads to considerable changes in its thermo-mechanical and fibrous properties.

ICI of England was the first industrial corporation to start commercial production of PHAs. In 1992, Zeneka Seeds and Zeneka Bio Products began producing polyhydroxybutyrate and hydroxybutyrate/hydroxyvalerate copolymers with the trade name of Biopol®. The polymers were synthesized by using the mutant strain of hydrogen bacterium *Alcaligenes eutrophus*, capable of utilizing glucose, in the process that was regarded as the best at that time. The cost of Biopol®, which was produced at 10–15 thousand tons per year, reached U.S.$ 16 000/t. That was an order of magnitude higher than the cost of polyolefins. This high cost prevents Biopol® from being used as common packaging material, limiting its application to special purposes such as biomedical ones, as suggested by specialists. In 1996, Zeneka transferred its rights to Monsanto and Astra (U.S.). These polymers have been extensively studied in all developed countries. Polyhydroxyalkanoates are manufactured by Monsanto Co., Metabolix Inc., Tepha, Procter & Gamble, Berlin Packaging Corp., Bioscience Ltd., Bio-Ventures Alberta Inc., and Merck, which produce polymers with the trade names Biopol®, Biopol™, TephaFLEX™, DegraPol/btc®, and Nodax™. Almost all developed countries are now engaged in manufacturing PHAs or plan to commercialize them, but the major factor that would allow wide-scale manufacture and use of PHAs is reduction of their cost.

In Russia, the Institute of Biophysics of the Siberian Branch of the Russian Academy of Sciences was the first to start research of PHAs. Studies of these promising polymers (PHAs with different compositions and special devices trademarked as BIOPLASTOTAN) have been conducted by the Institute of Biophysics SB RAS in close and fruitful cooperation with the Siberian Federal University and institutes of the Siberian Branch of RAS and other research institutes and medical clinics.

The fundamental studies of the synthesis of intracellular macromolecules by hydrogen-oxidizing bacteria revealed the major factors that determine polymer yields, chemical structure, and physicochemical properties. Our research team developed and tested the processes of high yield production of polymers with different compositions on a variety of substrates (hydrogen, acetate, alcohols, and sugars) and prepared Technological Procedures for polymer production. Polymer synthesis procedures were developed for different applications: for production of high-purity polymers for biomedicine – on sugar-containing media; on hydrolysates of plant wastes and hydrolysis lignin; on brown coal-derived syngas.

Since the late 1990s, the Institute of Biophysics SB RAS, in cooperation with the Federal Center for Novel Biomaterials at the V.I. Shumakov Institute of Transplant Surgery and Artificial Organs of the Russian Ministry of Health, has conducted biomedical studies of the PHAs synthesized in the Institute of Biophysics SB RAS. All studies have been conducted in accordance with the international standard ISO 10993 in the *in vitro* systems and in short-term and long-term experiments with animals. The polymers have been certified as suitable for medical use, including uses in contact with blood. Specifications of polymers with different chemical structure intended for use as a polymer basis for therapeutic systems, matrices of bio-artificial organs and implants, and packaging material have been developed and registered in Gosstandart of RF.

Different types of PHAs were used to fabricate bio-inert films, membranes, and 3D constructs, which were tested as cell scaffolds for cellular and tissue engineering; microparticles, including those loaded with drugs; constructs and hybrid composites with hydroxyapatite for bone tissue repair. In cooperation with a number of medical clinics, we have been conducting clinical trials of polymer constructs as implants for bone tissue repair in orthopedics and oral surgery; we have developed and are now testing biocompatible polymer-coated vascular stents. An approach to engineering and scale-up of polyhydroxyalkanoate synthesis facility has been developed. Our research team has constructed and put into operation the first Russian pilot facility for production of polyhydroxyalkanoates, which is capable of providing Russian research and development institutes with PHAs.

The goal of this book is to show current world trends in research of this type of biomaterials, the position and potential of polyhydroxyalkanoates, and to summarize results of experimental tests and clinical trials of PHAs obtained in the last few years at the Institute of Biophysics SB RAS, Siberian Federal University, and V.F. Voino-Yasenetsky Krasnoyarsk Federal Medical University. The book presents results of research in modern biotechnology aimed at creation of sustained-release drug delivery systems and constructs for cellular and tissue engineering and for reconstructive medicine. It also describes some of the authors' results of experimental tests and clinical trials of PHA effectiveness in reconstruction of skin and bone defects and in general and abdominal surgery.

These studies were supported by the budget allocations to the Institute of Biophysics SB RAS, the Development Program of the Siberian Federal University, the Russian Ministry for Education and Science, the Krasnoyarsk Regional Science and Innovation Foundation, the RF President's Program for Young Candidates and Doctors of Science, the Russian Science Support Foundation, the Program of the RAS Presidium "Fundamental Sciences to Medicine," the Program of Integration Projects of the Siberian Branch of RAS, the Program of the Russian Ministry for Education and Science "Development of the Potential of Universities," megagrants of the Government of the Russian Federation designed to support research projects implemented by leading scientists at Russian institutions of higher learning and to establish innovation centers at RF universities (Decrees of the RF Government No. 219 and 220 of April 9, 2010), and mega-project of the Federal Special-Purpose Program "Development of Pharmaceutical and Medical Industries in Russian until 2020 and further."

The authors gratefully acknowledge the assistance of their colleagues – researchers of the Institute of Biophysics SB RAS and the Basic Department of Biotechnology at the Siberian Federal University: Doctors of Sciences Galina S. Kalacheva, Svetlana V. Prudnikova; Candidates of Sciences Natalia O. Zhila, Anastasia V. Goreva, Elena D. Nikolaeva, Alexei G. Sukovatyi, Yevgeny G. Kiselev, Anatoly N. Boyandin; graduate students and Masters of Science Anna M. Shershneva (Kuzmina), Anna A. Shumilova, Daria A. Syrvacheva, Dmitry B. Goncharov; colleagues at

V.F. Voino-Yasenetsky Krasnoyarsk Federal Medical University Vladimir A. Khorzhevsky, Andrei V. Yakovlev, Ekaterina S. Vasilenya; surgeons Sergei V. Miller, Valentin A. Kukonkov, Igor I. Beletsky, and Alexei B. Mylnikov.

Special thanks to Elena Krasova, who translated this manuscript into English, and to Elena Nikolaeva, who prepared illustrations for this book.

PART I

REQUIREMENTS FOR BIOMATERIALS: THE POSITION AND POTENTIAL OF DEGRADABLE POLYHYDROXYALKANOATES

CHAPTER 1

CREATION AND USE OF ENVIRONMENTALLY FRIENDLY MATERIALS AS AN IMPORTANT PART OF CRITICAL TECHNOLOGIES OF THE 21ST CENTURY

CONTENTS

1.1 INTRODUCTION

Development and use of new, environmentally friendly materials, which can be involved in biosphere cycles, corresponds to the idea of environmentally safe, sustainable industrial development. In Agenda 21, adopted at the special UN session on the environment and development in 1991, the stress is laid on the need for the development and application of new, ecofriendly technologies and materials (Agenda 21). Environmental protection is an integral component of sustainable development. Human economic activities threaten all biotic and abiotic components of the environment. Moreover, as noted in Agenda 21, the increasing human population produces and consumes greater amounts of chemical substances, thus, exacerbating environmental problems. In spite of the increasing efforts to bring down accumulation of wastes and recycle them, more and more damage is done to the environment by enormous amounts of wastes and unsustainable land management.

One of the priorities stated in Agenda 21 is "to prevent, halt and reverse environmental degradation through the appropriate use of biotechnology in conjunction with other technologies, while supporting safety procedures as an integral component of the program." Specific objectives include the inauguration of specific programs with specific targets:

1) To adopt production processes making optimal use of natural resources, by recycling biomass, recovering energy and minimizing waste generation;

2) To promote the use of biotechnologies, with emphasis on bioremediation of land and water, waste treatment, soil conservation, reforestation, afforestation, and land rehabilitation;

3) To apply biotechnologies and their products to protect environmental integrity with a view to long-term ecological security.

Governments at the appropriate level, with the support of relevant international and regional organizations, the private sector, non-governmental organizations and academic and scientific institutions, should:

• Develop environmentally sound alternatives and improvements for environmentally damaging production processes;

• Develop applications to minimize the requirement for unsustainable synthetic chemical input and to maximize the use of environmentally appropriate products, including natural products;

• Develop processes to reduce waste generation, treat waste before disposal and make use of biodegradable materials;

• Develop processes to remove pollutants from the environment;

• Promote new biotechnologies for tapping mineral resources in an environmentally sustainable manner.

The increasing use of synthetic plastics has become a global environmental problem. Polymer materials have become an essential element of modern life, having replaced steel, wood, and glass in many applications. The term "polymeric materials" encompasses three large groups: polymers, plastics, and their morphological variety – polymer composite materials (PCM), or reinforced plastics. All these materials contain a polymeric constituent, which determines their basic thermal deformation and processing properties. The polymeric constituent is a high-molecular-weight organic substance produced by the chemical reaction between molecules of the starting low-molecular-weight substances – monomers. The term polymer usually defines high-molecular-weight substances (homopolymers) supplemented with stabilizers, plasticizers, lubricants, antirads, etc. Physically speaking, polymers are homophase materials, which retain all physicochemical properties of the homopolymers. Plastics are polymer-based composite materials, containing disperse or short-fiber fillers, pigments, and other free-flowing ingredients. Fillers are not present in the continuous phase. They are dispersed in the polymeric matrix. Plastics are heterophase isotropic (having the same properties in all directions) materials.

Plasticts are used in almost all areas of human activity (Table 1.1). About 60% of all synthetic plastics used for packaging are polyethylenes. They are low-cost materials and show excellent performance in various applications. High-density polyethylene (HDPE) has the simplest structure of all plastics, consisting of repeating units of ethylene, $(-CH_2-CH_2-)$ n. Low-density polyethylene (LDPE) has the same chemical formula, but it has a high degree of chain branching: $(CH_2CHR)n$, in which R may be $-H$, $-(CH_2)_nCH_3$, or a more complex structure with secondary branching.

Synthetic polymers (nylon, polyethylene, polyurethane) have revolutionized our way of life, but they have also created a number of problems. First, resources used to produce synthetic polymers are nonrenewable and, second, the application of polymers that cannot decompose in the natural environment and their accumulation lead to environmental pollution, presenting a global ecological problem. Outputs of synthetic plastics that do not decompose in the natural environment, mostly polyolefins (polyethylenes and polypropylenes), and are manufactured in processes of petroleum synthesis are huge: their annual production has reached 330 million tons,

TABLE 1.1   Application Areas for Plastics

Market sector	Product	Application area/ consequences
Packaging	Packs, films	Food package.
	Bottles, trays, plasters	Difficult recycling
	Hollow containers	Short lifetime of pack-
	Nets, bags	aged food
Fast Food	Dishes	Recycling may be
	Cutlery	impossible or expensive.
	Plugs, lids	Products are often bio- logically contaminated
	Drinking straws	through contact with
	Cups	food
Fibers/textiles	Clothes	"Breathing" cloth
	Technical textiles	Tactile properties
	Fabrics	Luster
Toys	Artificial materials	Educational advantages
	Bricks and blocks	Environment
	Golf tees	
Motor car con- struction	Car floor mats, car parts	
Everyday use	Trash bags	Short lifetime of the
	Fibers and nonwovens	product
	Personal hygiene products	Difficult recycling
	Cosmetic containers, containers for powders, lubricants, etc.	
	Golf tees	
	Adhesives, paints, coatings	
Gardening	Plant pots	Better compostable
	Supports	Very difficult recycling
	Peat bags	due to contamination
	Fertilizer strips	Cheaper manpower
	Bonding materials	
Agriculture	Film covers	Vegetable-growing, horti-
	Mulching films	culture
	Protective films	

TABLE 1.1 (Continued)

Market sector	Product	Application area/ consequences
Medicine	Implants	Safe residence and degradation in the body
	Medical packaging	
	Surgery materials	Short lifetime of the product
	Personal hygiene products	Disposable product
	Gloves	
Other	Filters	Specific advantages
	Assembly technologies	Cost reduction
	Burial	Compostablity requirements
	Desk sets	
Electronics	DVD, computer cases	
	Coatings for office equipment, video and audio equipment	
	Cases for mobile phones	

*Adapted from Fomin, Guzeev, 2001; the Data of Biodegradable Materials Interest Community Association; COPA Forecast.

increasing by about 25 million tons every year. In developed countries, no more than 16–20% of them are recycled, and they largely accumulate in landfills.

The main approaches of the plastic refuse policy are now landfilling and reclamation. Polymer landfilling is a time bomb, and this problem will have to be solved by the future generations. Moreover, up to 10,000 ha of the land, including agricultural fields, is annually occupied by new landfills. A possible way to reduce enormous amounts of plastic refuse is reclamation, which can involve several solutions: incineration, pyrolysis, recycling, and reprocessing. However, neither incineration nor pyrolysis of plastic containers and other plastic items can essentially improve the state of the environment. Moreover, incineration is a very costly process, releasing highly toxic and super-toxic compounds (such as furans and dioxins). Recycling is some solution, but it involves considerable labor and energy expenditures: plastic packaging items and containers have to be picked out of the household garbage; different plastics should be

separated, washed, dried and disintegrated; only then, they can be processed to fabricate a new end item.

Some countries have passed legislation that plastic waste, plastic packaging items and containers in particular, must be obligatorily collected
and recycled. For instance, the European Union Directives stipulate that
manufacture of plastic packaging must involve 15% recycled plastics; in
Germany, this quota is 50% and must increase to 60%. Landfilling and
incineration cannot solve the problem of reusing millions of tons of synthetic plastic refuse, and their accumulation in the biosphere can lead to a
global environmental disaster. A solution to this problem is to create and
use new types of materials that are naturally degraded in the environment
into harmless components. This solution grows in popularity as oil and
gas prices increase. In contrast to oil, polymers prepared from natural raw
materials or synthesized by microorganisms (the so-called biopolymers, or
bioplastics) do not contribute to greenhouse gas increase and global warm-
ing. Moreover, they can facilitate regeneration of the carbon cycle, or "carbon reincarnation." Petroleum-based polymers may also be regarded as
renewable materials, but it will take more than one million years for the
biomass to be converted into fossil fuels, which may be used to produce
plastics. As the level of consumption of plastics is much higher than the
level of renewal of fossil carbon resources, the carbon cycle is not in balance. By contrast, bio-based biodegradable polymers, which are produced
from renewable plant material and microbial biomass, can be produced
and recycled over comparable time periods. Therefore, countries developing this approach are exempt from emission quotas under the Kyoto
Protocol. The European Union undertook to reduce CO_2 emissions by 8%
relative to the level of 1990, by 2012. Japan undertook to reduce CO_2
emissions by 6%.

In 2000, the EU accepted the norm EN 13432, concerning biodegradable polymer products. EU Resolution No. 2001/524/WE brought
it in conformity with Directive No. 94/62/WE. The norm defines the
criteria for evaluation and procedures concerning composting of biodegradable synthetic materials and their anoxic treatment (i.e., recycling
rather than incineration of organic compounds). A radical solution to
the problem of "polymer garbage" is to create and use a wide range of
polymers that are, under the appropriate conditions, biodegraded into

components harmless to living and non-living nature. Biodegradability of high-molecular-weight compounds will be the basis for avoiding a large number of problems related to "plastic garbage" – discarded plastic packages and other polymer articles.

Economic situation also encourages the development of degradable plastics production from renewable sources: up to 98% of polymeric materials produced in the world are derived from nonrenewable fossil materials – oil, gas, and coal, whose resources are running out. As the production of synthetic plastics is increased, the costs of waste processing grow too. Furthermore, research over the past few years has shown adverse effects of food and drink plastic packages. Some materials used to produce plastic containers have been found to cause serious diseases, including cancer, because of the presence of bisphenol A in the synthetic package. Researchers of Boston Medical Academy announced that this substance tends to accumulate in the human body. A rough estimate shows that the global annual production of bisphenol A for the manufacture of food and drink packages amounts to 2.8 million tons. British manufacturers have stopped producing bisphenol-containing packages. Previous studies showed that phthalates (phthalic acid esters), components of vinyl food packages, medical pipes, and toys, impaired the development of some functions in newborn boys. Toxic 2-EHA was detected in polymer lids of the jars with baby food by scientists at Würzburg University (Germany).

Recent decades have seen a growth of interest in environmentally friendly materials, which are produced from renewable sources such as plant and microbial biomass. Biopolymers are synthesized from sucrose, starch, cellulose, lignin, vegetable oils, etc. During their "lifecycle" (from synthesis to complete degradation), biopolymers release much less carbon dioxide than petroleum-derived synthetic plastics do.

1.2 BIOPLASTICS: A NEW DIRECTION IN MATERIALS SCIENCE

Polymers produced in biotechnological processes or from the feedstocks synthesized by biotechnological techniques are named biopolymers. Bioplastics, according to IBAW, are biodegradable polymers that conform to the EN 13432 standard and polymers produced from non-biodegradable agricultural crops.

Production of polymers from plants is an energy-saving way. Production of one ton of bioplastics consumes 12 to 40 GJ less power than production of polyethylene. Disposal and composting of items prepared from biodegradable plastics derived from natural raw materials, especially containers and packages, which have a very short lifetime and comprise a considerable part of garbage, would be much simpler and more effective. Last but not least, development of "green" technologies is beneficial for agriculture. Thus, it is not surprising that biopolymers and biofuel have recently attracted worldwide interest. "Thinking outside the oil barrel" is a very popular phrase in the U.S. The market for various polymeric materials is growing rapidly, especially in China, India, and other countries in Southeast Asia. In Europe, plastic market is growing at a slower rate, and it is being restructured.

Analysis of the current situation regarding the production and use of biodegradable plastics suggests three main directions:
- production of plastics based on renewable natural sources,
- imparting biodegradability to high-molecular-weight synthetic materials, which are commonly used now, and
- synthesis of biodegradable polyesters of hydroxy acids.

Bioplastics based on natural biodegradable polymers such as starch, cellulose, chitosan or proteins are composite materials containing various additives. Water-soluble starch/pectin film is produced by adding glycerol or polyethylene glycol as plasticizers. Such composite biodegradable films are used in agriculture and for packaging. The costs of production of such materials are usually reduced by using unpurified starch mixed with polyvinyl alcohol, talc, and other additives. Water-resistant biodegradable composites are produced from blends of starch and polyoxyalkylene glycol. Absorbent biodegradable diapers and sanitary pads are produced from a hydrophilic blend of destructured starch impregnated with an ethylene/vinyl alcohol copolymer and aliphatic polyesters. Films made of this material are very strong and retain their properties exposed to 50°C during 3 months. Such films are used in agricultural mulch and as packaging materials for foodstuffs. Biotec GmbH produced starch-based compostable plastics (bioplast – granules for molding disposable items, foams for foodstuffs packaging, granules for fabricating Bioflex compostable blown and flat films, etc.). High ecological compatibility of these materials and their capacity to be degraded in compost at 30°C for 2 months, producing substances beneficial for plants, make them promising candidates for various applications. Other polysaccharides, such as cellulose and chitin, cellulose and starch, are also being investigated as a renewable biodegradable basis for the manufacture of thermoplastics. For instance, polymers generated by the interaction of cellulose with an epoxy compound and anhydrides of dicarboxylic acids are completely degraded in compost during 4 weeks. Bottles, disposable dishes, and mulch films are fabricated from these polymers by molding. To increase the biodegradability of materials based on cellulose esters, it is recommended to supplement the composition with ascorbic acid polyesters or cellulose acetate, partly interesterified with 6-hydroxycaproic acid. Composted materials manufactured from a blend of vegetable and natural initial products, with cellulose or its derivatives being the main component, are currently widely used to make disposable packaging items and indispensable things. Recently, researchers have specifically been attracted to compositions containing chitosan and cellulose. The blends containing 10–20% chitosan are used to make biodegradable plastics, strong and water-resistant films. The triple composition of chitosan, micro-cellulose fiber and gelatin is used to prepare films of enhanced strength, capable of being degraded by microorganisms when buried underground. Natural proteins are also used to make biodegradable plastics intended for packaging dry and wet foodstuffs etc. Methacrylated gelatin, casein, and derivatives of serine and keratin-containing natural products are used to make biodegradable packaging material for foodstuffs, perfume, and drugs. For instance, Showa (Japan) has devised a biodegradable thermoplastic polymer that can be used to make housings for television sets and personal computers. Generally speaking, the use of natural polymers to prepare biodegradable plastics is attractive because the resources of the raw materials are renewable (Gerngross, 1995).

The second approach is to modify synthetic polymers so as to make them biodegradable. Synthetic polymers – polyethylene, polypropylene, polyvinylchloride, polystyrene, and polyethylene terephthalate – are manufactured in large quantities. These polymers and items fabricated from them can persist "eternally" in landfills, and, thus, it is very important to render them biodegradable. There are various ways to attain this goal: by introducing molecules containing functional groups that make the polymer readily photodegradable; by synthesizing compositions with biodegradable natural additives that can, to a certain extent, initiate degradation of the basic polymer; and by synthesizing biodegradable plastics based on commercial synthetic products. Copolymers of ethylene and carbon monoxide are photodegradable. Vinyl-ketone monomers trigger photodegradation of polyethylene and polystyrene. If they are added (2–5%) to ethylene or styrene, the resulting plastics possess properties similar to those of polyethylene or polystyrene, but become photodegradable under ultraviolet radiation of 290–320 nm. Photosensitive additives – dithiocarbamates of iron and nickel – convert plastics into soil-degradable films. Cellulose pulp, alkyl ketones or fragments containing carbonyl groups are added to polyethylene, polypropylene, or polyethylene terephthalate to accelerate its photo- and bio-degradation. Such films persist for 8–12 weeks before their photo- and bio-degradation begins.

Recently, however, there has been little progress in the development of this approach: fewer publications have been devoted to the conversion of synthetic polymers into biodegradable plastics by adding natural components to them. The approach that receives the greatest attention now is the production of biodegradable synthetic plastics by synthesizing corresponding polyesters and polyester amides. Degradable copolyesters are produced by using aliphatic diols and organic dicarboxylic acids. In 1995, BASF manufactured Ecoflex F, a fully degradable plastic based on such polyester. It is used to make bags, agricultural film, sanitary film, and paper-coating material. The mechanical properties of Ecoflex F are comparable with those of low-density polyethylene. It is used to make film of high tensile strength, flexibility, water resistance, and water vapor permeability. It is processed by blown extrusion and cooling on rolls, like low-density polyethylene; owing to its capacity for plastic deformation it can be used to make thin films (less than 20 μm thick), which need not be specially treated. BASF is also producing biodegradable plastics based on polyesters and starch. Since the second half of the 1990s, BAYER AG has been producing new compostable aerobically biodegradable polyether-amide-based thermoplastics – BAK-1095 and BAK-2195. This material exhibits good adhesion to paper, thus, it is widely used to manufacture water- and weather-resistant packaging for food industry and agriculture. BAK-2195 is processed by injection molding. This material can contain such additives as cellulose, wood flour, and starch, which make it sufficiently stiff and tough. To reduce the cost of polyether- and polyamide-based materials, firms producing them use available production facilities and commercially available raw materials. Such compositions are processed into end products using conventional facilities. This is the way to speed up commercial production of new environmentally friendly polymers and to reduce the cost of biodegradable polymers, thus substantially decreasing the amounts of landfilled polymer containers and packages as well as the amounts of buried plastic wastes. According to BASF, the potential West European market for biodegradable polyether amides, copolyesters, and their starch blends is 200,000 t/y. Transparent moldable biodegradable copolyester for producing films and sheets is synthesized by ring opening polymerization and interesterification of lactide with aromatic polyesters based on tere(iso)phthalic acid and aliphatic diols. Recently, much attention has been focused on the development of biodegradable compositions containing both polyester-polyamide groups and urethane and carbonate groups and, especially, fragments of hydroxy acids. They are used to make a wide range of compostable items, which have good physical-mechanical properties and reasonable cost.

Lately, many large companies have been marketing modified biodegradable synthetic materials. Bayer offers new biodegradable polyetheramide, which has a semi-crystalline structure and which can be processed by jet molding. The feedstocks used to produce it are hexamethylenediamine, butanediol, and adipic acid. The film is translucent or transparent. The packaging film is biodegraded over 3 months of exposure to bacteria and soil fungi. This material is intended to be used to make garbage bags, foodstuff packages, and disposable dishes. Eastern Chemical (U.S.) has recently begun to produce Eastar Bio COPE polyester. It is also intended to be used as foodstuff packaging material and as material for manufacturing bags for horticulture and agriculture. The material is semi-crystalline and transparent; its oxygen barrier properties are better than those of polyethylene film. This material is degraded into carbon dioxide, biomass, and water; its degradation rate is equal to that of newsprint. DuPont (Switzerland) has announced the commercial production of Biomax – a hydro-biodegradable polyester. Its properties are similar to those of polyethylene terephthalate, and its production cost is just slightly higher than the cost of its chemical counterpart. A number of companies are marketing materials whose biodegradation parameters can be controlled. Symphony Environment Ltd (U.K.) has marketed a polyethylene-based biopolymer, whose degradation can be controlled by special additives. By choosing the additive and varying its quantity, one can change the degradation time of the package in the range between 3 months and 5 years (Volova, Shishatskaya, 2011; Prudnikova, Volova, 2012).

The third direction in production of degradable plastics is chemical or microbiological production of polymers based on hydroxy acids. Analysis of the latest literature on the development of biodegradable polymers shows that increasing efforts have been made to pursue this line of research. Polyesters based on hydroxy (glycolic, lactic, valeric, and butyric) acids are degraded in the environment by exodepolymerases of soil and water microflora, as its growth substrate. To produce polyesters of these acids, their dimeric derivatives are used: glycolides and lactides for glycolic and lactic acids and β-, γ- or ε-lactones for the other acids.

Complex polyesters of aliphatic hydroxy acids with various compositions have recently attracted increasing attention as materials for producing items for different applications, including medicine and related areas. Important advantages of these polymers and products fabricated from them are their biocompatibility and biodegradability that occurs via biodegradation of the macromolecular chain. Moreover, the end products of most of these polymers are harmless products such as carbon dioxide and water. As production processes are improved and the cost of these polymers is reduced, they may be used to produce biodegradable packaging materials and other items, which must be degraded when they reach the end of their lifetime.

About 100 polymers of aliphatic hydroxy acids or their copolymers have been discovered by now (Table 1.2). The lower aliphatic polymers are the best studied so far:

TABLE 1.2 Polyesters of Aliphatic Hydroxy Acids

Chemical structure	Nomenclature used by Chemical Abstracts Service	Other names
n = 0 R = H	Poly[oxy(1-oxo-1,2-ethanediyl)]	Polyglycolide, Polyhydroxy acetate, Polyhydroxy ethanoate
R = CH$_3$	Poly[oxy(1-oxo-2-methyl-1,2-propanediyl)]	Polylactide, Poly(2-hydroxypropinoate)
n = 1 R = CH$_3$	Poly[oxy(1-oxo-2-methyl-1,3-butanediyl)]	Poly(3-hydroxybutyrate), Poly(β-hydroxybutyrate)
n = 1 R = C$_2$H$_5$	Poly[oxy(1-oxo-2-ethyl-1,3-butanediyl)]	Poly(β-hydroxyvalerate), Poly(3-hydroxypentanoate)
n = 2 R = H	Poly[oxy(1-oxo-1,4-butanediyl)]	Poly(γ-hydroxybutyrate), Poly(4-hydroxybutyrate)
n = 2 R = CH$_3$	Poly[oxy(1-oxo-2-methyl-1,4-butanediyl)]	Poly(γ-hydroxyvalerate) Poly(4-hydroxyvalerate)
n = 3 R = H	Poly[oxy(1-oxo-1,5-pentanediyl)]	Poly(δ-hydroxyvalerate) Poly(5-hydroxyvalerate)
n = 4 R = H	Poly[oxy(1-oxo-1,6-hexanediyl)]	Poly(6-hydroxycaproate), Poly(ε-hydroxycaproate), Poly(6-hydroxyhexanoate)

$$\text{H}$$
$$[-\text{O}-\text{C}-(\text{CH}_2)_n-\text{C}-]_k$$
$$\text{RO}$$

The degree of polymerization of these polymers (n) usually ranges between 100 and 30,000.

Polymers of aliphatic hydroxy acids containing branched units exhibit optical activity, which largely determines the properties of these polymers and their biodegradability.

Polymers of hydroxy acids can be produced in a biotechnological process, by microorganisms capable of producing them, which will be the subject of this chapter, and by chemical synthesis.

In the latter case, polymers are synthesized via ionic or hydrolytic polymerization of their six-membered cyclic diesters (such as glycolide and lactide) or lactones such as ε-caprolactone (6-hexanolide):

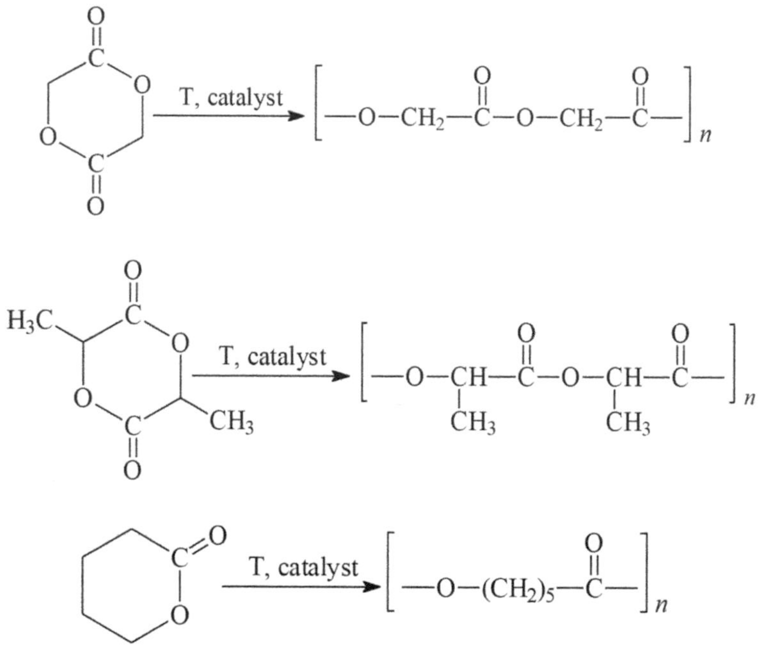

A polymer similar to hydroxy acid polymers is polydioxanone, produced synthetically, by polymerization of *p*-dioxanone, forming a polymer of 2-(2-hydroxyethoxy)propionic acid:

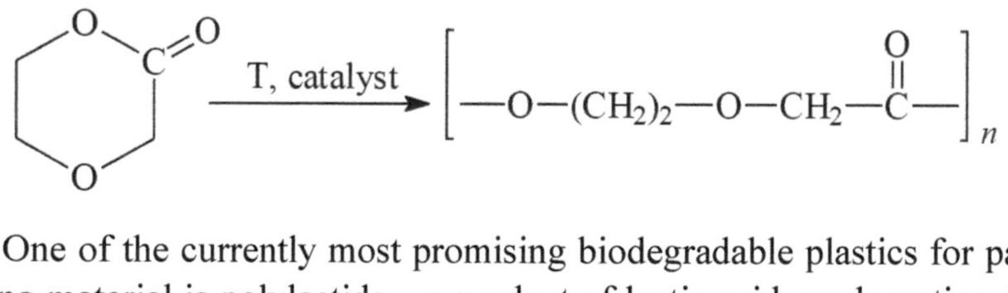

One of the currently most promising biodegradable plastics for packaging material is polylactide – a product of lactic acid condensation. The main reason for this is that lactide and polylactide can be produced both synthetically and through fermentation of dextrose or maltose, grain or potato mash. In compost, polylactide is biodegraded during one month; it is also utilized by seawater microorganisms. If plasticized properly, polylactide becomes flexible and imitates polyethylene, plasticized polyvinyl-chloride or polypropylene. The durability of the polymer increases as the proportion of the monomer in it is decreased and as a result of alignment, which enhances polymer toughness, Young's modulus, and thermal stability. In spite of all these advantages of polylactide, it has not been widely used for household and commercial applications so far. The reasons are low outputs, low efficiency of processing lines, and, as a consequence, the high cost of the product. That is why, much attention is currently focused on the reduction of the cost of the biodegradable product. Cargill Inc. (U.S.) has been actively developing the technology of producing lactic acid. Sheets made of Eco-Pla, a biodegradable polymer manufactured by that company, show impact resistance comparable to that of polystyrene. Eco-Pla coating materials and films are very tough, transparent and lustrous; their extrusion temperature can exceed 200°C; and they have a low friction coefficient. Cargill Inc. produces up to 6000 t/y of polylactide through corn dextrose fermentation; the firm plans to increase its capacity to 50,000–150,000 t/y and reduce the cost of polylactide from US$/kg 250 to 2.2. To reduce the cost of the lactic acid polymer, Mitsui Toatsu (Japan) has installed a pilot facility for one-phase production of polylactide. The resulting product is a thermostable polymer with better properties than the plastic produced in a two-phase process. The cost of the new material is US$/kg 4.95.

Polyesters of microbial origin (polymers of hydroxy fatty acids, or polyhydroxyalkanoates, PHAs) currently occupy a special position among biodegradable polyesters. Interest in polyhydroxyalkanoates has been growing since the late 1980s. This is a new class of natural polyesters, which are not hydrolyzed at a high rate by non-biological processes and whose properties – molecular weight, crystallinity, mechanical strength, and degradability – may vary substantially. Polyhydroxyalkanoates are promising materials to be used in food industry (as packaging and antioxidant materials), agriculture (as coating materials for seeds, fertilizers, and pesticides; as degradable films and containers for hothouse gardening), and in other areas, including medicine and pharmacology. The physicochemical properties of PHAs, their diversity, and the feasibility of producing PHA-based composites with various materials make them most promising materials of the 21st century.

Biodegradable polymers, especially those of biological origin, are too costly, and, thus, do not play an important part in the global plastics market today (Figure 1.1).

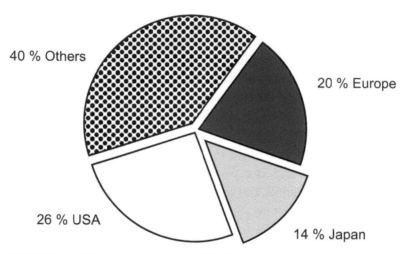

FIGURE 1.1   Global market for degradable bioplastics (the early 21st century).

The greatest consumers of plastics in Europe are Germany, the U.K., France, Italy, and the Netherlands. Belgium, Norway, Austria, Spain, and Switzerland are also developing production of bioplastics. IBAW (Biodegradable Materials Interest Community Association) estimates that European consumption of bioplastics nearly doubles every year. The market for biopolymers is increasing owing to environmental concerns and the endeavor to replace fossil raw materials in polymer industry. The greatest increase in the global market for biodegradable polymers is expected in the next five years. The fastest growth of this market can be seen in the U.K., Italy and the Netherlands. By 2010, Japan had increased the production of degradable "Toyota's" (starch-based) plastics to 200,000 tons, and the annual production of bioplastics may be increased to 30 million tons – one fifth of Japan's plastic consumption. In the U.S., the interest in bioplastics is also growing fast. The Freedonia Group reported a 20% annual increase in demand for bioplastics; in 2010, annual production of bioplastics reached 185,000 tons. Between 2005 and 2020, global polymer production is expected to increase from 180 to 258 million tons. Hence, the amount of polymer waste will increase almost proportionally, and only part of it will be recycled, while the remaining part will be landfilled or dumped along the roads, on the banks of the rivers, and in ravens. Over the same time period, the proportion of bioplastics is expected to increase from 1.5% to 4.8%, or from 4 to 12.5 million tons. Toyota forecasts are more optimistic, predicting that in 2020, the world plastics market share of bioplastics will reach 25%, or about 30 million tons, owing to the increasing interest in renewable resources. Between 2001 and 2003, global consumption of biodegradable materials had doubled, reaching 40,000 tons. This trend is expected to persist, especially because of the steadily growing oil prices. The European Bioplastics association's forecast is that in 2020, bio-based degradable polymers will occupy 5% of the European polymer market. However, areas of potential applications of bioplastics and the scale of their production are mainly determined by their cost. COPA (EU Committee of Professional Agricultural Organisations) and COGECA (EU General Confederation of Agricultural Cooperatives) have estimated potentials of bioplastics and their applications in Europe (Figure 1.2). Certain drawbacks of biopolymers also hinder their wide use. Cellulose-based products are brittle; starch-based polymers are humidity sensitive; polylactides are not thermoplastic and polylactide products are permeable to water vapor and oxygen. This, however, cannot stop rapid development of bioplastics production.

Biopolymers and composite materials are not unique products anymore; they are regarded as attractive and available goods. This change has been facilitated by integration of agriculture and industry, advances in

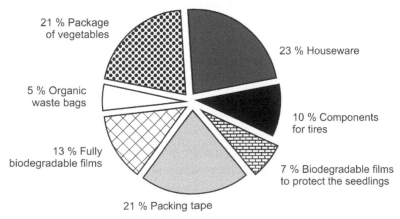

FIGURE 1.2 Potential applications of bioplastics.

biotechnology, genetic engineering and selection, and growth of production facilities (Table 1.3).

Research on biopolymers has been developing rapidly recently. Biopolymers are divided into two categories: polymers produced by biological systems (such as microorganisms) and polymers synthesized by chemical means but from biological feedstocks: organic acids, sugars, fats. Their primary application is in packaging. The lifetimes of most products of packaging industry are very short: a few months or even days. At the same time, bottles, containers, boxes, blister packs, and bags comprise the greater part of household garbage. Degradable bioplastics are very promising materials for light and food industries, agriculture, and municipal engineering: bags, films, bottles, trays, plasters, nets, disposable dishes, clothes, technical textile, fabrics, cosmetic bottles, toys, plant pots, plant supports, peat bags, fertilizer strips, protective and mulching films, cases for computers, office equipment, video and audio equipment, mobile phones, etc.

In the past 10–15 years, designing of biopolymers has become a major interdisciplinary research area, whose main goals are: (i) to discover and study new biopolymers; and (ii) to develop a basis for constructing

TABLE 1.3   Factors Influencing Biopolymers Market

Technological	Advances in molecular biology, fermentation techniques, genetic engineering and plant selection
	Advances in designing of composites and compounding
	Implementation of large-scale economic projects
	The use of "organic" recycling instead of mechanical one
Economical	Increasing costs of natural resources
	Higher waste disposal costs
	Enhanced competitive ability of biopolymers owing to increasing pollution taxes
Political	Laws
	Support of the governments and local administrations
Social	Favorable public attitude to biopolymers
	High level of environmental education among consumers

biological systems capable of synthesizing polymers with tailored properties. At the present time, the most widely used and actively developed biodegradable polymers are aliphatic polyesters, polyamides, segmental polyester urethanes, polymers of lactic and glycolic acids (polylactides and polyglycolactides), silicone, and polyethylene terephthalate, which have recently been joined by polymers of hydroxy fatty acids, the so-called polyhydroxyalkanoates (PHAs). Biodegradable polymers are attractive materials for manufacturers and environmentalists because they can be degraded rather quickly (within a few months or a few years), yielding environmentally safe products such as water, biomass, carbon dioxide or methane (depending on the type of the degradation process – anaerobic or aerobic one).

In their physical and technical properties, bioplastics compare well with traditional plastics, but are also environmentally safe. Production of bioplastics is a budding industry, and there are still certain misunderstandings. Some people think that all polymers prepared from plant materials are biodegradable, but this is wrong: the ability to be degraded in natural conditions is not determined by the natural origin of the basic material but rather by the molecular structure of the material (and many other parameters). Moreover, not all biodegradable polymers are fabricated from natural components. Synthetic materials are also used to prepare biodegradable plastics. One example is biodegradable, carbohydrate-derived, polymer Ecoflex (BASF, Germany). This material was used as a basis for the composite polymer Ecovio, produced from both synthetic and renewable (polylactide) materials. Research conducted by European Bioplastics (the European association of manufacturers, suppliers, and consumers of bioplastics and other biodegradable materials) shows that bio-packaging is mainly demanded by food industry. In 2000, the EU accepted the norm EN 13432, concerning biodegradable polymer products.

The market and the leading manufacturers of plastics are given in Table 1.4; application areas for bioplastics and their advantages are listed in Table 1.5. The current global capacity for the production of biodegradable materials is about 300,000 t. Most of the industrial facilities have been in operation since 2000.

TABLE 1.4   The Major Manufacturers of Biodegradable Plastics and the Market for Them

Company	Product	Production capacity	Notes
Nature Works	Polylactic acid	140,000 t/y	Produces the greatest amounts of bioplastic
Novamont, Italy	Vegetable starch- and biodegradable polyester-based materials	5,000 t production facilities to be upgraded in 2006	Has been over 10 years on the market
BIOP Biopolymer Technologies AG, Denmark	Starch-based materials	10,000 t/y	Production facility in "Schwarzheide," Germany, under contract with "BASF." Facility cost – € 7 million
Stanelco, U.K.	Starch-based materials		
Plantic, Australia	Starch-based materials		
BASF	Composite biodegradable plastics	Increasing from 6,000 t/y to 14,000 t/y in 2006	The most significant manufacturer of composite biodegradable plastics
Metabolix and ADM	PHAs	50,000 t/y	To be started in 2008
DuPont and Tate & Lyle	Polyester containing about 40% 1,3-propane-diol (PDO), produced biotechnologically	50,000 t/y	Started in 2007
Toyota, Japan	Polylactic acid	1,000 t/y	Pilot plant. If production costs are reduced, a 50,000 t/y plant will be constructed
Hycail, the Netherlands	Polylactic acid	50,000 t/y	In the final stage of modernization
Procter & Gamble Chemicals, U.S.	PHAs		Planning to establish production in Europe
Tianan, China	Polylactic acid		Started
Rodenburg Bio-polymers	Bioplastic Soyanyl	47,000 t/y	The largest plant in Europe

*Adapted from Biodegradable Materials Interest Community Association; COPA forecast.

TABLE 1.5 Application Areas for Bioplastics and Their Advantages

Application area	Advantages
Packaging (foil, film, bottles, blister packs, nets, bags)	Ideal as food package for foods with a short shelf life
Restaurants, fast food (dishes, cutlery, drinking straws)	Disposable dishes are economic and do not produce adverse effects on food contacting with them
Fiber production, textiles (clothes, technical textile, fiber)	"Breathing," pleasant to touch, lustrous fabrics
Toys	Environmentally safe
Everyday products (bags for organic garbage, personal hygiene items)	"Natural," readily degradable materials
Agriculture (various films, protective materials, plant pots, seed packages)	Economic materials; waste management is not costly
Medicine (implants, surgical materials, mouth hygiene items, gloves)	Clean and disposable
Technological installations, stationery	Various processing techniques and low-cost waste management by composting

Tables 1.4 and 1.5 list application areas for biodegradable polymers and their advantages. The majority of biodegradable polymers are manufactured from vegetable materials. Several attempts have been made to investigate and produce bioplastics from potato, wheat, legumes, helianthus, sugar beet, and poplar and aspen wood. As a result, some of these feedstocks were found to be useless, while others – wheat, corn, and sugar beet – proved to be very productive. Such natural polymers as cellulose, natural rubber, polysaccharides, polypeptides, chitin, epoxidized oils, lignin, pullulan, polyesters, etc. are widely used now. Starch has attracted great attention as a relatively cheap material, extracted from potatoes, wheat, corn, and rice. However, starch-based polymers are not water resistant. This disadvantage can be overcome by fabricating composite materials consisting of starch and natural or synthetic components enhancing water resistance. Rodenburg Biopolymers was the first in the world to sell Solanyl at a price comparable with the prices of traditional synthetic plastics. Solanyl is prepared from industrial potato wastes; this plastic is degradable by microorganisms. Since 2001, the largest bioplastic-manufacturing plant in Europe has been producing Solanyl; the annual outputs

of this bioplastic reach about 47,000 t. A Solanyl-manufacturing facility operates in the Netherlands. Farmers annually supply about 400,000 t potato wastes to the plant. Novamont SpA (Italy) and Environmental Polymers Group (EPG) (U.K.) market packaging bioplastics based on starch. Novamont SpA manufactures four compositions of nontoxic starch-based polyacetal material – Mater-Bi. Mater-Bi is a fully biodegradable and biochemically degradable bioplastic, whose stability and mechanical properties are comparable with those of traditional plastics. This plastic is based on natural material, and its greenhouse gas releases and power consumption are very low. Novamont SpA has marketed several grades of Mater-Bi®. Some major British supermarkets have been using Mater-Bi® films to pack organically grown vegetables and fruit. Traditional plastic bags are replaced by the bags made of NatureFlex (Innovia Films) and Mater-Bi® layers. This combination has excellent mechanical and isolating properties of biodegradable packaging, which can be used to pack dried food for long-term preservation. In Austrian and Swedish McDonald's, customers may use corn-based forks and knives. Goodyear has produced the first bio-tires, Biotred GT3. Carrefour (France), Esselunga (Italy), and Co-Op (Norway) shops provide Mater-Bi shopping bags. Italian companies plan to produce bioplastic from sugar beet, which would be degraded by aquatic microorganisms over 10 days at room temperature. In Australia, bioplastic made from cornstarch has been developed by the CRC for International Food Manufacture and Packaging Science over the past six years. The material will be used for dry foods packaging. Product engineers claim that "corn syrup" can be used to manufacture polymer bags, plates, cups, spoons, and forks. Toyota (Japan) and Cargill Dow (U.S.) also produce bioplastics. Until recently, green plastic production has been below 100,000 t/y. Toyota is going to increase this amount dramatically. In 2007, the company constructed a plant producing 50 000 t bioplastic annually, and by 2020, the annual output will be increased to 20 million t. This must earn Toyota an annual profit of US$ 38 billion. The price of one kilogram of the polymer will be decreased from $5–10 to $2, i.e., it will be equal to the price of synthetic plastics. Toyota's bioplastics are produced from sugar cane, corn, and tapioca. Now one kilogram of the bioplastic costs 500–1000 yens, and this is 5 times more expensive than the price of usual, petroleum-based, plastics. Japan consumes about 14 million tons plastic materials a year (about one tenth of the world's plastics output), but only 10,000 tons are bioplastics. Another giant automaker in Japan – Mazda Motor Corp. – plans to produce bioplastics from nonfood feedstocks. They emphasize that by using nonfood materials to produce bioplastics, they will not cause losses for food industry. They propose to use wood industry wastes: sawdust, etc.

Polylactide (lactic acid polymer) has been a leader among commercially produced degradable bioplastics until recently. Lactic acid can also be manufactured by biotechnological means – by fermentation of dextrose, sucrose, or maltose, corn or potatoes. Then lactide is polymerized. To enhance PLA thermal stability and improve its mechanical properties, it is usually blended with polyglycolide. This plastic can be prepared by chemical synthesis or by fermentation of sugar feedstocks. Commercial lactic acid is produced by hydrolysis of chloropropionic acid and its salts at 100°C or lactonitrile, $CH_3CH(OH)CN$ (100°C, H_2SO_4), which yields lactic esters; the esters are recovered and hydrolyzed to form a high-quality product. Other ways to produce lactic acid are oxidation of propylene by nitrogen oxides (15–20°C) followed by treatment with H_2SO_4; interaction of CH_3CHO with CO (200°C, 20 MPa). Polylactide is prepared by polymerization of optically active lactide in a solution at 100–150°C or in a melt at 140–200°C. Polylactide is used for various applications (Table 1.6). It is processed into disposable dishes, candy wraps, films, packaging, and medical devices. Cargill annually produces up to 6000 t polylactide prepared by fermentation of corn dextrose. The corporation plans to increase annual outputs to 50–150 thousand tons and reduce the price of polylactide from US$250 to US$2.2 per kg. The most successful large project has been implemented by Cargill Dow – a joint venture of agricultural Cargill Corporation and Dow Chemical, a leader in chemical production. Cargill Dow plays the leading role in the production of polylactic acid. NatureWorks, a subsidiary of Cargill, is the largest private company in the United States making polylactides at a plant in Nebraska (U.S.). The annual capacity of the plant being 140,000 tons polylactide, it now produces only half as much. In 2008, Cereplast, producing plastics from vegetable feedstocks, started production of PLA in Indiana (U.S.). The Biotechnology Department at Toyota has established a pilot plant annually producing 1000 t PLA. Hucail (the Netherlands) is completing the construction of a plant with an annual capacity of 50,000 t polylactide. In China, there is a Tianan plant producing PLA. Solvay (Belgium) intends to invest into the production of polyvinylchloride from bio-ethanol (sugar cane based renewable feedstock), with the 60,000 t/y capacity. The facility was supposed to begin operating in 2010.

Bioplastics are also represented by polyesters of alkanoic acids – poly-hydroxyalkanoates (PHAs), degradable linear polyesters of microbial origin. The first PHA, poly-3-hydroxybutyrate (P3HB), was discovered by Maurice Lemoigne, a French microbiologist, in 1925. P3HB is now one of the best-studied PHAs. P3HB is a thermoplastic high-crystallinity polymer. The discovery of the ability of microorganisms to synthesize PHA heteropolymers gave an impetus to PHA-related research. By now, about 100 structurally different naturally occurring and genetically engineered polyhydroxyalkanoates have been identified. Even an insignificant alteration of the fractions of monomer units in PHAs can change their properties significantly. ADM and Metabolix (U.S.) have announced construction of a plant for production of biodegradable PHAs with a 50,000 t/y capacity. Procter & Gamble Chemicals plans to start production of PHAs in Europe (Shishatsky, Shishatskaya, 2010; Shishatsky et al., 2010; Volova, Shishatskaya, 2011).

The use of biodegradable packaging materials and creation of favorable conditions for developing biopolymer industry is a global issue. In some EU countries, e.g., in Germany, companies that sell goods in biodegradable packaging materials are exempt from waste management taxes. Denmark and Ireland levy taxes on the use of polyethylene bags by retail networks. The use of non-degradable packages is partially or completely banned not only in the U.S. and EU countries, but also in Bangladesh, Singapore, Taiwan, and some states in India. Knesset in Israel has been discussing a law that would prohibit free distribution of PE bags in shops. In China, production of thin synthetic polymer bags and their use in shops have been prohibited since June 1, 2008. A similar ban has been introduced in Australia. In 2010, Italian government obligated manufacturers and retailers to produce and use only biodegradable packaging. In Los Angeles, 2.3 million PE bags used to be distributed in shops every year. The average usage time of one bag is 20 min, and, finally, it is buried in the landfill, where it will stay undegraded for several hundred years. To improve the environment in the city, since July 1, 2010, distribution of plastic bags has been prohibited in Los Angeles supermarkets, grocery stores, and clothes shops. Paper or other biodegradable bags should be used instead.

In Russia, however, there are no biodegradable polymers in commercial production. At the same time, several federal programs such as "Integrated Program for Development of Biotechnologies in Russia 2020," "Development of Pharmaceutical and Medical Industries in Russia: 2020 and Further," "Industrial Biotechnologies" (a subprogram of "Development of Industry and Enhancing its Competitive Ability 2020"), "BioTech 2030," and "Medicine of the Future" are aimed at establishing production of biodegradable materials for various applications, primarily, for medicine, in Russia.

Thus, although the market share of biodegradable polymers is still insignificant, things are changing. Growing environmental awareness does not allow increasing the global plastic garbage heap. In addition to that, processes of production of biodegradable polymer materials are being improved, leading to their cost reduction and making them more attractive to consumers.

1.3 THE NEED FOR NOVEL FUNCTIONAL MATERIALS IN MEDICINE

Development and testing of novel materials for medical applications, which are intended for interacting with the living environments and which are necessary for producing surgical elements and repairing damaged tissues and organs, pose a difficult challenge. To achieve medical breakthroughs, researchers need to have novel materials and advanced functional devices, including systems capable of reproducing biological functions of living systems. Biotechnological processes that can be used to synthesize a wide range of tailored biomolecules with various structures have enormous potential. In order to improve the effectiveness of treatment and quality of life, reconstructive medicine needs high technologies and novel, highly functional and specific, materials, including devices capable of reproducing biological functions of living systems.

One of the most recent approaches is creation of bio-artificial organs and tissues, which requires novel functional materials. It is especially important to have biocompatible materials that can be resorbed in biological media without releasing toxic products. Development of biocompatible and biodegradable materials for temporary devices is an even more urgent and complicated task, as after these materials and devices have fulfilled their task of repairing tissue defects, they must be replaced by biological structures in a definite time. The goal of global science today is to create and test novel biomaterials, whose production amounts are approaching those of pharmaceuticals.

Production of novel biocompatible materials and special biomedical devices based on these materials has recently become one of the leading lines of research and commercialization. Considerable achievements of medicine are based on the application of a wide spectrum of devices constructed using novel materials (artificial heart valves and ventricles, blood vessel prostheses, implants for the locomotive system, etc.). New biomaterials have been successfully used to design extracorporeal devices and implants, which have improved the quality of life and saved the lives of millions of people. The issue of creating novel medical-grade materials is growing in significance because of the alarming tendencies of the past few decades: population aging and the increasing incidence of chronic diseases and disabilities among working-age people. The treatment, rehabilitation, and social support of these people place a heavy burden on the federal budget, annually consuming a considerable part of health care financing. Thus, more effective and less costly approaches are needed to restore patients' health.

Development of novel medical technologies aimed at improving the effectiveness of treatment and quality of life needs high technologies and novel, highly functional, formulations, devices, and materials, including systems capable of reproducing biological functions of living organisms. The progress of medical technologies and health services suggests considerable advances in this socially oriented field.

Medical industry is a science-based high-tech industry. Its achievements are largely associated with results of technological change. The one-way interaction between medicine and technological change that was observed in the past has been gradually replaced by a more complex pattern of interaction. Now, medicine induces technological change through feedback mechanisms.

The major segments of medical industry are:

- the chemical pharmaceutical industry,
- the medical device industry,
- manufacture of medical products based on glass, ceramics, and plastics; materials for dentistry and prostheses, and
- cultivation of medicinal plants.

Between 2000 and 2010, the average annual increase in the globally marketed medical devices, instruments, and drugs was about 10%. The amounts of costs in the global market of medical equipment have exceeded US$ 240 billion, and this is higher than the costs in the global market of machine tools by a factor of 6. However, just a few countries – U.S., Japan, and EU states – purchase 82% of the global medical products (Figure 1.3). The same countries are the main manufacturers of medical devices and health products. In the U.S. alone, this industry employs over 411.4 thousand people. At the same time, all BRIC countries (Brazil, Russia, India, and China) annually spend only US$ 9.1 million on purchasing medical equipment, instruments, and drugs.

This disproportion is caused by the huge difference in healthcare investment on a per capita basis. In 2006, the U.S., EU countries, and Japan spent US$ 287, 250, and 273, respectively, per capita. Other countries spent about US$ 6 per capita, and African countries – not more than US$ 2.5. Russia ranks in the middle, purchasing medical products for more than US$ 20 per capita a year. It is the first among the BRIC countries (Figure 1.4) regarding investments on a per capita basis but not the total production and marketing of medical items (Figure 1.5).

The Russian market for medical and pharmaceutical products on a per capita basis is smaller than the corresponding markets of Hungary, Czech Republic, and Italy by a factor of 5–8, and it is even smaller – by a factor of 60–100 – than the markets of the leading countries. Moreover, the conditions for marketing medical products in Russia are not equal for Russian and foreign manufacturers. If a foreign manufacturer comes to the Russian market, they only need to register their medical product. If, however, licensed Russian manufacturers want to market their medical devices, they need to register them and have their documentation examined by experts, which usually takes at least 5 or 6 months.

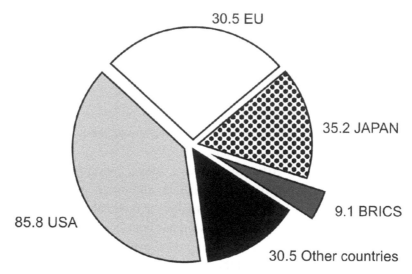

FIGURE 1.3 Medical product global market shares *(US$ billion)* (The data found at http://minpromtorg.gov.ru).

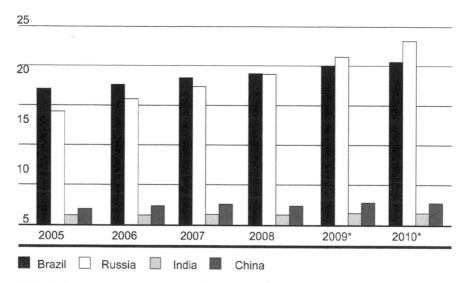

FIGURE 1.4 Average annual expenditures on medical products of the BRIC countries on a per capita basis, US$ (The data found at http://minpromtorg.gov.ru).

The structure of medical products manufactured in Russia is shown in Figure 1.6.

The main objective of Russian industry is to expand the share of products manufactured in Russia at home and abroad. The Russian Ministry of

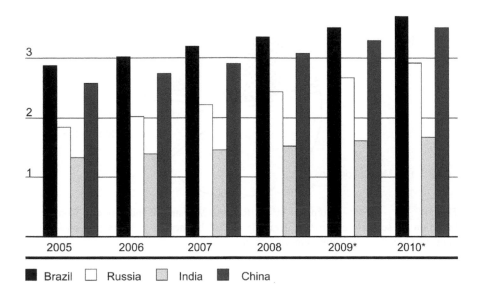

Brazil ■   Russia □   India ▨   China ■

FIGURE 1.5   Expenditures on medical equipment in the BRIC countries (US$ billion) (The data found at http://minpromtorg.gov.ru).

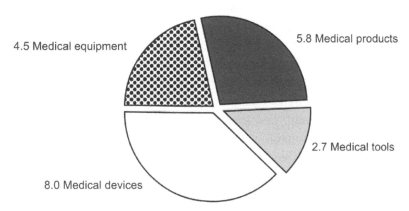

4.5 Medical equipment

5.8 Medical products

2.7 Medical tools

8.0 Medical devices

FIGURE 1.6   Manufacture of medical products in Russia (billion Rubles) (The data for 2010 found at www. minpromtorg.gov.ru).

Industry and Commerce has worked out a long-term strategy for developing medical industry and facilities for manufacturing medical products by other industries (Figure 1.7).

Federal control of the market for medical products is justified by the need for healthy working-age people, as the health of the nation is one of

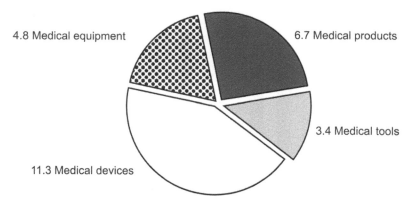

FIGURE 1.7   Forecasts prepared by the Russian Ministry of Industry and Commerce based on the scenario of strong federal support of the medical industry (billion Rubles) (The data found at http://minpromtorg.gov.ru).

the main indicators of the level of development of the country and state concern for people's welfare. The state should pursue flexible fiscal policies regarding this economic sector. At the same time, medical market is based on the general business laws, and it strives to get profit – by selling more medical devices, drugs, etc.

Development of biotechnology is a way to achieve considerable progress in manufacturing medical products, including novel materials and new-generation high-tech devices. The goal of biotechnology as a field of knowledge and a dynamically developing industry is to solve many of the major contemporary problems and, at the same time, maintain the balance in the "humans-nature-society" interactions. Biological technologies (biotechnologies) are based on using the potential of the living systems and are aimed at friendly and harmonic interactions between humans and their environment. Biotechnology has become one of the major lines in national economies of all industrially developed countries, and it is very important to increase competitive ability of biotechnological products in the global market.

Nowadays there exist five main groups in biotechnological applications, which have been identified by a color system: "white," "green," "red," "gray," and "blue" biotechnologies. White biotechnology is industrial biotechnology, which designs products that have been manufactured by chemical industry – alcohol, vitamins, amino acids, etc. – making them more energy efficient and less polluting than traditional ones. Green biotechnology is focused on agriculture as working field. Green biotechnological approaches and applications include producing pesticides to control pests and pathogens causing diseases in cultivated plants and domestic animals, creating biofertilizers, and increasing plant productivity by various methods, including genetic engineering. *Grey biotechnology* includes all those applications of biotechnology directly related to the environment: land reclamation, removal of pollutants from wastewater and gas emissions, industrial waste processing, and degradation of toxicants by using biological agents and biological processes. *Blue biotechnology* is based on the exploitation of sea resources to create foodstuffs, technical and bioactive products, and medicinal substances.

Red (medical) biotechnology is the most significant branch of modern biotechnology: production of this sector of global biotechnology reaches about 60%. Biotechnology makes a great contribution to the development of novel medical materials and devices. The global market for biotechnologies and its medical segment are growing at a very high rate. A few years ago, its cost in the U.S. was predicted to reach US$ 6 billion in 2010 and 10 billion in 2015. Leaders among pharmaceutical companies are Merck & Co, Astra Zeneca, Pfizer, Glaxo Welcome, Novartis, Bristol-Myers Squibb, Johnson & Johnson, Warner-Lambert, American Home Products, and Eli Lilly.

Biomedical sciences include fundamental, experimental, and clinical research, based on achievements of microbiology, molecular genetics, and proteomics. The main branches of biomedicine are:

- pharmaceutical nanotechnologies, including nanotechnological drug delivery systems;
- designing polymers and novel drug formulations;
- targeted drug delivery;
- synthesis of peptides and nucleic acids, technologies of extraction and transformation of bioactive compounds;
- biophysical technologies in medicine;
- new methods of analysis and diagnostics; and
- processes of cultivation of microorganisms and tissues producing physiologically active substances.

The global pharmaceutical market is growing very fast: 7–10% per year. This market is dominated by North America, Europe, Japan, and Latin America (up to 85%). As reported by IMS Health, North America earned US$ 170 billion, Europe – 101, Japan – 46, Latin America – 31, Southeast Asia (including China) – 20, East Europe – 7, Middle East – 11, Africa – 5, India – 7, Australia – 5, and FSU – 3 (IMS Health Annual Report, 2003). In Germany, at the end of the 20th century, there were about 300 biotechnological companies, whose sales returns reached about US$ 400 million (German Biotech, C&EN, 12 June 2000, http://www.gly-cotope-bt.com/en/services/Biotechnological_Service.php). In 2000, the market of European biotechnological products, including the sales of the leading pharmaceutical companies, reached US$ 160 billion (Risks and Rewards; European Chemical News, 25–30 April 2000).

Progress achieved in different branches of medicine is associated with the development and use of a wide range of new-generation products, devices, and drugs, based on novel natural and synthetic materials and composites, including bio-artificial systems and devices for biomedicine. Novel biomaterials have been used to design various extracorporeal devices and implants. Pump oxygenators have been employed in thousands of surgeries; several hundred thousand artificial heart valves and ventricles, up to one million small-diameter blood vessel prostheses and vascular stents, several dozen million implants for the locomotive system, hundreds of thousands of various implants for reconstructive surgery have been implanted. These devices have improved the health and saved the lives of millions of people (Shtilman, 2006; Hench and Jones, 2005; Biosovmestimost, 1999).

Injured or diseased tissues and organs can be repaired or replaced in one of the following ways:

- by transplanting organs and tissues;
- by implanting artificial organs and tissues;
- by tissue engineering methods.

Each of these ways has both advantages and disadvantages. The use of *autografts* – organs or tissues that are transplanted within the same person's body – are regarded as the "gold standard." Some disadvantages of this approach are that the amount of the person's own material is limited and that a separate surgical procedure is conducted to remove an organ or tissue. The amount of the material that can be transplanted from another person's body (*allografts, or homografts*) is also limited because of its restricted availability, high cost, and the necessity to take immunosuppressants for the rest of the patient's life. *Xenografts* (transplants of organs or tissues from one species to another) must be genetically or chemically modified to prevent transplant rejection.

Transplantation of organs is an effective and economically sound method of treating life-threatening diseases affecting the heart, kidneys, and other organs. However, because of the limited availability of the transplant organs, this approach cannot be used to help all potential recipients. As reported by the United Network of Organ Sharing, UNOS, in the U.S. alone, over 87,000 patients are waiting for organ transplants.

Organs and cells of animals (pigs, cattle, and other mammals) may be regarded as donor organs and sources of therapeutic cells (*xenografts*). The main obstacle to transplanting xenografts is immune response – hyperacute rejection. Genetic modification is the most promising approach to overcoming this obstacle. One way is to remove, e.g., the porcine gene encoding the enzyme responsible for rejection. Another possible approach is to add human genetic material to the porcine genome in order to disguise the pig cells as human cells. The potential spread of infectious disease through xenotransplantation is another limitation of this method.

Thus, there is not enough transplant material to help all patients that need new organs or tissues. Artificial organs are currently used only as temporary devices for maintaining the functions of vital organs for relatively short time periods. Therefore, it appears impossible to enhance the effectiveness of treatment and improve the quality of life without introducing new revolutionary technologies into medicine.

Implants, prostheses or artificial organs are used to repair or replace body parts much more often than transplants because they are more readily available and are seldom rejected by the body's immune system. However, it is important to remember that all implants and artificial organs are made outside the body, by a technological process (or processes). They have a number of limitations: no ability to self-repair and no ability for self-adaptation; their mechanical and biological properties are a compromise for the living tissues and organs being replaced. Thus, no artificial spare parts are as good as the living parts they replace. There will always be relative success and relative failure of implants, just as there is for transplants. There is a risk that an implant or transplant will fail and need to be replaced during the lifetime of the recipient.

Analysis of research in the field of transplantation and artificial organs that was performed at the end of the 20th and beginning of the 21st centuries suggests emergence of a fundamentally new approach to restoring the functions of damaged tissues and vital organs – *cell and tissue engineering*. The latest achievements in molecular and cell biology have provided a wide range of opportunities for creating fundamentally new effective biomedical technologies, which can be used to restore injured tissues and organs and treat serious human metabolic disorders (bio-artificial liver, kidneys, pancreas) (Sevastyanov, 1990; Ratner, 1993; Pişkin, 1997; Glick, Pasternak, 2010; Jones, 2005). Thus, a third way to circumvent the limitations of transplants and implants is to engineer tissues and organs. The objective is to use the patient's own cells or an immunotolerant 'universal' cell source to grow replacement tissues or organs *in vitro* (*tissue engineering*) followed by transplantation into the patient. Alternatively, tissue-engineered cells, with or without genetic modification, can be placed into a patient and stimulate regeneration of tissues and organs.

However, the development and improvement of the methods of regenerative medicine based on tissue engineering is impossible without creation of new, highly functional and specific, materials and construction of systems capable of mimicking biological functions of a living organism. Interaction of new materials with body tissues is both a subject of fundamental research and an issue of practical medicine. Studies of medical-grade materials are among the currently important lines of research, corresponding to the high level of development of science and technology, including the critical technologies. This is integrated research, uniting medicine, chemistry of high-molecular-weight compounds, biotechnology, biophysics, molecular biology, and cell biology. It has the following interrelated objectives:

- to develop novel materials and methods to modify and process them into biomedical items;
- to study the mechanism of interaction between biomaterials and blood and tissues;

- to investigate physicochemical and biomedical properties of bio-materials and products fabricated from them;
- to conduct experiments and clinical trials and use new materials and items.

Novel biomedical materials and biomedical devices fabricated from them have been increasingly studied and commercialized recently. The emphasis is placed on the search for technologies for designing bioartificial materials and organs – a system of materials of artificial or biological origin, which either include functioning cells of organs and tissues or induce regeneration of the corresponding cells at the implantation site. The need for resorbable materials with high biocompatibility is particularly acute. In spite of considerable investments into this branch of research in developed countries, scientists have not managed so far to create, for example, an artificial surface whose properties will be identical with those of the natural biological tissue. Artificial organs are currently used only as temporary devices for maintaining the functions of vital organs for relatively short time periods. Material that would be fully compatible with the living organism has not been created yet. Many branches of medicine today need biocompatible materials: general and cardiovascular surgery (prostheses of blood vessels, artificial heart valves, and cardiovascular assist systems), orthopedics, and dentistry. They would be also useful for designing new formulations of drugs, sorbents, etc.

Modern cardiovascular and transplant surgeries need materials suitable for long-duration interaction with blood. *Hemocompatibility* is the most important aspect of biological compatibility of biomaterials. Analysis of the literature suggests that materials and devices intended for interaction with blood must have the following biological properties. Hemocompatible medical devices must not have toxic, allergic and inflammatory effects; activate enzyme systems; adversely influence blood protein, blood cells, organs, and tissues; cause antigenic and carcinogenic responses and metabolic disorders; provoke infections and cause electrolyte imbalance; and change their medical and technical properties because of undesirable calcification and (or) biodegradation. Formation of thrombi and thromboembolisms induced by the surface of medical materials and devices indicates that they are not hemocompatible.

The necessity of creating biodegradable materials capable of mimicking the properties of biological structures is related to revolutionary changes that have recently occurred in medicine, which has seen new developments in transplant surgery and implantation of artificial organs, based on the fundamentally new approach to recovery of functions of vital organs. These are the use of processes of genetic, cell, and tissue engineering to design bioartificial (hybrid) organs and tissues and *in vitro* cloning of organs and tissues from the patient's own stem cells. The emphasis is placed on the search for technologies for designing bioartificial materials and organs – systems of polymer materials with functioning cells. Bioartificial systems are intended to completely or partially, temporarily or permanently, replace the functions of vital organs and tissues. The end goal is to recover the 3D structure of the tissue at the site of the defect. To direct organization of cells and facilitate their growth, proliferation, and differentiation during regeneration of the damaged tissue, cells are grown on scaffolds. This line of research is named biomimetics, and it uses self-monitoring, smart or intelligent materials. These materials are used in a wide range of applications. These are transdermal or implantable devices with controlled release of biologically active substances for pharmacological, cell or genetic therapies; devices with shape memory for orthopedic applications and cardiovascular surgery; biosensors; biodegradable devices for reconstructive surgery; biotechnological devices for separation, purification, and identification of biological structures at the molecular and cellular levels, etc.

Surgery is among the medical sciences that especially need novel biocompatible biomaterials for clinical use. For reconstructive procedures, researchers investigate not only materials and devices for cardiovascular system but also various surgical elements, endoprostheses, strong suture materials, devices and tools for abdominal surgery, traumatology, orthopedics, and maxillofacial surgery. New materials are needed to improve results of surgical intervention in abdominal surgery, to prevent postoperative ileus, to construct anastomosis and suture the intestine, and to reconstruct bile ducts. Traumatology, orthopedics, maxillofacial surgery, and dentistry face a serious problem of plastic surgery to correct defects of bone tissue that may form during surgical treatment of some bone diseases and injuries. Results of surgical reconstruction of bone tissue defects are largely determined by the effectiveness of reparative osteogenesis. Long-term clinical observations show that reparative osteogenesis in post-traumatic bone defects occurs slowly – for months and years, and that sometimes bone defects remain unfilled. Taking into account the high incidence of traumatic injury and the great number of orthopedic manipulations, traumatology, orthopedics, maxillofacial surgery, and dentistry need new, effective, methods, materials, and drugs for repairing bone defects. Proper materials are needed for plastic surgery of large bone defects that originate during chronic inflammations and after removal of cysts, chondromas, and tumors. Thus, it is important to continue searching for new effective osteoplastic materials for improving the methods of bone defect repair.

Biodegradable materials are also useful in pharmaceutical applications. The most promising and effective method of drug administration today is controlled sustained-release drug delivery. Such systems are especially necessary in transplantation medicine, when donors' or bioartificial organs are transplanted, for treating chronic infections, including posttraumatic and postoperative infections. The matrix used as a carrier for a drug must be made of proper material. Such materials must be perfectly harmless to the body, have certain physical and mechanical properties, and be biodegradable without forming toxic products; they must not inactivate drugs or mix with them in different phase states.

Thus, research and clinical use of novel materials and devices fabricated from them will contribute to progress in medicine and, ultimately, provide improved quality of life for patients.

1.4 REQUIREMENTS FOR MEDICAL GRADE MATERIALS

The range of materials used in medical applications is very wide, including natural and artificial materials: metals, ceramics, hydroxyapatites, synthetic and naturally occurring polymers, various composites, etc. A biomaterial is material that interacts with living systems. Biomaterials are used to fabricate medical items and devices. In spite of the considerable progress of biomaterials science, biomaterials are not produced in sufficient quantities, and no fully biocompatible material has been created yet.

Materials necessary for the development and improvement of reconstructive and transplant surgeries include various materials and composites with different functional characteristics and basic properties:

- materials biocompatible with the living system (materials that do not cause any adverse response being implanted in the body and staying in it for extended time periods. Today, these are silicon, Teflon, polycarbonates, polyglycolides and polylactides, polyethylene, titanium, etc.);

- materials with antithrombogenic properties (materials that can stay in contact with blood for long time periods, which are used to produce vascular prostheses, heart valves, artificial pericardium, and thoracic diaphragms);

- adsorbents (materials used in artificial exchange systems for organ replacement – kidney, lung, heart. Today, these are activated carbon, zirconium, ion-exchange resins, etc.);

- oxygen-carrying substances (fluorinated hydrocarbons, used to dissolve high oxygen concentrations, and systems based on encapsulated animal and human erythrocytes or chemical binding of high-molecular-weight substances with the erythrocyte heme);

- diffusion dialysis films (used to produce dialysis membranes, which selectively remove urea, creatinine, and other metabolic products from the body);

- fibrous materials (microporous materials with high exchange effectiveness, used in artificial exchange systems, e.g., vinyl acetate fibers (artificial kidney), silicon capillaries of artificial lungs);

- materials for microencapsulation (used to fabricate microcapsules of micron diameter for delivering drugs carrying oxygen);

- resilient abrasion-resistant materials (materials used to fabricate artificial bones and joints, heart valves. These materials must have mechanical and physical properties that would enable them to meet the mechanical load applied to them);

- bioadhesives for connecting living tissues (used to connect intestine fragments, blood vessels, bile ducts, etc.; these substances must take instantaneous effect, be resistant to the aggressive environments in the human body, release no heat or toxic substances);

- composite materials, including reusable ones (these materials may be constructed by varying combinations of homologous polymers and combinations of synthetic materials with metals and biopolymers with synthetic polymers or metals). This approach is used to produce materials with fundamentally new functional properties (Chench, 2000; Ratner, 1993; Shtilman, 2006; Hench, Jones, 2005; Biosovmestimost, 1999; Volova et al., 2013).

Thus, requirements to be met by biomaterials and the meaning of biocompatibility vary depending on the application of a material.

Modern cardiovascular and transplant surgeries need materials suitable for long-duration interaction with blood, i.e., not only generally biocompatible but also hemocompatible ones. Research addressing bio- and hemo-compatibility of materials has been going on for quite a short time – since the early 1960s, as part of research aimed at creating artificial heart. Studies of the end of the 20th century focused on creation of materials with tailored physicochemical and medical-technological properties.

A branch of medical materials science is biomaterials science – research of interactions between materials and living tissues and, based on this, construction of devices for different branches of surgery, including transplantation. One of the major objectives of biomaterials science is to develop and test novel, highly functional and specific, materials and construct material systems capable of reproducing biological functions of living organisms. A new class of biocompatible materials can mimic certain properties of biological structures. These materials are named self-monitoring, smart or intelligent materials. They are capable of changing their properties in response to changes in the environmental parameters (temperature, pH, osmotic pressure).

Self-monitoring materials are used in various applications, including devices with shape memory for cardiovascular surgery and orthopedic applications; biodegradable sutures, surgical elements, drugs; implantable devices enabling controlled release of bioactive substances for pharmacological and genetic therapies; biotechnological systems for identification, separation, and purification of various compounds at the molecular and cellular levels.

Construction of self-monitoring biomaterials is based on chemical and biotechnological methods and different approaches:

- synthesis of materials capable of modifying their properties in response to changes in the environment and containing bioactive compounds;
- construction of hybrid materials by methods of cell and genetic engineering (tissue engineering);
- creation of materials with surfaces modified specifically for direct contact with blood and tissues of human body;
- construction of materials based on processed and modified human and animal bio-tissues;
- development of biodegradable materials and composites with controlled biodegradation rates, including biopolymers produced by microorganisms.

The use of novel materials that are not only biocompatible and functional but also degradable *in vivo* is a special challenge – more serious than the difficulties arising during construction of materials and systems intended to function *in vivo* for long time periods or permanently. The temporary implants fill the defect of an organ or tissue in the body, and, then, in a definite time, they must be biodegraded and replaced by the new tissue.

The main obstacles to the wide use of biodegradable polymers in medicine are the rather limited variety of these materials and the problem of controlling the degradation processes in the body. Biodegradation products of these materials may be substances natural for the body and involved into cell metabolism, such as monosaccharides, lactic, glycolic, and β-butyric acids, or substances that are not metabolized by cells and tissues. The latter products must not be toxic, and their concentrations in the blood flow must be below the maximum permissible level. Therefore, the biocompatibility of the biodegradable material may be either passive, when degradation products are removed from the body without inflicting any harm to it, or active, when degradation products are involved into cell metabolic cycles.

Biodegradable materials and devices are used in various applications, including repair of tissue defects, sealing of parenchymal organs with polymer films, dressing of wounds and burns, production of sutures and elements for urology, orthopedics, dentistry, vascular and tissue engineering, and prolonging of drug release.

Among the new polymer materials that have already been developed or are being developed now for medical applications are aliphatic polyesters, polyamides, segmented polyester urethanes (PURs), polymers of lactic and glycolic acids (polylactides (PLA) and polyglycolides (PGA)), silicon, polyethylene terephthalate (PET), poly-β-hydroxybutyrate (PHB), and other polymers of hydroxy derivatives of alkanoic acids, the so-called polyhydroxyalkanoates, PHAs.

Any material intended to be used in biomedical applications must be harmless to the human body and functional. The mechanical/physical and biocompatibility properties of the medical-grade polymers vary rather widely, depending upon the specific functions of the material, its implantation site, lifetime, etc.

Materials implanted into the body must not:

- cause poisoning or allergic reactions,
- injure living tissue,
- be carcinogenic,
- be antigenic,
- cause destruction and degradation of proteins and enzymes,
- cause electrolyte imbalance,
- cause metabolic disorders.

Biocompatible materials to be used in contact with blood are named *hemocompatible materials*, i.e., materials compatible with blood. All hemocompatible materials are biocompatible, but not all biocompatible materials can be used in contact with blood. For example, metal ceramic composites or alloys used to prepare pins, nails and orthopedic or dental prostheses are biocompatible, but this does not imply that they would not induce changes in the blood cell composition.

Hemocompatible medical devices must not:

- cause toxic, allergic, and inflammatory responses;
- activate enzyme systems (the coagulation, fibrinolytic, and complement systems);
- produce adverse effects on blood protein, blood cells, organs, and tissues;
- be antigenic and carcinogenic;
- cause metabolic disorders;
- induce infections;
- cause electrolytic imbalance;

- change their medical and technical properties because of undesirable calcification and (or) biodegradation.

Formation of thrombi and thromboembolisms induced by the surface of medical materials and devices clearly indicates that they are not hemo-compatible. Therefore, sometimes, hemocompatible devices (materials, coatings) are named thromboresistant (athrombogenic), and non-hemo-compatible devices are named thrombogenic, although these are not syn-onymous terms. There is no exact correspondence between foreign body reactions of different components of blood. For example, a thrombore-sistant surface may not activate the blood coagulation system, but, at the same time, the complement system or the neutrophil response may be acti-vated. Thus, thromboresistance is only one aspect of hemocompatibility of the device contacting with blood, which is determined by its specific inter-action with the human body at the molecular, cellular, and systemic levels.

The relatively new multidisciplinary branch of biomaterials and transplantation sciences – tissue engineering – acutely needs specialized biocompatible materials. The goal of tissue engineering is the design of constructions that can be used to recover, strengthen, and enhance the functions of tissues. There are several approaches to the use of tissue-engineered constructions: infusion of the isolated cells, creation of materi-als inducing tissue growth, and implantation and culturing of cells on the scaffold – a bio-implant model. Materials used in tissue engineering must have a number of special properties. Products of construct degradation must not be toxic; the construct must retain its shape and be sufficiently strong until completely replaced by the host tissue. Material used to fab-ricate the construct must not elicit an immune response from the host; it must maintain the growth of cells and their organization into tissues. The implant must not prevent removal of cell metabolites.

The material implanted in the body must not:
- be mechanically destroyed or abraded,
- change the surface structure and configuration,
- be chemically transformed and decayed,
- be adsorbed and sedimented,
- be extracted.

Unfortunately, materials that would meet all of these requirements have not been designed yet. Nevertheless, consistent research will finally produce materials that will be literally biocompatible. No multifunctional and biocompatible materials can be created, though, without understanding the mechanism of material – body interaction at the molecular level. Therefore, the fundamental objective of biomaterials science is to gain insight into molecular compatibility of the material with biological structures of the body. The following questions need to be answered:

- What materials should be used?
- What factors will influence them?
- How will they change *in vivo*?

Devices fabricated from polymer materials and intended for medical applications must meet a number of biological safety requirements. First, physicochemical properties of the material are investigated; then, biological tests of the material and extracts of the material are conducted *in vitro* and *in vivo* in the following order: *in vitro* and *ex vivo* screening, short- and long-duration *in vivo* tests, and, finally, clinical trials. ISO 10993 standards are currently used for safe use evaluation of medical devices in the U.S., Russia, and some other countries (ISO 10993).

Thus, to increase life expectancy of the patients and enhance the quality of treatment, it is necessary to design new materials and improve the construction of devices and technologies of their use. Prior to clinical trials, the safety of all materials, devices, and therapies needs to be evaluated. The regulatory process required to ensure product safety consists of several steps. This process is called *technology transfer*. Technology transfer involves a series of paths, from research to market assessment and production. Each path has a different output and must have a separate budget. Effective development of a new healthcare product requires completion of all of the paths. Profitability often depends on the time and cost invested in each path. Time and expense of the paths are additive and must be accurately projected and followed if the technology transfer process is to succeed.

1.5   PERMISSION TO USE NOVEL BIOMATERIALS AND DEVICES IN CLINICAL PRACTICE

Medical-grade materials should meet very high safety requirements. Special approaches have been developed to achieve regulatory approval of a medical material.

In the U.S. and EU countries, there are special laws controlling production and marketing of medical devices. Classification of medical devices is based on the risk presented by each given device; it also determines the scale and level of their testing. A medical device is an instrument, apparatus, implement, machine, contrivance, implant, *in vitro* reagent, or similar or related article, including a component part, or accessory that is intended for use in the diagnosis of disease or other conditions, or in the cure, mitigation, treatment, or prevention of disease or intended to affect the structure or any function of the body, and that does not achieve its primary intended purposes through chemical action within or on the body and that is not dependent upon being metabolized for the achievement of any of its primary intended purposes.

Devices assigned to different classes are subject to different levels of control. Registration of medical devices is controlled by Food and Drug Administration (FDA) in the U.S., Medicines and Healthcare Products Regulatory Agency in the U.K., the Ministry of Health, Labor and Welfare in Japan; in the EU countries, the CE marking is obligatory for medical devices.

The definition of a medical device encompasses a wide variety of devices, from *in vitro* diagnostic reagents and simple bandages to very complex life support devices such as pacemakers and heart valves. The U.S. Food and Drug Administration has recognized three classes of medical devices based on the level of control necessary to assure the safety and effectiveness of the device. Class 1 devices are usually low-risk ones, and they may not present a potential risk of illness or injury; examples of Class 1 devices include *in vitro* diagnostic reagents and non-invasive products like bandages. Class 2 devices pose a higher risk to the patient and may include noninvasive or invasive, short-term devices, like catheters that are only placed for a short term, or some of the electronic devices that are used to test or to analyze a patient's health. EKG (electro cardiogram) and radiography equipment are Class 2 devices. Class 3 devices are high-risk ones; they are intended for long-term implantation. According to the laws of the U.S., a device cannot be commercially distributed until it is proved to be substantially equivalent to one legally in commercial distribution in the United States before 1976 [Notification 510(k)] or a premarket approval is obtained after the submission of clinical data to support claims made for the device.

The classification of medical devices in the European Union is outlined in Directive 93/42/EEC regarding medical devices. There are basically four classes, ranging from low risk to high risk. The European classification depends on rules that involve the medical device's duration of body contact, its invasive character, its use of an energy source, its effect on the central circulation or nervous system, its diagnostic impact or its incorporation of a medicinal product. Medical devices are part of the general legislation that regulates the quality and safety of all devices marketed in Europe. Certified medical devices should have the CE mark. The European Council issues the directives aimed at coordinating the rules relating to the safety and performance of medical devices in the EU. These directives contain the classification and other significant requirements that must be met by manufacturers and EU member states. The special directives also define the processes that must be used by those engaged in manufacturing, marketing, use, and control relating to medical devices. Class 2 devices in the EU are very similar to the FDA Class 2 devices in the U.S., but the EU classification has Class 2b, which includes implantable devices containing resorbable materials and other types of long-term implants. Class 3 devices are high-risk ones, whose failure could result in immediate danger to the patient, such as life-supporting long-term implantable devices. Examples of Class 3 devices include pacemakers, total hip and knee joint replacement systems, bone transplants, and artificial heart valves.

Until 1999, the Russian legislation relating to medical devices had been based on the laws of the USSR. State Standard R ISO 10 993.99 "Biological Evaluation of Medical Devices" was issued in 1999, and it was based on the international standard ISO 10993–99 on Biological Evaluation of Medical Devices, which is accepted in the U.S. and EU member states. In 2007, the Ministry of Health and Social Development of Russia issued Order No. 488 "On Approval of the Administrative Regulations of the Federal Control Service in Health and Social Development for Certification of Novel Medical Technologies." In 2013, new acts regulating permission and performance of clinical trials and certification of medical devices were issued in Russia.

The classification necessary for medical device regulation is based on potential risks presented by medical technologies:

- Class 1 – low-risk medical technologies, which are not included in Classes 2 and 3;
- Class 2 – medium-risk medical technologies, which immediately (surgically) affect skin, mucous membranes, and body cavities; therapeutic, physiotherapeutic, and surgical manipulations in cosmetic dermatology;
- Class 3 – high-risk medical technologies, which immediately (surgically) affect body organs and tissues (except Class 2 technologies); plastic reconstructive surgery; medical technologies involving cell technologies and genetic manipulation, organ and tissue transplantation.

Certification of novel medical technologies is performed by the Federal Control Service in Health and Social Development based on results of studies, tests and evaluations that confirm their effectiveness and safety. The effectiveness of a technology is defined as the achievement of the intended aim; safety is defined as the potential risk to patients, users, devices or the environment presented by the properly used medical technology weighed against its benefits. The Federal Control Service in Health and Social Development controls the observance of the acting regulations of clinical and biomedical research of new medical technologies.

The steps required to achieve regulatory approval of new medical technologies and devices include sanitary-chemical and toxicological-sanitary research and sanitary evaluation of the device; technical approval tests and medical tests of the device.

In Phase 1, material and extracts are evaluated by using state-of-the-art physicochemical methods (chromatography, mass spectrometry, etc.). Assays are performed (with isolated animal organs, tissue and cell cultures, etc.) to reveal possible toxic effects on cells and tissues. Toxicological studies are performed on laboratory animals. Material is selected based on results of the sanitary-chemical, cytotoxic, toxicological, and pathomorphological studies. These studies are carried out in certified laboratories and institutes.

Phase 2 includes toxicological-sanitary tests of prototypes of commercial products, taking into account processing conditions, sterilization conditions, storage conditions, etc. The same methods are used as in Phase 1.

If a device or material has been approved in this phase, it may be subjected to clinical trials.

Clinical trials also consist of several steps. In Step 1, the new medical device and technology are tested on 20–30 volunteers who have given their informed consent in the so-called local trials. Then, if approved by the supervisory bodies, limited clinical trials, which involve 300–500 volunteers, are performed. If successful, they are followed by the clinical trial involving at least one thousand individuals and performed in several clinics. Only after a reliable proof of the beneficial effect has been obtained, certification documents are prepared.

The Food and Drug Administration (FDA) was created by an Act of the United States Congress in 1906 through the passage of the Food and Drug Act. Over the years, the U.S. Congress has continually revised and refined the scope of the enforcement and oversight of the FDA. In 1938, the Congress passed the Food, Drug and Cosmetic Act, which set up and gave specific guidelines for what cosmetic and drug manufacturers were required to do to prevent the misbranding and adulteration of their devices. The Act in 1938 established the role of the FDA in overseeing and allowing the sale and distribution of drugs and cosmetics and established their authority to prosecute companies acting outside of the law. In 1976, a critical piece of legislation was passed known as the Medical Device Amendments. This established, in the U.S., clearer definitions as to what medical devices were, defined the classification process, and established ways for the medical device manufacturers to get products into the market. In 1990, the Congress went further in enacting the Safe Medical Devices Act. The 1990 Act increased control of the FDA by allowing the agency to require companies to submit performance data so that the FDA could better assess performance of devices. Manufacturers of Class 3 high-risk devices are now required to ensure a method of tracking a device from the start of manufacturing to implantation. In this way, if something goes wrong with a device after implantation it can be tracked to the patient and, if possible, retrieved and changes made. This revision in the medical device legislation also reclassified many of the devices that had been Class 3 devices to Class 2 devices.

In 1997, as a result of the EU issuing the Medical Device Directives, and with the recognition that the two largest markets for medical devices were operating under two different systems, Congress passed the 1997 FDA Modernization Act. This Act has moved FDA rules and regulations towards standardization throughout the world. With the creation of the EU in the 1980s, there was a move to allow for uniform regulations within its member states so that medical devices could be sold and distributed following one set of regulations. Landmark legislation and agreement was in the form of Medical Device Directive 93/42 EEC that allowed for the conformity, or the CE marking of medical devices. The EU Medical Device Directive created a mechanism that established notified bodies, i.e., organizations that would be responsible for granting the conformity mark, the CE mark, to medical device manufacturers. In this Directive, it was established that the manufacturers needed to comply with ISO 9001 quality system regulations and ISO DIS 13485 quality systems for medical devices.

The CE mark is the EU conformity marking that provides a basis for ensuring that a product will conform to certain essential requirements. The essential requirements for a CE mark as stated in the Medical Device Directive 93/42 are 40 items written into this directive that cover all known sources of danger to the patient. These 40 essential requirements are broken down into six major classifications. Essential requirements for a CE mark are: internal production control, EU type examination, conformity of type, production quality assurance, product verification, and full quality assurance.

Figure 1.8 summarizes the five paths required for successful commercialization of biomedical technologies and new materials.

The research path is often much longer than 2–10 years (minimum), due to the need for cell-based or animal testing of new materials. Results of research may be expressed not only as scientific publications and theses but also as patents. If it is possible to pursue the five technology transfer paths in parallel, then a cumulative time for a successful technology transfer process is between 8 and 10 years. However, if it is necessary to complete each path prior to commencing the next, the cumulative time is nearly doubled, to 14–16 years. Often, this is the case because the costs associated with patent protection and technology demonstration are usually considerably larger than research costs. New layers of management become involved in the decision-making process as one moves from

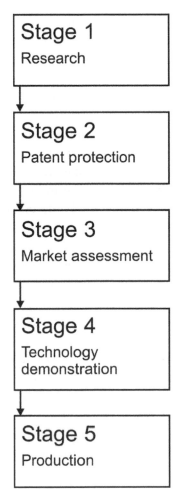

FIGURE 1.8   Paths necessary for the transfer of biomedical technology, involving the use of new materials and devices (Volova's figure).

research to patent protection, market assessment, technology demonstration, and production. The number of decision makers increases. Since the project costs increase from research to production, they finally become substantially higher than usually budgeted in universities or small companies. The major impedance step in the serial process is the transfer from path 3 (market assessment) to path 4 (technology demonstration). The level of financial commitment goes up by a factor of 10 at this point.

However, cost is not the only barrier in moving from path 3 to 4; additional personnel, management and facilities are equally important factors.

Usually, funds to pursue a Technology Demonstration project may be approved without completion of a marketing and business study. The marketing and business analysis will attempt to project cost/benefit ratios, capital required, size of market, time to reach the market, percentage of market penetration, competitive position of the new technology, lead times over the competition, etc. University and academic science does not have the staff or experience to make this analysis. Usually, a licensing agreement must be in place to move from path 2 to 3. Delays often occur because funds are seldom available to initiate path 3 until patents are issued, the end point of path 2, and license agreements are signed. This takes time and money.

The greater the advance of the new technology, the more difficult it is to make marketing and business assessments. Therefore, the greater the potential of the new technology, the greater is the risk and the longer is the time required to pass the judgment that it should be supported to enter path 4 and become a technology demonstration program.

The scale up from pilot plant to production (path 4 to 5) requires even more capital and market assessment. It is critical to know the size of the production facility required to achieve projected profit margins; however, production rates must be targeted towards sales projections. Thus, identification of profitable niche markets in the first years of scale-up is a key requirement for projecting the profit/risk ratio for the new technology. Large corporations have the expertise to make these assessments, but their large overheads inflate the required margins of profits to be a successful business. Small companies, conversely, have low overheads, but often lack the ability to assess accurately the multiple factors involved during the transition to production.

Rapid technology transfer depends upon several key factors:
- The researchers responsible for creating the technology (path 1) and the patents (path 2) must participate in the early steps of path 3 (market assessment) and path 4 (technology demonstration).
- Transitions between paths 1–2–3 must be quick and efficient.
- License agreements need to be flexible to allow rapid pursuit of paths 2, 3 and 4 simultaneously even though information to conduct the market assessment (path 3) is incomplete.

- Milestones with specific timelines need to be agreed upon by the research, management, marketing and product development teams responsible for paths 3 and 4. These timelines need to be coordinated with budgets and reward incentives need to be built in to ensure the effort is made to meet the milestones.
- Sufficient capital must be available to pursue paths 3 and 4. The capital should be available in stages that correspond to performance milestones for each path. Tying budgets to performance is one way to ensure that the capital will be properly targeted towards the output of the paths instead of being used to continue research.

One of the easiest ways to delay the technology transfer process is to continue to pursue research in path 1 with limited capital resources instead of moving the program into paths 3 and 4. This often happens in universities and academic institutes because salary and promotion depend on publications, the primary output of path 1.

When technology is created within a corporation, there is usually a management structure in place to make the decisions and budgets for paths 2, 3, 4 and 5. Milestones and timelines are often imposed as part of the job requirements of the teams involved. In contrast, when the technology is created within a university or a government laboratory, there is seldom a management structure or budget for paths 3 and 4. The approach to technology transfer is typically a license agreement with a company. Each path of technology transfer has a different degree of risk, time and personal capital investment associated with it.

1.6 MODERN MEDICAL GRADE MATERIALS

Materials intended to come into contact with living systems and used to fabricate medical devices are named biomaterials. In spite of the considerable progress of biomaterials science, biomaterials are not produced in sufficient quantities, and no fully biocompatible material has been created yet. Medical grade materials must be biocompatible with the living system. They must not cause any adverse response being implanted in the body and staying in it for extended time periods (Biosovmestimost, 1999; Shtilman, 2006; Hench and Jones, 2007; Sevastianov and Kirpichnikov, 2011).

Traditionally, most of biomaterials for defect repair have been derived from tissues of humans and animals (bones, skin, tendons, meninges, etc.). However, these routinely used technologies have very many limitations: from the immunological barrier and shortage of material to the declines in the quality of life. The preferred transplants are autografts, which are prepared from a patient's own tissue, thus circumventing the major immunological complications and most of the infections that may be caused by transplantation. However, harvesting autograft tissue creates a second surgical site, increasing the time of surgery, blood loss, and the time of the patient's hospital stay. In about 20% cases, harvesting autograft bone causes such complications as vessel and nerve injuries, hematomas, chronic neuropathic pains, and infections. The size of the autograft must not exceed 20 cm^3. Should more donor material be harvested, the risk of complications would increase. Moreover, autografts are resorbed too fast, before complete bone defect repair takes place. Fresh-frozen, freeze-dried, demineralized, formalized, and poorly differentiated bone tissues are the most widely used autografts in clinical practice. These materials must be prepared immediately prior to transplantation, or the hospital must have a bone bank. Actually, only very large hospitals can afford to have such banks because of the high cost of preparing and keeping bone graft materials. However, there is often an insufficient supply of autograft bone available, and the patient is subjected to serious surgery. These are the limitations for the wide use of autografts.

An alternative source of the donor material for bone replacement is the use of allografts (often called homografts): soft or hard tissues harvested from another patient or human cadaver. The benefits of allografts are shorter surgery time, the possibility of pre-storage and modeling of the implant, and the practically unlimited availability of the grafted material. The disadvantages are that bone allografts are slowly incorporated into the bone; there is a risk of transfer of bacterial or viral diseases from the donor to the recipient; allografts may precipitate tissue incompatibility immune response and development of granulomatous inflammation; bone allografts are costly; there may be religious barriers to the use of allografts. To minimize the risks, allografts are subjected to severe pretreatment, which significantly decreases the osteoinductive properties of the implants (weak osteoconductive properties are usually revealed only in the frozen and freeze-dried allografts of spongy bone) and their mechanical properties (almost by 50%). The risk of infection, though, is not eliminated altogether.

Xenografts are prepared by using special tissue harvesting techniques and by carefully selecting animals to ensure the patient's safety; the xenografts are also subjected to chemical and physical treatment, which increases the cost of such materials and reduces their availability. The risk of prion diseases, however, cannot be eliminated by preserving the animal biomaterial in formalin, and, thus, the use of xenografts is forbidden in a number of countries. Allografts and xenografts have the following common limitations: slow incorporation into the bone structure, the risk of transmitting different diseases from the donor to the recipient, the possibility of tissue incompatibility immune response and development of chronic granulomatous inflammation, the high cost of bone allografts, and religious considerations. To minimize the risks, allografts, and xenografts are subjected to severe pretreatment, which significantly decreases the osteoinductive properties of the transplants and their mechanical properties (almost by 50%), without completely eliminating the risk of infection. Therefore, in spite of successful experiments and clinical trials, these materials are not able to meet the needs of practical medicine and do not hold considerable promise.

Metal, ceramic, and polymer implants are more widely used in reconstructive surgery.

1.6.1 METALS

Metals are used in orthopedic applications, where their high mechanical strength is an important property, in internal electrical devices, and in artificial organs. The following factors influence the choice of metals or alloys for medical applications: (i) biocompatibility, (ii) good physical and mechanical properties, and (iii) the absence of adverse reactions caused by material aging. The most widely used metals are stainless steels and titanium and its alloys with other metals. Nitinol (nickel titanium), the shape memory metal, is also beginning to find applications in different devices and implants. Small quantities of noble metals, such as platinum and gold, are also used to fabricate chemically inert prostheses.

Metals are the material of choice for orthopedic applications because of their high yield strength and toughness. Metallic implants and devices, which are widely used to consolidate the bones by intramedullary and extramedullary metal-osteosynthesis techniques, ensure a favorable outcome of the treatment in just 80% cases. The use of compression-distraction systems is not always effective either. When the system stays for a long time on a limb, one of the usual complications is inflammation of soft tissues along the pins, which may lead to osteomyelitis, phlegmon, and even sepsis. Joint stiffness is another complication, and the most frequent is extension contracture of the knee and ankle joints. Aneurisms and arteriovenous fistulas sometimes, although not very often, form as a result of vessel injury by pins or vessel wall erosion. Although external fixators have been greatly improved over the 100 years of their history, they cannot fully support the mechanics of osteogenesis. Metallic (steel) pins, rods, plates, etc. interact with the surrounding tissues, and this causes such complications as metallosis, autoimmune responses, aseptic inflammation, release of toxic ligating components such as nickel, chromium, etc. from the metallic construct. Recently, trauma surgeons have preferably used titanium/nickel systems with a decreased nickel content. Many researchers report that nickel causes aggressive response of the local cellular immunity. Surgical steel has been used widely and successfully, but, as many patients are allergic to the metal, researchers have tried to coat the metal constructs with bio-inert materials.

The goal of the new (30–35 years old) branch of materials science is to enhance the biocompatibility of metals by modifying their surface, as a way to circumvent unfavorable effects of the use of metallic implants and systems in orthopedic applications. Titanium constructs are coated with a protective oxide film to reduce tissue response at the implant site. However, the use of such constructs for healing the fractures in the cortical bones or skull flat bones has a number of limitations, because of the insufficient shear strength of the oxide layer and because it does not interact with the bone tissue. Thus, the implant is not properly integrated into the bone, causing various complications.

The use of metallic implants and devices does not reduce the risk of infection, which is a serious challenge. In the case of open fracture, the likelihood of bacterial infection increases to 5–33%. It is very difficult to treat chronic osteomyelitis and so-called implant-associated osteomyelitis, which develops at the orthopedic implant site (joint prostheses, connecting rods, pins, screws, etc.). The weak point of using metallic orthopedic implants in reconstructive surgery is that it is necessary to prevent infection at the tissue/implant interface. Anti-inflammatory and antimicrobial drugs need to be delivered to this site and their concentrations need to be maintained at a specified level over a long time.

1.6.2 CERAMICS

Ceramics have been increasingly used in tissue regeneration applications because they combine bioactivity and mechanical strength. Their major advantages are compatibility with bone tissues, mechanical strength, and suitability for fabricating construction implants for the skull and spine. Ceramics used in medical applications are called bioceramics. Examples of bioceramics, which are mainly used for bone tissue regeneration, are tri-calcium phosphate, hydroxyapatite, calcium aluminates, bioactive glasses and glass-ceramics. Depending on the type of response in the body, bioceramics can be classified as bio-inert, bioactive and resorbable.

Materials with good load-bearing capacity – synthetic and natural calcium-phosphate materials (hydroxyapatite and di- and tri-calcium phosphates) – are widely used to repair bone defects. The use of hydroxyapatite-based implants in the form of cement is indicated if the size of the defect is below 30 cm^2. To repair defects of larger areas, implants are reinforced with titanium mesh. One of the advantages of hydroxyapatite-based implants is their nearly complete biocompatibility. In the case of small defects, hydroxyapatite is completely resorbed and replaced by the bone tissue in 18 months. In the case of large defects, the edges of the implant grow into the bone, and part of the material becomes resorbed, while the central part of the implant remains intact. The use of hydroxyapatite reduces the risk of infection. The composition of this material is similar to that of the human bone tissue, and it induces biological reactions similar to those in bone remodeling. During resorption, degradation products of calcium-phosphate materials (calcium and phosphate ions) are naturally metabolized without inducing an increase in the level of calcium or phosphates in urine, serum, or any organs. The main properties of ceramics are biocompatibility, high hardness, insulating properties of heat and electricity, and their heat and corrosion resistance. A common characteristic of all ceramic materials is that they are subjected to high temperatures (above 500°C) during manufacture or use. One of the disadvantages of ceramics, which limits their use in medicine, is their brittleness. Ceramics consist of inorganic and organic compounds. Ceramics are usually polycrystalline (granular) structures or mixtures of two or more crystalline phases. The size of crystallites in different types of ceramics varies widely (1–1000 μm). Porosity may be fine or coarse, open or closed. The elastic properties of ceramics determine their mechanical behavior and are closely related to the crystal structure. Glass-ceramics represent a special class of ceramics. They are polycrystalline materials having fine ceramic crystallites (of size < 1 μm).

Hydroxyapatite is used in the form of powders with different particle sizes, which can be processed into paste, granules, porous blocks, and composites with different substances. Powdered hydroxyapatite is not the best form for clinical use because hydroxyapatite particles are distributed in the bone wound non-uniformly and may be partly removed, during the postoperative period, together with the blood flow and wound exudate. The present-day research is focused on preparation of hydroxyapatite for filling bone defects in the form of ceramic granules or porous implants. Hydroxyapatite has the following advantages: it is easy to sterilize; it has a long shelf life; it is highly biocompatible and relatively slowly resorbed. The major disadvantage of granular hydroxyapatite is that being implanted into the bone tissue, it does not induce sequential remodeling and strengthening of the bone. Therefore, granular hydroxyapatite is generally used as a filler, while mechanical stability is achieved in other ways.

One of the modern lines of research is design of hydroxyapatite composites with other materials and substances. One of them is a composite of granular hydroxyapatite and water-soluble gelatin. The latter, however is resorbed in a week after implantation; it does not optimize osteogenesis but serves as shape generating component. Successful experiments have been performed with hydroxyapatite/glycolic acid and hydroxyapatite/fibrin composites. Hydroxyapatite has been blended with calcium sulfate and the osteoinductive component of the growth factor isolated from bovine bone. Antibacterial formulations have been prepared by linking antibacterial drugs with ceramics, e.g., Lincogap – a composite of the ultrafine hydroxyapatite powder and lincomycin and gentamycin. A number of fundamental studies proved that hydroxyapatite has osteoinductive properties (Urist, 1965). However, osteoinduction and osteoconduction of calcium-phosphate materials remains a complex and insufficiently studied process. What calcium-phosphate material should be chosen to repair a specific defect of bone tissue is not clear yet. The main common drawback of these materials is a decrease in the activity of calcium phosphates in the physiological environment because of the high temperatures (over 1200°C) necessary for their synthesis. Moreover, despite the high compressive strength, ceramics are very brittle and have low tensile strength and elasticity.

Synthetic osteoplastic materials such as calcium sulfate, bioactive glass, cyanoacrylate esters, polyacrylamide gel, tricalcium phosphate, sodium alginate formulations, etc. have been tested in experiments and clinical trials. Porous glass-ceramic bioimplants have been produced and studied. Bioactive glasses are among the best-studied bioceramics used in bone tissue engineering applications. They are divided into two main classes: glasses based on silica and glasses based on calcium phosphate. Silica-based glasses, such as bioglass, consist of SiO_2, Na_2O, CaO, and P_2O_5, with SiO_2 below 60 mol.%, high Na_2O and CaO concentrations, and the Ca/P ratio higher than 1, in order to decrease oxidation and solubility of the material (Ulery et al., 2011). The biological activity of these materials is accounted for by the presence of a highly hydrated silicate layer, which forms when the material interacts with human plasma. There are literature data supporting the high osteogenic potential of bioglass. However, its brittleness and low viscosity are obstacles to using it for making fixing systems and regeneration of large bone defects. Modern osteoplastic materials based on bioglass and calcium phosphates, such as Interpore-200 and Interpore-500, Calcitite-2040, Ostrix NR (U.S.), Ceros-80 (Switzerland), Osprovit 1,2 (Germany), BAK-1000 (Russia), and Bioapatite (France), are both bioactive and mechanically strong materials. However, their use is limited to dentistry applications because although they are highly osteoconductive, their osteoinductivity has not been proved yet. Natural corals and silk, whose structure and mechanical properties make them suitable for carrying bone marrow cells, have been studied as materials for bone tissue regeneration (Velema, 2006; Sviridova, 2010), but it is still unclear whether these materials are osteoconductive.

Developments of bioresorbable ceramics present complications, mainly due to problems related to matching the rate of resorption with the replacement by the natural host tissue. The rate of tissue growth varies from patient to patient and with tissue type. The mechanical properties of bioceramics, in particular their low fracture toughness, are disadvantages for their direct use in medicine. Usually, bioceramics are combined with metals, effectively forming composite materials with enhanced mechanical properties and elastic constants matched to those of bone.

1.6.3   POLYMERS

Polymers are used widely for medical devices and implants. The diversity of synthetic and natural high-molecular-weight polymer compounds used in medical applications is growing steadily. The widely varying stereoconfiguration and molecular weight of the polymers and the feasibility of producing various composites with different materials provide the basis for designing a wide range of novel materials that would have new valuable properties. Early medical applications of polymers dealt primarily with their use as materials enhancing the properties of medical items: glassware was replaced by elastic and unbreakable polyolefin plasticware, and nonwoven materials came into wide use. Present-day medical applications of polymers are very wide. They are used to produce portable devices, clinical equipment and instruments, items for personal hygiene, equipment for medical analysis, artificial organs (kidneys, blood vessels, valves, pacemakers, heart-lung machines), materials for dentistry, etc. Eucomed, a European medical technology organization, estimates the total market of medical technology at €184 billion, of which about 80 billion is the U.S. market size and 20 billion – the Japan's market size. In Europe, medical device market grows at the highest rate, reaching €55 billion. European countries annually spend about 8.4% of their GDP on healthcare. Germany spends 11% of its GDP on medical services. Plastics are used in medical devices together with metals, glass, ceramics, etc. The use of polymers in medical applications opens up great profit possibilities. Specific properties of the materials, their flexible designs, and, what is very important, low processing costs, are obvious advantages of polymers over conventional materials. Development of new technologies in reconstructive medicine has given an impetus to the broadening of the range of polymers to be used for constructing various implants and prostheses. Many of the polymers used in medical applications are produced on a large scale by the chemical industry. Requirements that the polymers must meet to be used in medical applications are different from those for general-purpose plastics. Medical grade polymers must be high-purity materials, which must not contain even trace amounts of substrates, catalysts, and technological additives. Medical grade polymers may be synthetic and naturally occurring materials, bio-inert (non-degradable) and degradable (bioresorbable) polymers; highly crystalline thermoplastic polymers and rubber-like elastomers.

Bio-inert synthetic polymers are neither hydrolyzed in liquid media nor degraded by blood and tissue enzymes and cells. They are intended to produce implants and long-term devices. They include polyethylene, polypropylene, polyethylene terephthalate, nylon, polytetrafluoroethylene, polymethylmethacrylates, etc.

The portion of plastics used to manufacture medical devices is growing. Investigation conducted by the commercial magazine Kunststoff Trends showed that 1.8 million tons of plastics had been used globally for this purpose. General-purpose plastics such as polyethylene, polypropylene, polystyrene constitute 80% of the total plastics used for constructing medical devices. The other 20% are special-purpose plastics such as polycarbonates, polyamides, and polyurethanes. The amounts of polycarbonates used in medical industry approach 100,000 t/y. They are used to manufacture lenses for eyeglasses, inhalers, pressure-resistant ampoules and other containers for medical substances, and parts for dialysis machines and blood oxygenators.

Polyethylene (PE) $[-CH_2-CH_2]_n$ is a hydrophobic and bio-inert material. The low yield strength of polyethylene limits its applications. There are three commercial types of PE: low density, high density, and ultra high molecular weight PE. Low-density polyethylene (LDPE) is a wide-spread material, exhibiting high biological inertness, with a molecular weight of 50–200 Da. High density polyethylene (HDPE) is a more crystalline material. Porous HDPE produced by Porex Surgical, Inc. (U.S.) under the trademark MEDPOR® is used to manufacture implants. LDPE and HDPE are readily moldable. Ultra high molecular weight polyethylene (UHM-WPE) has a molecular weight over 10^6 Da; it is non-moldable and has to be sintered prior to processing; then the material is processed mechanically to get a desired shape. UHMWPE is a partially crystalline polymer with a degree of crystallinity of 45–60%. Because of its high molecular weight, this PE has enhanced mechanical properties, which are in many respects better than those of LDPE. The tensile strength of UHMWPE is 30–44 MPa and its hardness reaches 4 MPa, while LDPE is characterized by values that are 2.0–2.5 times lower. The wear rate of UHMWPE amounts to 18 min/mm³. UHMWPE is produced under different trademarks (Hostalen GUR, Hercules H, RCH-WOO, Hifax-1900, Spectra 900, Hylamer M, etc.). UHMWPE is used as a basis for constructing composite materials with enhanced mechanical properties. These materials are used to fabricate vascular grafts, to regenerate bone tissue, and as fixing systems.

Polyamides are polymers that contain a –CONH– linkage. Traditional polyamides [poly(ε-caproamide) and poly(hexamethylene adipamide)] have been used to prepare fibers, films, and meshes for medical applications. Lower aliphatic polyamides are capable of biodestruction, and products prepared from them are biodegradable. Nylon 6, Nylon 6,6, and Nylon 6,12 are polyamides used in medicine. Polyamides are currently being replaced by polymers of hydroxycarboxylic acids and polyurethanes. Another polyamide used as a construction material for a long-term stay in the body is polyamide-12, which is produced in Russia and prepared by polymerization of the corresponding lactam: it shows low water uptake and higher resistance to biodegradation than poly(ε-caproamide) and poly(hexamethylene adipamide). Poly(p-phenylene terephthalamide) is synthesized from p-phenylene amide and dichloroanhydride of terephthalic acid in a hexamethylphosphoramide solution. This polymer is manufactured in the U.S. by DuPont de Nemours & Co., Inc. under the trademark Nomex, in Japan – as Conex, and in Russia – as Fenilon. Poly(p-phenylene terephthalamide) fiber under the trademark Kevar is produced in the U.S. This fiber shows good mechanic properties, chemical stability, and heat resistance. Polyamides that may be suitable materials for fabricating implants are polyvinylpyrrolidone (–NH–(CH$_2$)$_3$–CO–) and polyamide polyamines, which are prepared by various techniques: they can be used as components of the inner layer of non-thrombogenic vascular implants and as binders of different drugs.

Poly(methyl methacrylate) is a bio-inert, hard, rigid, glassy, brittle polymer, with glass transition temperature of about 100°C. Methacrylates are thermosetting plastics of medium strength and hardness, which were originally used only as dental cement or as cement for fixing joint replacements (hip, knee, etc.). However, these replacements were often unstable and misshaped. Until recently, polymers based on acrylic acid (acrylates, methacrylates) were used in surgical regeneration of skull bones. Methacrylates were used in 73% of all cranioplasties. These materials are easy to mold into any shape and have relatively low cost, but the thermal effects during polymerization cause additional defects of bone tissue and may induce infection (e.g., in the case of open paranasal sinuses). The use of polymethacrylates has serious limitations. For instance, Protacryl is prepared by mixing a monomer powder and a liquid (plasticizer), which form clayey mass, which hardens during polymerization reaction. Protacryl and other methacrylic acid-based polymers are potent allergens (they must not be used in children and individuals with allergies). Moreover, it is quite often resorbed in a number of years, and the presence of this polymer in the body may induce inflammations (purulent fistulas, empyema). The surgeon needs to shape a graft very quickly, before the mass has hardened, and constantly pour liquid on the implant, to cool it, as the polymerization reaction releases heat, which may both damage the adjacent brain and injure the bone tissue. Furthermore, the process may release toxic and carcinogenic substances. In summer 2014, the approval of clinical applications of these materials in Russia was withdrawn.

Poly(tetrafluoroethylene) (PTFE), in contrast to other linear polymers, is a most inert polymer, and, thus, is can be used in reconstructive surgery, in spite of its poor mechanical properties. The drawbacks of PTFE include its hydrophobic properties, poor cell adhesion to its surface, and complete inability of the material to be integrated with the bone without forming a connective tissue interlayer.

Polypropylene is a high-crystallinity homopolymer, brittle at low temperatures; it has good thermal insulation properties; polypropylene thin films are translucent. Polypropylene is a thermoplastic polymer with the melting point about 180°C. Polypropylene is resilient, bend-tolerant, and durable. It is used as a basis for preparing surgical sutures and parts of bone grafts. This polymer is prone to oxidative destruction, and the products yielded in this process exert an adverse effect on the adjacent tissues; therefore, polypropylene implants are surrounded by fibrous capsules. Porous polypropylene with pore size between 125 and 350 μm is produced by Porex Surgical, Inc. (U.S.).

The best known poly(p-xylylene) (Parylene) is 4-fluoro-poly(p-xylylene) (Parylene AF-4). It has unique properties and can be used to prepare strong, water-resistant, and bio-inert coatings for metals, polymers, fabrics, etc. Parylene coatings are used to protect and seal technical and medical products. Protective properties of a 20-μm layer of parylene are equal to those of a 100–200 μm layer of lacquer. Parylene N, produced by ParaTech Coating, Inc., Vitek Res. Corp, U.S., and exhibiting high biocompatibility, is used as a coating for metallic implants.

Polysiloxanes (silicones) are widely used in medical applications. Silicones can be elastomers, gels, lubricants, foams, and adhesives. Polysiloxanes are very stable and inert chemical substances. They are low-absorption materials intended for long-term use. Polydimethylsiloxane usually needs to be modified to enhance its properties. Polysiloxanes are used in breast implants; polysiloxane adhesives cure and crosslink on contact with water, e.g., Silastic® Medical Adhesive type A, and they can be used to affix materials to the skin.

Polyurethanes constitute one of the main groups of polymeric materials used to fabricate various implants and medical devices. Most polyurethanes for medical use are two-phase block copolymers. Commercial segmented polyurethanes may be of different chemical structure, including biodegradable ones. The polyurethane under the trademark Cardiothane is manufactured in the U.S., and its Russian analog is Siluram. Tecoflex™ (Thermedics, Inc., U.S.) is synthesized from polyether and cycloaliphatic dicyclohexylmethane-4,4'-diisocyanate; Biomer (Ethicon, U.S.) is prepared from diisocyanate, poly(tetramethylene oxide), and ethylenediamine. Segmented polyurethanes are produced by CardioTech International, Inc., U.S. These are biodegradable polyurethane elastomers ChronoFlex® AL, C, and AR, which contain flexible polycarbonate blocks; Chrono Thane™, consisting of aromatic fragments and polyether blocks; Hydrothane™ –

a hydrophilic thermoplastic polyurethane. Some of these polyurethanes have varieties differing in hardness. Segmented polyurethanes Gemotan are produced in Russia. They are synthesized from poly(tetramethylene oxide), ethylenediamine, and aromatic diphenylmethane-4,4'-diisocyanate. Polyether urethanes have good hemocompatibility and are one of the preferred polymer types for blood contacting devices.

Polyvinyl alcohol is seldom used in reconstructive processes because it is quickly degraded *in vivo*, being a highly hydrophilic and water-soluble material. In some cases, PVA crosslinked by glutaraldehyde (marketed under the trademark Gelevin) can be used alone and in composites as sorption material for absorbing wound exudate.

The cost of synthetic materials is usually lower than that of naturally occurring polymers; they are easy to mold into 3D structures; raw materials are readily available; polymer properties such as mechanical strength, degradation rate, and microstructure can be tailored to particular applications. Major disadvantages of synthetic materials are their unpredictable interactions with cells and immune system components of the body and uncontrollable time of *in vivo* biodegradation. The main reasons of complications developing in the case of using synthetic biodegradable materials in medicine are possible adverse responses of the body (inflammation and allergy) to degradation products and carcinogenicity of the materials.

Designing bioresorbable polymers, both synthetic and natural, is a special challenge – more serious than the difficulties arising during construction of materials and systems intended to function *in vivo* for long time periods or permanently. The temporary implants fill the defect of an organ or tissue in the body, and, then, in a definite time, they must be biodegraded and replaced by the new tissue. Hence, the rate of polymer degradation must match the rate of tissue or organ repair. Degradation products must be removed from the body, without adversely affecting it.

When certain synthetic polymers are exposed to liquids, they may swell or dissolve. Polymers that are produced by a condensation route are prone to hydrolysis, which is the reaction with water to produce –OH bonds. In addition, polymers may contain side groups that are capable of being hydrolyzed. The rate of hydrolysis is dependent on the water absorption of the polymer and is often limited by the diffusion of water through the polymer. Diffusion of water in polymers is related to their solubility parameter, their glass transition temperature and their degree of crystallinity. Polyesters based on $(-R-COO-)_n$ are often susceptible to hydrolysis. Degradation is pH dependent.

Polyesters of monocarboxylic acids are regarded as the most promising biodegradable materials. Chains of these polymers consist of repeating units of short-chain-length acids, the most widespread of which are lactic (PLA) and glycolic (PGA) acids. Polymers of lactic acid are often used in combination with polyglycolic acid and poly(ε-caprolactone) (PCL). Polymers of lactic and glycolic acid passed the regulations of the United States Food and Drug Administration (FDA) in 1970. These polymers are widely used as sutures, reinforcement structures, meshes for hernia repair, as materials for reconstructive and regenerative medicine – for repairing bone and cartilage defects, as wound dressings.

Polyglycolide (PGA) is based on $(O-CO-CHR)-n$ where R = H. PGA is a hard, tough crystalline polymer melting at about 228°C with a glass transition temperature of 37°C. The molecular weight of the polymer is determined by the conditions of synthesis and processing, ranging between 2×10^4 and 1.45×10^5. PGA is synthesized from glycolide under the influence of metal salt catalysts by a ring opening polymerization. The polymer cannot be dissolved in most of organic solvents. As PGA is both mechanically strong and elastic, it is used to prepare various constructs: from nonwoven spongy materials to plates and screws for fixation of bone fragments. The material is not cytotoxic, favoring cell attachment and facilitating cell proliferation. However, PGA implants retain their mechanical strength for a short time only, losing up to 40–50% of their mass in about 20 to 30 days.

Polylactides are based on $-(O-CO-CHR)-n$ where R = CH_3. Replacing H by CH_3 leads to a more hydrophobic polyester and results in lower water uptake and lower hydrolysis rates. Having R = CH_3 results in a chiral center and leads to D and L forms as well as DL racemic forms where there is a random arrangement of chiral centers. D and L forms can crystallize while the racemic form cannot. The PLA monomer unit – lactic acid (lactide) is prepared by chemical synthesis (from lactone and chloropropionic acid) or by toluene recrystallization. Lactic acid can also be manufactured by biotechnological means – by fermentation of saccharides, which are hydrolyzed and extracted from natural materials (corn, starch, etc.) Then lactide is polymerized. In its chemical, physical, and biological properties, polylactide is similar to polyglycolide. PLA is inferior to many synthetic polymers in its heat resistance: when heated to temperatures above 50°C, PLA products lose their shape, and, hence, they cannot be sterilized by heating. Polylactide is a resorbable polymer, which is degraded at high rates in biological media. It is resorbed *in vivo* in 10 to 12 days. Degradation is often autocatalytic. Degradation of thick sections can occur faster than thin sections due to the build-up of a localized low pH (3.2–3.4) accompanying degradation within the section and the formation of lactic acid. This can result in the rapid release of lactic acid and polylactide oligomers, causing a toxic response of surrounding tissues. The reduction in molecular weight results in a marked reduction in strength.

Polylactide/polyglycolide composites show better heat resistance, degradation rates, and mechanical strength parameters. The properties of the polymer can be varied considerably by varying the lactic/glycolic acid ratio. *In vivo* degradation mainly occurs by non-enzymatic hydrolysis. Hydrolytic degradation of these polymers is controlled not only by the diffusion of water through the polymer but also by the release of water-soluble acid oligomers – polymer degradation products. PLA and PLA/PGA degradation rates are faster in the bulk of the polymer than on its surface due to autocatalytic effect. This decreases the strength of the products.

Many studies showed good potential of PLA/PGA copolymers as cell scaffolds, plates, sponges, and fixers intended for repairing bone defects. However, polylactic acid-based material is quickly degraded *in vivo*. Also, PLA/PGA implants cause various adverse responses, usually after between 2–4 months and 5 years of implantation. These are tissue acidification, local inflammation, and osteolysis. Histological examination of tissues at the site of the polylactic acid-based implant site revealed dense fibrous capsules surrounding the remaining material.

Natural biodegradable polymers, such as polysaccharides chitosan and alginate, which are structurally similar to glycosaminoglycans, are also used in medical applications. Glycosaminoglycans are found in the connective tissue and are linked to proteins through intermolecular interactions. In practical medicine, these are films, sponges, and hydrogels used as wound, burn, and healing dressings; they are also used in systems for cell, drug, and growth factor delivery, as sutures and tissue adhesives. Chitin and its derivatives and their combinations with other natural materials (gelatin, heparin, alginate) are used rather widely. The drawback of chitosan is its brittleness and structural changes under sterilization. Chitosan is a product of the natural polysaccharide chitin, which is extracted from the shells of crustaceans. The availability, biodegradability, and biocompatibility of chitosan make it a suitable material for use as dressings, burn covers, and as cell scaffolds in cellular and tissue engineering.

Alginate is a linear polysaccharide derived from seaweed or produced by biotechnological fermentation. It has been approved by the FDA as material for preparing wound dressings. Alginate fibers (prepared from a blend of sodium and calcium alginates) and films are used in the initial treatment of wounds and burns. Since alginates are able to be polymerized and gelated in the presence of divalent metals (calcium, magnesium, etc.), they are used as cell scaffolds. Alginates containing high percentage of hyaluronic acid are used as matrices for drug delivery and as material for constructing cell scaffolds. Alginate stabilized by polylysine is used to immobilize islets of pancreatic cells (hybrid pancreas), genetically modified fibroblast cells, hematopoietic cells of bone marrow, parathyroid cells, various cells releasing monoclonal antibodies, and complex biomolecules (antibiotics, hormones, vaccines, enzymes, etc.).

Hyaluronic acid is a glycosaminoglycan, a component of intercellular substance of vertebrates' soft tissues. This is one of the most promising materials in reconstructive surgery and tissue engineering. One molecule of hyaluronic acid is able to bind up to 1000 water molecules. This glycosaminoglycan is used in medicine as a viscoelastic material in ophthalmologic surgery, in the treatment of joints in orthopedics, as barrier membranes, burn covers, and in cosmetic applications. Hyaluronic acid and its composites with other materials are used as barrier and anti-adhesion material in abdominal surgery. Films prepared from hyaluronic acid polyesters have been successfully tested as cell scaffolds.

Collagen is a natural polymer that is found abundantly throughout the body (in connective, bone, and cartilage tissues). Collagen in used in various medical applications: as material for soft tissue and fluid duct implants, wound dressing components, including ones with hemostatic effects. Collagen has been successfully tested as material for arterial vessel implants, ligament prostheses, and nervous system components. To improve the properties of collagen-based materials and to impart increased strength to them, collagen is blended with ceramics and synthetic polymers (polyethylene, polyvinyl alcohol, polysiloxanes). Collagen/hydroxyapatite composites are considered as a bone replacement material for repairing bone defects in maxillofacial surgery and dentistry. Limitations of using collagen are its uncontrollable biodegradation rate and short life (no more than one month) in the body.

Polyesters of microbial origin, polyhydroxyalkanoates (PHAs) are also very promising resorbable biopolymers. Polyhydroxyalkanoates are a relatively new class of natural polymers synthesized by microbes, which has been given much consideration recently, but has not been sufficiently studied yet. Interest in these linear polyesters has been growing since the late 1980s – early 1990s. These polymers offer a variety of potential applications, including medical and pharmaceutical ones (Amass et al., 1989; Williams et al., 1999; Sudech et al., 2000). The most widespread and the best studied representatives of the PHA family are a polymer of β-hydroxybutyric acid (poly-3-hydroxybutyrate, P3HB) and copolymers of hydroxybutyric and hydroxyvaleric acids (P3HB/3HV), which are stronger and more elastic materials. PHAs have been extensively studied for medical applications as material for surgical elements, tissue engineering, bio-artifical organs, and drug delivery systems (Shum-Tim et al., 1999; Stock et al., 2002; Williams, Martin, 2002). PHAs, which can be processed by various methods from different phase states and which are slowly degraded in biological environments, can be used to solve some pharmacological problems; e.g., they are promising candidates as materials to be used in sustained-release drug delivery systems. As *in vivo* bioresorption rates of PHAs are several times lower and their mechanical properties better than those of other known biodegradable biomaterials (polylactides, polyglycolides), these polyesters can be used for the long-term regeneration of large and complex bone defects and injuries (Jones et al., 2000).

Although natural biodegradable polymers have been known for decades, their market share is still insignificant. Their main limitations are the high production cost, processing difficulties, and insufficient mechanical strength. Things are changing, however. Thanks to the establishment of large-scale production facilities, the cost of biodegradable polymers is reduced, and complex polymerization and blending methods make these materials stronger.

A great variety of medical-grade polymers with different compositions are known now. The data based on the analysis of the literature and data-bases on commercially produced biodegradable polymers have been sum-marized in Table 1.6.

The European Bioplastics association's forecast is that in 2020, bio-based degradable polymers will occupy between 5 and 10% of the Euro-pean polymer market.

1.7 APPLICATIONS OF MEDICAL GRADE MATERIALS

Efforts are being made to classify biomaterials and compile a database on them. Biomaterials can be broadly classified as materials for cardio-vascular surgery, tissue surgery, urology, orthopedics, dentistry, wound covering, and drug delivery systems. Improved technologies and novel biomaterials and devices for temporary or permanent replacement of the functions of diseased tissues and organs broaden the possibilities of recon-structive medicine, thus improving the quality of life and increasing life expectancy. Biomedical materials science has considerably contributed to the progress in designing bio-artificial organs and tissues. Prostheses fab-ricated from novel—highly functional and biocompatible—materials are being increasingly used in various branches of reconstructive medicine.

1.7.1 MATERIALS AND PROSTHESES FOR RECONSTRUCTION OF THE PARTS OF THE CARDIOVASCULAR SYSTEM

The cardiovascular system (blood circulation) is essential for all vital functions of the body. The three main components of the cardiovascular system are the blood, the blood vessels and the heart. They are respon-sible for the delivery to the tissues of all the nutrients that they require, including oxygen, the removal of waste products, the control of multiple functions through the endocrine hormone system, and for thermoregula-tion. Failure of any of the components of the system can have catastrophic effects, including irreversible damage to organs, tissues, and brain, cardiac arrest and death. In spite of the achievements in the therapeutic and surgi-cal methods of treatment of cardiovascular pathologies and the decline

TABLE 1.6   Trademarks and Manufacturers of Biodegradable Polymer Materials

Trademark	Chemical characterization	Producer
Bellfree, Lactron	PLA (granules, filaments, threads, films)	Kanebo
Biophan	PLA (films)	Trespaphan
Biotech Color	PLA- and starch-based pigments	Dainippon Ink & Chemical
Ecoloju	PLA (films)	Mitsubishi Plastics
EcoPla, Nature-Works	PLA (granules, films)	Cargill Dow Polymers
PLA, DL-PLA, etc.	Biodegradable polymers (PLA/PGA: lactide/glycolide/ε-caprolactone melts)	Birmingham Polymers
Terramac	PLA (films, threads)	Unitika
Biomax	Polyester (PET modification)	DuPont
BioMer	PHA (Polyhydroxybutyrates)	Biomer
Bionolle	PHA (Polybutylene succinates, granules)	Showa Highpolymer
Biopar	PHA (polyhydroxybutyrate/valerate copolymers)	Biopolymer Technologies
Bioplast	Composites based on starch, cellulose, and synthetic binder	Biotec
Biopol	Polyhydroxybutyrates, polyhydroxybutyrates/valerates	Monsanto (before 1999), Metabolix
Celgreen	Polycaprolactone and modifications thereof	Daicel Chemical
Cellophane, Cel-Green, Celluflow, Grafix, Mylar, Rotuba H,	Cellulose acetate (cellophane), cellulose propionate (films, labels)	-
Eastar Bio	Copolyester (films, nonwovens, etc.)	Eastman
Eco-Excell, Eco-Foam, Eco-Maize, Eco-Plus, Eco-Shape	Starch-based thermoplastics for different applications	National Starch & Chemical
Ecoflex F	Copolyester of terephthalic acid, adipic acid, and butane-1,4-diol (granules, films)	BASF
Ecoplast	Composites based on wood dust, starch, and binder	Novamont, Groen Granulaat

TABLE 1.6 (Continued)

Trademark	Chemical characterization	Producer
EverCorn Resin	Modified starch-based plastic	Japan Com Starch
Greenfill	Composite based on starch and poly-vinyl acetate	Green Light Products
GSPIa	Polybutylene succinates	Mitsubishi Chemical
Highcelon	Polyvinyl acetate (film)	Nippon Synthetic Chemical Industry
Luncare	Polyalkylene succinates and copolymers thereof	Nihon Shokubai
Mater-Bi	Composites of starch (40–85%), cellulose, and polycaprolactone and polyvinyl acetate (packaging and nonwoven materials, foam)	Novamont
Nodax	Polyhydroxyalkanoates (nonwoven materials, films, granules for spinning)	Procter & Gamble
Novon	Composite (43% starch, 50% synthetic material, 7% other)	Novon Intl.
Plantic	Polyhydroxyalkanoates	Plantic Technologies
Solanyl	Biopolymer based on starch production waste for molding	Rodenburg Biopolymers

*Adapted from BioWorld – http://www.cbio.ru; German Biotech, C&EN. 2000, http://www.glycotope-bt.com/en/services/Biotechnological_Service.php – online-k. www; Hench and Jones (2007); Sevastianov and Kirpichnikov (2011).

in their rate in recent years, diseases of the cardiovascular system still account for over 40% of all deaths in the world. Coronary heart disease is the most prevalent cardiovascular condition.

Thus, materials for cardiovascular surgery constitute the largest group of medical-grade materials. They are used to fabricate containers for blood, needles, and syringes, vascular catheters, blood vessel prostheses, replacement heart valves, coronary artery bypasses and assist systems. The most stringent demands are placed on the hemocompatible materials used to produce blood vessel prostheses. Hemocompatibility of vascular prostheses is determined by the material used and technology employed to produce the prostheses. Other important factors are the design and diameter of the prosthesis, the hemodynamics at the implant site; physical and mechanical properties, etc.

1.7.2 MATERIALS FOR VASCULAR PROSTHESES

The first vascular prostheses were tubes of metal or bone used to repair the damaged artery in 1882. In the late 19th century, the first vascular prostheses were made from metals. Then, the range of materials used to prepare prostheses was increased to include aluminum, silver, paraffin-coated glass, polymethylmethacrylate. In the mid-20th century, prostheses were also made from synthetic fabrics, primarily, parachute silk (Biosovmestimost, 1999; Shtilman, 2006). For 70 years, all attempts to replace arteries by artificial vessels that had various structures and were made from various materials (tubes of ivory bone, paraffin-coated glass, aluminum, silver, methylmethacrylate, polyethylene, etc.) failed at the stage of experiments. In 1952, porous synthetic cloth tubes were tested in the U.S. In 1955, the U.S. Society for Vascular Surgery established the Committee for the Study of Vascular Prostheses. The prostheses were constructed from synthetic polymers (Vynion-N, Orlon, Dacron, Teflon, Ivalon) and stainless steel. A considerable improvement of vascular prostheses was the use of crimped synthetic tubes. Polymeric materials – polyethylene, polyvinyl alcohol, polyacrylonitrile, etc. – have come into use since the mid-1950s. However, vascular prostheses constructed from these materials were not sufficiently strong. The greatest achievement of that time was construction of knitted Dacron (polyethylene terephthalate) prostheses. These prostheses were used widely and produced by many companies. Researchers studying synthetic materials for vascular prostheses finally chose polyester (Dacron, Terylene) fibers and, then, polytetrafluoroethylene (Teflon) fibers. Dacron vascular prostheses were first constructed in the U.S.; those were woven seamless bifurcated tubes prepared from commercial taffeta Dacron. Teflon (polytetrafluoroethylene) is the most suitable material for constructing vascular prostheses. Production of Edwards seamless arterial grafts of Teflon was established in the U.S. by C. R. Bard, Inc. Since 1974, U.S. researchers have been studying new synthetic materials for arteriovenous prostheses – expanded microporous polytetrafluoroethylene. There are currently three main types of synthetic textile fabric prostheses: knitted, woven, and braided ones. Present-day cardiac surgery uses various types of vascular prostheses: xenografts made of bovine blood vessels, autografts prepared from the patient's femoral veins, prostheses made of naturally occurring and synthetic materials, biodegradable prostheses of bioresorbable polymers, prosthesis constructed from composite materials, hybrid prosthesis of synthetic materials covered by biological coatings (fibrin or collagen) or seeded with cells (tissue-engineered vascular grafts).

The main requirement for the materials to be used in the construction of vascular prostheses is that they must show high hemocompatibility, which depends on a number of factors:

- the type and properties of the material used,
- the fabrication technique employed, and the shape and size of the prosthesis,
- physical and mechanical, elastic properties of the material,
- surface properties,
- implantation site,
- hemodynamics at the implant site,
- the patient's state and the postsurgical healing of injured tissues,
- the patient's blood coagulation and the tactics of anticoagulant therapy, and
- possible postsurgical infections and other complications.

Vascular prostheses are elastic tubes of various diameters and lengths. Their inner diameter may range between 3 and 10–12 mm; the walls are 0.2–0.4 mm thick. Construction and implantation of small-diameter vascular prostheses presents the most serious challenge. It is very difficult to connect such prostheses to the native vessels, and they are prone to frequent thrombosis. The vascular prostheses that are currently used or designed may be solid, porous, knitted, braided, and woven. Knitted tubes are the most popular choice of the prostheses, as they are compliant and flexible. Crimped grafts, which have been developed more recently, are more compliant. Advances in the use of vascular grafts are associated with successful development of medical materials science. The availability of the material with adequate properties is the main factor determining successful application of vascular grafts. The most widely used vascular prostheses are made of porous PET – Gore-Tex®, commercially produced by W.L. Gore & Associates, Inc. (U.S.). Gore-Tex® modifications include bifurcated, tubular, ringed, conical grafts, etc.

Biodegradable polymers with high hemocompatibility are relatively new materials for constructing vascular prostheses. As the implant is gradually degraded, it is replaced by the new tissue, causing less trauma to blood cells and decreasing blood vessel wall response. Such prostheses with tissue ingrowth are able to grow longer with time, which is important if they are implanted to children. Biodegradable (temporary) vascular grafts are fabricated from PET, polylactide, polyglycolide, and fibrin. One of the challenges in vascular graft surgery is the lack of reliable small-diameter (below 5.0 mm) grafts. It is very difficult to connect such prostheses to the native vessels, and they are prone to frequent thrombosis because of the low blood flow rate in them. Such polymers as modified segmented polyurethanes, polycarbonate urethanes, polyethylene oxide, and silicone-coated polymers are being evaluated as materials for constructing small-diameter prostheses. The most promising approaches are to use microporous segmented polyurethanes and hemocompatible coatings with enhanced thromboresistance and antimicrobial coatings. A new approach is to create hybrid prostheses, employing the methods of cellular and tissue engineering. Chemists in Germany came up with an invention that may have considerable influence on the development of vascular surgery. They synthesized a polymer whose molecules can return to their original shape after deformation. A mixture of polymer materials contains photosensitive molecules that can retain their shape when irradiated with the ultraviolet of a certain wavelength. In the experiment that proved the possibility of using this invention in medicine, a straight plastic filament was inserted into a tapered vessel. Irradiation performed by using fiber optics caused the filament to coil, thus preventing the vessel walls from collapsing.

1.7.3 SUBSTITUTE HEART VALVES (SHV)

First surgeries to replace heart valves were performed in the 1960s. Now, heart valve replacement is a common clinical surgery. Heart valve disease is one of the causes of death of cardiac patients. Tens of thousands of replacement valves are implanted each year because of acquired damage to the natural valve and congenital heart anomalies. There are two main types of valve replacements – mechanical and biological substitute heart valves (Table 1.6). Aortic valves are normally repaired with an artificial mechanical or biological prosthesis and, while mitral valves can be replaced, many surgeons prefer to repair the existing valve by reshaping it. The life of biological valves is lower than the life of mechanical valves so, despite their rather less satisfactory flow characteristics, modern mechanical heart valves tend to be placed in young patients and may last for their whole lifetime. The life of biological valves is about 10 years, and they should not be implanted to patients with life expectancy over 10 years; otherwise, they need to be re-implanted. Mechanical valves are, however, thrombogenic, and patients with such valves always need anticoagulation treatment with such drugs as warfarin and heparin. Other drawbacks of mechanical valves are hemolysis (the rupturing of erythrocytes) and the high noise level. Thus, neither mechanical nor biological heart valves are perfect, and new-generation valves need to be designed.

Bio-artificial heart valves must provide endothelialization of the surface contacting with blood, be able to synthesize extracellular matrix, have stable shape and potential for growing in the patient's body, cause no undesirable immunological response and other inflammatory processes, be stable to calcification and excessive growth of tissues (cells), have stable mechanical properties, have a large effective area of the valve orifice and tight closure of valve leaflets, be resistant to infections and chemically inert, induce no hemolysis, and be easy to implant.

Biological valves were first used about 25–30 years ago. Occasionally, autologous grafts are inserted, when, for example, the pulmonary valve is used to replace a defective aortic valve, with a less critical mechanical valve being inserted into the pulmonary artery. Xenografts have been made from porcine or bovine tissue, which is 'fixed' with materials such as glutaraldehyde to eliminate immunological reactions. Devices include whole valve systems, single valve leaflets attached to a metal ring, or valves shaped from pericardial tissue. All these valves have a much more normal anatomical geometry than any of the mechanical valves. However, the xenograft tissue may be calcified and be susceptible to fatigue damage. The life of xenograft valves is often shorter than the life of mechanical valves.

Mechanical valves were first implanted in the 1950s, but their design has been revised continually since then. More than 100 types are known now. The solid components of mechanical valves are usually manufactured from stainless steel alloys, molybdenum alloys, and increasingly pyrolytic carbon and polyester urethanes are used for the valve housings and leaflets. The earliest artificial valves were ball-in-cage devices. The mechanical valve designed in the 1980s is a bileaflet valve in which two approximately hemispherical plates pivot on hinges. Such valves cause the least damage to blood. In modern models, the plates are made of silicone rubber reinforced with a titanium ring and of polyoxymethylene and UHMWPE. To decrease the incidence of complications caused by using mechanical valves (thrombus formation and embolism), new materials with enhanced properties need to be used to fabricate them. A trileaflet valve is a physiological analog of a natural valve. This design creates fewer obstacles to the blood flow (compared with the bileaflet valve, in which the leaflets and fixers divide the blood flow). Since 2001, the Roscardioinvest company (Russia) has been developing a new model of the heart valve prosthesis – a unique trileaflet valve, 'Tricardiks.' The trileaflet design ensures central blood flow, decreasing the risks of thrombosis and malfunctioning of the valve. Patients with a 'Tricardiks' valve may stop taking or take substantially lower doses of anticoagulants for the rest of their lives. This will significantly improve the patients' quality of life. Moreover, the new trileaflet heart valve is physiologically very close to the natural heart valve. The new trileaflet mechanical 'Tricardiks' heart valve developed by Roscardioinvest combines the best characteristics of biological and mechanical valves. The present-day heart valve market has several leaders (Table 1.7). The largest producer of bileaflet heart valves (the dominating type of mechanical valves) is St. Jude Medical, which is followed by Carbomedics and Edwards Life Sciences.

1.7.4 STENTS

Today, one of the most frequently performed interventions in cardiovascular surgery is intravascular arterial stenting (Sosudistoye i vnutriorgannoye stentirovaniye: rukovodstvo). Stenting decreases the number of

TABLE 1.7   The Leading Producers of Heart Valve Prostheses (Data of Roscardioinvest – www.roscardioinvest.ru)

Company	Mechanical valve	Biological valve	Notes
Medtronic, Inc., Minneapolis, Minnesota	Medtronic Hall® mechanical valve (single-leaflet)	1. Mosaic® aortic and mitral bioprosthesis	Mosaic bioprosthesis was approved for use in the U.S. in 2000
		2. Hancock® standard mitral bioprosthesis	Hancock biological valve was approved for use in the U.S. in 1999
St. Jude, Medical St. Paul, Minnesota	1. SJM Regent Valve (bileaflet) 2. SJM Masters Series Valve (bileaflet)	1. SJM Toronto, SPV Valve (porcine) 2. SJM Toronto, Root Valve Regent	Regent Valve was presented in 1977; Masters Series, based on the same design, with improved properties, was presented in 1997
Edwards Life-Sciences Irvine, California Sorin/Carbomedics	1. Starr-Edwards Silastic Ball Valve 2. Edwards MIRA Mechanical Valve (bileaflet)	1. Carpentier-Edwards PERIMOUNT, PERIMOUNT Magna, and PERIMOUNT plus 2. Carpentier-Edwards Porcine Bioprosthesis	PERIMOUNT (pericardial valves) was presented in 1991. Starr-Edwards Silastic Ball Valve was presented in 1968, has not been modified, obsolete technology. Edwards MIRA – modified Sorin/Carbomedics valve, clinical trials stage
CarboMedics Division of Sorin Group, Italy	1. Standard aortic and mitral valves 2. Top HatTM Supra-Annular Aortic Valve 3. OptiForm™ Mitral Valve	Mitroflow Pericardial Heart Valve	Titanium-carbon housing
Medical Carbon Research Institute (MCRI) Austin, Texas	ON-X Valve (bileaflet)	None	Entirely carbon bileaflet heart valve, approved for use in the U.S. in 2002

TABLE 1.7 (Continued)

Company	Mechanical valve	Biological valve	Notes
Cryo-Life, Inc. Kenesaw, Georgia		The CryoLife-O'Brien® Stentless porcine aortic bioprosthesis	Only available outside the U.S. – CryoLife purchased the right to produce and sell the porcine valves designed by Australian surgeon
			Mark F. O'Brien, M.D., F.R.C.S. in 1995; the prosthesis is a trileaflet valve and is implanted with custom tools

acute complications in balloon catheter angioplasty (such as thrombosis or intima dissection). In addition, stent implantation is the only method of preventing late restenosis. Stents are usually cylinders fabricated from metal (stainless steel, nickel and titanium alloy) meshes. In EU states and in the U.S., 0.3–0.4% of the population is annually subjected to the stenting procedure; in 2010, e.g., about 4 million stent implantations were performed there. The major complication associated with stenting is restenosis – renarrowing of the lumen of a vessel as a biological response of the vessel wall to the implantation of a foreign body. Restenosis develops in 10–58% of the patients with implanted coronary stents up to 6 months after surgery. Intravascular implantation of a metal stent triggers a cascade of complex histopathologic processes – the response of the body to the invasion of a foreign body. The outcomes of these processes depend on the structure, physicochemical properties, and the degree of chemical purity of the material, on the shape and surface properties of the implants, and on the stent implantation technique employed. Decreasing of the risk of thrombosis remains the major issue of the postoperative period. There may be several ways to enhance the biocompatibility of stents: to modify the surface properties of the metal implants, to coat them with biocompatible materials ("coated stents"), or to prepare stents of bioresorbable polymer materials ("biodegradable stents"), including stents loaded with antiproliferative drugs ("drug-eluting stents"). The major issue here is to find an optimal material that would be ideally bio- and hemo-compatible, possess certain physical and mechanical properties, and be bioresorbable *in vivo* without releasing toxic products or causing adverse response of the vessel wall.

The world leaders in stent production and the types of the stents are as follows: balloon-expandable grafts – Palmaz Crown (Cordis, U.S.), Itias Bridge (Easy Wallstent, AVE, U.S.), VIP (Medtronik, U.S.); self-expandable stents – Easy Wallstent (Schneider, Switzerland), Memotherm (Bard, Ireland), Symphony (Boston Scientific, U.S.), Instent Vascucoil (Medtronic, U.S.). The most popular uncoated coronary metal stent is MULTI-LINK VISION – the first in the new class of cobalt-chromium alloy stents used to treat ischemic heart disease. The cobalt-chromium alloy – a stronger and more radiopaque material than stainless steel – was used to fabricate a thin-walled structure whose strength and visibility by X-ray were comparable to those of metallic stents with thicker walls. The cobalt-chromium technology and thin-walled structure (0.0032") ensure excellent deliverability and conformability of the stent and minimize the injury of the vessel wall, thus decreasing the incidence of restenosis, as confirmed by clinical trials. The MULTI-LINK VISION® stent system was marketed in Europe in December 2002 and in the U.S. in July 2003 by the Guidant Corporation – a leader of the U.S. market of metallic stents since 1997, when they marketed their first stent. In April, Abbott Vascular Devices merged with the endovascular department of the Guidant Corporation, and now, the Abbott Company is the leader in the international market of metallic stents. Johnson & Johnson has a considerable stent market share (U.S.). The BX VELOCITY stent mounted on the Raptor Rapid Exchange Stent Delivery System is intended to increase the lumen of the coronary artery in the treatment of acute or impending vascular occlusion in the patients after the failure of the treatment intervention, when the occlusion length is no more than 30 mm and the initial diameter of the artery is between 2.25 and 4.00 mm.

The second-generation stents are the stents coated with biocompatible substances. Among the coatings proposed and developed for increasing biocompatibility of stents are impregnation of the stent surface with heparin ("Jomed," Sweden), diamond-like carbon coating ("BioDiamond," Germany), gold coating ("InFlow Dynamics," Germany; "Medinol," U.S.), phosphocholine coating ("Biocompatibles" U.K.), and silicone-carbide coating ("Biotronic," Germany). Investigations are being conducted with biodegradable polymer materials: polyglycolic and polylactic acids, polyurethanes, etc. However, the problem of restenosis has not been resolved yet, and the coating that would not affect the vessel wall remains to be found.

In recent years, attempts have been made to enhance biocompatibility of vascular stents by adding biological agents inhibiting hyperplasia of a neointima (dexamethasone, rapamycin, paclitaxel, sirolimus etc.) to stent coatings. Coronary stents Hercules III (Intek, Switzerland) are metallic constructs coated with the drug that contains paclitaxel – the active substance preventing restenosis [data found at the sites of Intek – http://intek/koronarnye_stenty_intek; ROSSLYN Medical® http://rosslynmedical.com]. Hercules III was based on the Hercules II stent. The polymer coating of the stent is biologically stable and bio- and hemocompatible, enabling controlled release of paclitaxel. The active component is concentrated on the outer surface and is directly delivered to the restenosis initiation site. Therefore, Hercules III coronary stents not only mechanically widen the artery but also suppress cell growth in its tissues, thus perceptibly increasing the effect of angioplasty. The active component – paclitaxel – interrupts the cascade development of restenosis; the service life of the drug coating is 18 months. The material used to fabricate Hercules III stents is both mechanically strong and highly pliable. Thanks to this combination, coronary stents are capable of assuming the natural shape of the problem region of the artery and effectively maintaining the width of the lumen sufficient for the normal blood flow. Drug-eluting stents reduce the incidence of restenosis by over 90%, making unnecessary repeated transluminal angioplasty.

Cypher stents (Cordis) are world leaders. Numerous investigations of long-term effects, however, have shown that premature termination of antithrombotic treatment after implantation of stents coated with this cytostatic drug is a predictor of acute late DES thromboses [sirolimus-delivering (Cypher™) and paclitaxel-delivering (Taxus®) stents]. Moreover, acute late and very late DES thromboses [sirolimus-delivering (Cypher™) and paclitaxel-delivering (Taxus®) stents] can develop even in the patients receiving double antithrombotic medication. Thus, new biocompatible coatings for such stents and new coating drugs need to be found.

A third approach is based on the idea that implanted stents must be gradually replaced by normal vascularized tissue. This approach could solve the problem effectively, but it has not been sufficiently developed yet. Experimental stents have been prepared from fully degradable materials such as polylactides, polyurethanes, and silicone.

1.7.5 SUTURES

Practical surgery involves the use of great amounts of suture materials, which are prepared from natural or synthetic polymers (the use of metal sutures is very limited). Chemical fibers are fibers produced by using chemical or physicochemical processing of natural and synthetic high-molecular-weight compounds (polymers). Depending on polymer origin, chemical fibers are broadly divided into artificial fibers (from natural polymers) and synthetic fibers (from polymers produced by chemical synthesis of low-molecular-weight units – monomers). Chemical fibers are divided into two major classes: carbon-chain fibers, whose polymer chain solely consists of carbon atoms, and hetero-chain fibers, in whose polymer chain there are both carbon-carbon bonds and bonds between carbon atoms and the so-called heteroatoms (e.g., nitrogen or oxygen atoms). The hetero-chain fibers comprise fibers that differ in the type of the bond between the elementary components of the macromolecule of the fiber-forming polymer: polyamide fibers, with –CO–NH– bonds, polyester fibers, with –CO–O– bonds, etc.

Cellulose, the basic component of the majority of plants, and its derivatives were the first fiber-forming polymers used in commercial production of chemical fibers. Production of synthetic fibers began in the 1930s. At the end of the 1970s, the global production output of synthetic fibers exceeded 3 million t/y, but in the late 1980s and early 1990s, this value decreased somewhat. Polyester fibers constitute up to 60%, polyamide fibers – 21%, and polyacrylic fibers – slightly less than 15%. Much attention is currently attracted to elastomeric fibers (Spandex, Lycra), which are produced by processing synthetic hetero-chain polyurethanes. Very strong fibers are fabricated from flexible-chain polymers such as polyethylene. The mechanical properties of very strong synthetic fibers are greatly superior to those of the fibers intended only to be used as textile fibers.

Suture materials should have a number of different properties, including mechanical strength, the ability to hold the edges of the sutured wounds and tissues together, biological safety, and ease of use (ease on hands, ease of tying, and ease of sterilization). In recent years, much research effort has been focused on designing biodegradable surgical sutures. Biodegradable surgical suture material must be able to hold the tissue together following separation by surgery, be sufficiently elastic, and be gradually resorbed at rates corresponding to the rate of tissue regeneration. Material degradation products must be easily eliminated from the implant site, be completely harmless to the body, causing no unfavorable responses of the surrounding tissues or the body as a whole. The most stringent demands are placed on the suture material used to construct anastomoses and intestinal sutures, particularly in the surgery of the colon, because it has thin walls and because of the presence of aggressive microflora and active enzymes in the colon. The most common technique of connecting intestinal loops is ligature, and the most common anastomosis is double layer anastomosis. The availability of synthetic suture materials (Vicryl, Maxon, Polysorb) has encouraged the common use of single layer suturing, which is more convenient for the surgeon and less traumatic for the patient. There is great activity in developing biocompatible polymer coatings for synthetic suture materials and fully resorbable polymer sutures. Surgical sutures are made of natural and synthetic polymers. Non-degradable sutures made of synthetic polymers like polypropylene (Prolene, Surgilene, Deklene), polyamides (Surgilin, Dermalon, Nylon 66, Polyamide 6), and halogen-containing polymers (Gore-Tex, FUMALEN®) are used to construct deep buried sutures. They remain indefinitely intact when placed in the body, with fibrous tissue encapsulating them. If these materials are used for surface sutures, the thread is pulled out after the tissue heals.

Table 1.8 lists the data on the commonly used suture materials, showing their properties and lifetimes *in vivo*. Recently, a considerable research effort has been directed towards development and use of degradable suture materials. As the tissue is healing, the threads are gradually replaced by the new tissue, enhancing the strength of the suture. The "oldest" biodegradable suture material is catgut, which was first mentioned in the scientific literature in 1885. Catgut is prepared from sheep intestines. German companies produce twisted catgut of different thickness under the trademarks CATGUT PLAIN and SMI-AG and metal-reinforced catgut (Catgut 0,41101). Catgut is quickly (in 10–15 days) biodegraded in the implant site and causes rather strong inflammation. Catgut has been replaced by stronger and more durable monofilament sutures prepared from resorbable synthetic polyesters of glycolic acid (Dexon) and copolymers of glycolic and lactic acids (Vicryl); 80% of the strength of these sutures, however, is lost in 21 days after implantation. Such sutures were regarded as very promising surgical materials, but, recently, it has been proved that they cannot be used in surgical operations that require the sutures to remain mechanically strong for a long period of time. Also, these sutures are not elastic enough, causing unnecessary trauma (i.e., they show a "wicking" tendency and high tissue drag). The more recently developed monofilament sutures prepared from glycolic acid/trimethylene carbonate copolymer (Maxon) and polydioxanone are more long-lasting (30–50% of their strength is lost in 30 days post implantation), but they need to be tied with very complicated knots. A common drawback of absorbable synthetic sutures is their unpredictable behavior on interaction with cells and components of the patient's immune system. The main reasons of complications associated with the use of synthetic biodegradable materials (such as polylactides and polyglycolides) are tissue acidification and inflammatory and allergic reactions of the host organism to their degradation products. Therefore, an important task of materials science is to find materials for preparing biocompatible, mechanically strong, and bioresorbable sutures.

The largest producers of surgical materials include Tyco Healthcare Group AG, Davis & Geck and U.S. Surgical Corporation, Chirmax (Czech Republic), Siemens Medical Solutions (Germany), GE Healthcare Technologies (U.S.), Listem (Korea), Storz Medical (Sweden), Nihon Kohden (Japan), Ulrich (Germany), Mediprema (France), Merivaara (Finland), Medtronic (U.S.), 3M (U.S.), Trumph (Germany), Johnson & Johnson (U.S.). In Russia, suture materials are also produced on a commercial scale. The leading Russian producers of suture materials are BALUMED, LINTEX, VOLOT, Optikum, AKMED, the MERCURY Group, Lizoform-SPb, VIPS-MED, Ekran, MedInzh.

TABLE 1.8 Major Commercially Available Sutures (data found at http://www.lintex.ru/index.php.; http://www.volot.ru)

Suture trademark	Composition, structure	Tensile strength, absorption rate	Applications
Absorbable			
PROXIL® twisted coated polymer filament of glycolic acid Other trademarks: Vicryl, PHA, Dexon, Polysorb, Safil	Synthetic braided coated absorbable material of the glycolic acid polymer	About 75% of the initial strength is retained over the first 2 week after implantation and about 50% in 3 weeks. Holds the wound edges together during the critical period of wound healing. Predictable hydrolytic absorption is complete in 60–90 days.	General surgery, abdominal surgery, gynecology, plastic surgery, ophthalmic surgery, and other applications where absorbable materials should be used
PROXIL – Rapid® Other trademarks: Vicryl-Rapid	Synthetic braided coated absorbable material of the glycolic acid polymer (PGA-Rapid)	Short-term wound closure, loss of strength over the first 9–11 days after implantation. Predictable hydrolytic absorption is complete in 40 days.	In applications where rapid suture dissolution is desirable: skin sutures, minor surgery, dentistry, urology, abdominal surgery, gynecology
MONOPRO® Other trademarks: Monocryl	Synthetic monofilament suture material of the glycolide/caprolactone (PGA-PCL) copolymer	Short-term wound closure. Retains 60% of its initial tensile strength over the first week and 30% in 2 weeks. Hydrolytic absorption is complete in 90–110 days.	The smooth surface of the monofilament minimizes the risk of trauma to tissue and postoperative infection. Pediatric surgery, abdominal surgery, general surgery. Subcutaneous suture.
PROLONG® Other trademarks: PDS II	Synthetic monofilament suture material of polydioxanone (PDO)	Long-lasting wound closure. Retains 75% of its initial tensile strength over the first 2 weeks and 60% in 4 weeks. Hydrolytic absorption is complete in 180–210 days.	In applications where the edges of the slowly healing, contaminated wounds need to be held together over long time periods.

TABLE 1.8   (Continued)

Suture trademark	Composition, structure	Tensile strength, absorption rate	Applications
Polished catgut	Natural absorbable material of purified collagen made from cattle intestines	Holds the edges of the wound over about 10 days. Complete enzymatic degradation in up to 70 days.	General surgery, gastrointestinal surgery, obstetrics and gynecology, urology, mucous membranes, subcutaneous sutures, ophthalmology
Polished chromic catgut	Natural absorbable material of purified collagen made from cattle intestines. Treated with chromium salts	Holds the edges of the wound over about 20 days. Complete enzymatic degradation in up to 90 days.	General surgery, gastrointestinal surgery, obstetrics and gynecology, urology, mucous membranes, subcutaneous sutures, ophthalmology
Nonabsorbable			
Virgin silk and braided silk	Natural material of the fibers produced by the silkworm, virgin or braided, impregnated in and coated with natural wax or silicone	Superior manipulation properties, but inferior in strength to synthetic materials with corresponding conditional numbers. Tensile strength is nearly entirely lost over the first year; complete degradation occurs in 2 years.	General surgery, gastrointestinal surgery, dentistry, skin sutures, ophthalmic surgery, plastic surgery. Not used in bile duct surgery, urology, and infected tissues
Nylon monofilament Other trademarks: Nurolon	Synthetic nonabsorbable material, monofilament, polyamide polymer	Gradual elimination from the body through hydrolysis (15–20% strength lost per year)	Ophthalmic surgery, microsurgery, skin sutures, plastic surgery
Braided nylon Other trademarks: Nurolon	Synthetic nonabsorbable material, braided polyamide polymer coated with silicone or natural wax	Gradual elimination from the body through hydrolysis (15–20% strength lost per year)	General surgery, skin sutures, gastrointestinal surgery, ophthalmic surgery, plastic surgery. Used instead of silk, being superior to it in strength and knot sliding.

TABLE 1.8   (Continued)

Suture trademark	Composition, structure	Tensile strength, absorption rate	Applications
Polyester-silicone Other trademarks: Dacron	Synthetic nonabsorbable braided suture material coated with silicone	Nonabsorbable. Gradually encapsulated	General surgery, cardiovascular surgery, plastic surgery, orthopedics, ophthalmic surgery
Polyester-Teflon Other trademarks: Ftorest, Ftorlan	Synthetic nonabsorbable braided suture material coated with Teflon	Nonabsorbable. Gradually encapsulated	General surgery, cardiovascular surgery, orthopedics
PROXILEN® (Polypropylene) Other trademarks: Monofil, Polypropylene monofilament, Prolene	Synthetic nonabsorbable monofilament, propylene polymer, a polyolefin	Nonabsorbable. Encapsulated in body tissues	Cardiovascular surgery general surgery, orthopedics, ophthalmic surgery, plastic surgery, microsurgery, skin sutures
Polyvinylidene fluoride (PVDF)	Synthetic nonabsorbable material, fluoropolymer, monofilament	Nonabsorbable. Encapsulated. Stronger than polypropylene	Cardiovascular surgery general surgery, ophthalmic surgery, plastic surgery, microsurgery, skin sutures, purulent surgery

1.7.6   SURGICAL ADHESIVES AND COMPOSITES

Surgical adhesives and sealants are often used instead of suture materials to connect wound edges during surgery. This application of polymer materials in medicine has not been sufficiently developed yet, but it may be effective during surgeries of the liver, kidneys, and lungs. Surgical glues must be able to quickly form elastic adhesion in the presence of moisture, which is characteristic of tissues of internal organs; they must be elastic; no heat must be released during healing; they must remain functional for long time periods; and no toxic compounds must be released during degradation of adhesives. The most commonly used surgical glues are isocyanate-, polyurethane-, and polyurea-based ones. The first commercial medical glues were based on polycyanoacrylates; such glues are still being produced. These are "Ciacrin SO-4" in Russia and Istman-910 in the U.S. These glues have short lifetime – 6–9 months. Polyurethane adhesives have been successfully used in neurosurgery, urology, muscle tissue bonding, and partial gastrectomy. Composites of polycyanoacrylates with caprolactone and polymers of monocarboxylic acids (polylactides/polyglycolides) are used for aneurysm repair and embolization of cerebral aqueduct. Fibrin glues, which are based on protein systems, are difficult to prepare, but they firmly join the edges of the tissues of internal organs. Another adhesive is gelatin crosslinked by formaldehyde and glutaraldehyde. The main research effort is focused on the development of safe and very strong materials and compositions.

Injectable compositions are used to fill postsurgical cavities, to fill and strengthen aneurysms, fistulas, and channels in internal organs (such as liver or pancreas ducts). These liquid compositions, derived from organic-silicon compositions and low-molecular-weight polysiloxanes, are administered with a syringe and then structured. These are Elastosil MI, Embosil, Pankreasil (Russia). Depending on the viscosity, crosslinking density, and curing time *in vivo*, the composition may be injected subcutaneously (e.g., the silicone Elastosil MI, with curing time of 20–24 h) or into the organ (Pankreasil, which is cured within between 5 and 60 min, depending on the proportions of components, when injected into the pancreas in the case of pancreatitis). Embosil, whose composition is similar to that of Pankreasil but also includes small inert Teflon balls, is used to speed up embolization. Subcutaneously injected polymer compositions are used to repair cosmetic soft tissue defects and defects caused by tumor removal in plastic surgery. These are hydrogels based on hyaluronic acid, collagen formulations, and colloidal dispersions containing, e.g., polylactide particles distributed in the polyvinylpyrrolidone gel or a suspension of polymethyl methacrylate particles in hyaluronic acid. These formulations are used in corrective plastic surgery of soft tissues and for filling the cavities.

1.7.7 MATERIALS USED TO REPLACE DEFECTS OF SOFT TISSUES AND SKIN

Postoperative cavities in soft tissues are filled with curable polymer compositions or pre-shaped prostheses. Such prostheses may be fabricated as porous sponges or sacks filled with bio-inert harmless gels or liquids. Mammary prostheses are used for cosmetic purposes and after cancer surgery. These are oval-shaped or spherical constructs – elastic sacks with soft fillers (salt solutions, oils, silicone-based gels, polyacrylamide gel). One of the currently used fillers is polyacrylamide gel Formakril produced by "Gel Cosmetic Technology" (Moscow, Russia). The skin of mammary prostheses is made from methylvinyl siloxane rubber (SKTN-M and SKTN-MED, Russia).

Hundreds of thousands of such prostheses are used every year, and materials scientists and biotechnologies are trying to improve their biocompatibility in order to reduce the risk of adverse effects. The main complication associated with the use of prostheses filled with liquid polymer materials, which are the most common implants, is that the shell may burst and a large amount of material may leak out of the prosthesis into the body. Salt solutions are used as an alternative to polymer fillers. Such prostheses are widely used in the U.S. The most common saline-based implants are Spectrum® and BIO-CELL®. In the late 1990s, U.S. and U.K. companies produced soybean oil-filled implants, but, later, because of toxicity of soybean oil, production of such implants was stopped. Thus, the search for the optimal material for soft tissue prostheses is still going on.

Pre-shaped prostheses based on organic-silicon materials are used to repair soft tissues of the face and cartilage. These are usually foams or porous sponges, which can be explanted at any time without injuring tissues, thanks to their surface properties. Such prostheses are fabricated from silicone rubber, such as SKTV-1-MED or stronger compositions, based on chlorinated polyethylene. The cartilage of the nasal septum or the pinna of the ear is replaced by the constructs prepared from polymers and copolymers of monocarboxylic acids (PLA and PGA).

The use of prostheses to reconstruct ligaments and muscles is a new challenge. More experience has been gained in construction of tendon prostheses. Ligaments and tendons can be replaced by xenografts and constructs prepared from synthetic and natural materials. Another option is to use prostheses fabricated from only synthetic materials or to use them together with the grafts based on natural tendons to enhance the strength of the latter. In this application, polyethylene terephthalate is used as a synthetic material; braided PET grafts are strong enough to be used in load-bearing sites. As the bulk of the grafts is porous, new tissues immobilize the grafts by growing through them. Polyethylene terephthalate-based ligament prostheses are produced by Phoenix Bio-Medical Corp. (U.S.) under the trademark GORE-TEX. Ligament grafts based on polymers of lactic and glycolic acids and carbon fibers were successfully tested in experiments with primates. Materials for constructing such prostheses must be biocompatible and very strong. Elastic and porous polypropylene cords strengthened with carbon fibers are used to fasten and reinforce ligament xenografts. Biodegradable ligaments are constructed from collagen fibers. Carbon powder is blended with polysulfone and PTFE as a binder. More effective implantation is achieved by introducing bioactive substances (growth factors) or functioning cells into the grafts prior to implantation. Prototype ligament grafts have been constructed from collagen meshes seeded with fibroblasts.

Reconstruction of muscle tissue defects with prostheses is still experimental. The first grafts for this application were made of very strong and elastic composite of ethylene dimethacrylate and a 2-hydroxyethyl methacrylate copolymer filled with polyethylene terephthalate. However, the grafts fabricated from this composite were unable to mimic muscle contraction. There are attempts to mimic functional properties of the muscle tissue by using polymer gels, which have the ability to reversibly swell or shrink due to changes in their environment such as pH or electric field. An example of polymer systems that are able to contract under electrochemical impact are gels with ionic groups. These gels can be used to fabricate ionogenic membranes, reinforced with inert metals. Porous gels based on mixtures of polyvinyl alcohol with polyelectrolytes (e.g., polyacrylic acid and polyamine hydrochloride) and prepared using repeated freeze-thaw cycles are able to swell or shrink under the impact of electrical pulse.

The most common need for artificial skin is a result of burns. First-degree burns destroy the epidermis. Second-degree burns destroy both layers but leave epidermis around hair follicles, allowing some scar-like regeneration. Third-degree burns destroy all skin, exposing the fat and muscle. Severely burned skin must be removed immediately before bacteria can breed. The preferred covering for burns is an autograft; however, a patient may not have enough to transplant. Animal or cadaver skin is immediately rejected by the body. The alternative is the use of artificial skin dressings.

Depending on the degree of injury and the stage of skin regeneration (inflammation, regeneration, and formation of new epithelial skin), the artificial skin must meet different requirements. In the first stage, artificial skin must ensure wound cleansing; in the second, it must create conditions favoring the growth of new connective tissue; and in the third stage, it must ensure proliferation and migration of epithelial cells and formation of new epidermis. Artificial skin is made from various natural and synthetic materials in the form of gels, fabrics, nonwovens, films, powders, fibers, porous covering, etc. Traditional textile bandages, which are generally used in first aid procedures, are currently being replaced by materials and systems with enhanced cleansing, sorbing, and hemostatic properties. Polymer-based sorption materials are produced as fibers, sponges, and powders. When wetted, powders form a film on the wound surface. Natural polymers used to prepare sorption skin covering include collagen compositions in the form of sponges. For example, Digispon is a system based on collagen and crosslinked polyvinyl alcohol, containing an antiseptic. Agarose-based gel systems are also used. Materials designed in Russia (Algipor and Algimaf) are porous sponges loaded with furacilin. Sorption skin materials have been prepared from synthetic polymers, including materials based on polyvinyl alcohol, polyvinylpyrrolidone, cross-linked polyethylene glycol, and block-copolymers of ethylene oxide and propylene (Pluronik F-127).

Multilayered functional wound dressings have been introduced into clinical practice rather recently. They may consist of the upper hydrophobic layer, which is impermeable to microbes but permeable to oxygen and water vapor, and underlying hydrophilic layer(s), which absorb the liquid and contact with the wound surface. Op-Site® (Acme United Corp.) is a clear oxygen-permeable polyurethane film that is suitable for first-degree burns. Biobrane® (Woodroof Inc.) is a silicone rubber–nylon compound coated with collagen extract used for more serious burns. These skin bandages can be used for up to 2 months but then a permanent covering is required to reduce scarring. Water vapor permeability of Op-Site® is 20 g/$m^2 \cdot h$, and this is comparable with this water vapor permeability of healthy skin (10 g/$m^2 \cdot h$).

Loading of drugs into the hydrophilic layer closest to the wound facilitates regeneration of the epithelium and wound healing. The search for the effective wound-healing materials is in progress.

1.7.8 MATERIALS FOR REPAIRING BONE TISSUE DEFECTS

Bone tissue repair and facilitation of healing of bone tissue defects are challenges facing reconstructive medicine. Orthopedic traumas as well as dental and other socially significant bone diseases are among the most commonly occurring disorders. Based on their causation and pathogenesis, bone tissue defects can be divided into several groups: bone defects caused by cancer, inflammatory processes, atrophic processes in bone tissue, and traumas.

Diseases and traumas of musculoskeletal system are among the most frequently occurring disorders that impair the quality of life, and their percentage in the total morbidity grows steadily; about 30% of these disorders are caused by injuries. More than 1700 injuries per 100,000 population are recorded annually; 37.5% of them are upper limb fractures and 31.1% – fractures of lower limbs. Up to 20 million traumas are annually recorded in Russia. Fractures of long bones constitute between 27 and 88.2% of all fractures. Patients with femoral shaft and tibial shaft fractures usually require treatment in hospital. Large and deep open wounds and fat embolism pose a threat to the lives of the patients with such traumas. Moreover, this severe disability requires great material costs. In spite of the use of modern constructs and treatments, unfavorable clinical outcomes still reach 37%.

New plastic materials and technologies would also be useful for maxillofacial surgery and restoration of the destroyed periodontal tissue. The occurrence of diseases accompanied by the loss of tooth supporting tissues is very high. Inflammatory processes in the periodontium are recorded in 86% of the people in the 15 – 19 age group, reaching 100% in the 35–44 age group (Data found in the WHO oral data bank – www.medlinks.ru/sections.php/1994). There is high demand for materials that could be used to construct bone grafts to repair cranial bone defects, which heal very slowly. Up to 60,000 open head injuries are annually recorded in Russia, and up to 50,000 patients are qualified as disabled.

There are various materials to be used as transplants for bone tissue replacement, which differ in composition, form, and properties. None of them, however, fully conforms to the requirements of modern reconstructive surgery. Thus, new materials need to be found and the available ones should be improved.

Historically, biomaterials for repairing bone defects have been mainly derived from cartilage and/or bone tissues of humans and animals. Natural biomaterials are the most favorable materials for cell adhesion, wound healing and tissue regeneration, providing a matrix for the new growing tissue. For example, Bio-Oss® – a natural bone mineral – provides an optimal scaffold, into which new blood vessels grow from the peripheral host bone, and osteoblasts migrate into the graft and form a new bone at the graft site (Data found at http://www.mt21.ru/bio_oss1.shtml). The natural bone is chemically and physically treated to achieve high purity of the material. The material comes from the limbs of the animals. Calf bones, unlike their brain or spinal cord, belong to tissues free from prions as even in case of infected animals no contagious material has been found in bones. Bio-Oss® grafts are manufactured from U.S. cows only. The United States of America is considered to be a country without BSE-cases. The bone material is harvested from healthy animals only. During Bio-Oss production, the organic scaffold of bovine bone is removed leaving intercrystalline microtunnels and microcapillaries between apatite crystals. The remaining mineral matrix is similar to that of human bone, and, therefore, Bio-Oss® is better integrated into the bone than other natural deproteinized materials, which consist of larger crystals. The internal surface area of Bio-Oss® ($79.7 \text{ m}^2/\text{g}$) is more than 5 times as great as any of the bone replacing materials, due to the spaces retained between crystals and microtunnels. This facilitates revascularization and formation of new bone cells, increasing graft incorporation rate. Bio-Oss® has natural rough surface, which favors bone cell attachment and proliferation and synthesis of organic matrix. Bio-Oss® has a high osteoconductive property. Bio-Oss® resorbs itself slowly, being integrated in the process of natural remodeling, while synthetic preparations are broken down as a result of chemical decomposition.

Bioceramics are both bioactive and sufficiently strong materials. Bioceramics are used to fabricate replacements for teeth, bones, and joints. The main requirement for these materials is biocompatibility. Porous materials are materials of choice for osteoplastic surgery, as they are quickly incorporated into body tissues. The new bone grows into the pores, ensuring mechanical stability of the interface. The main advantages of porous biocomposites are that their mechanical strength is comparable to that of the bone and that they can be used to fabricate construction implants for the skull and the spine. The physical/mechanical properties of biocomposites are similar to those of spongy bone tissue. The commercially produced biocomposite (calcium-phosphate) materials are Interpore-200, Interpore-500, Calcitite-2040, Ostrix NR (U.S.), Ceros-80 (Switzerland), Osprovit 1,2 (Germany), BAK-1000 (Russia), and Bioapatite (France).

Hydroxyapatite and its composites with other materials are commonly used in maxillofacial surgery. Calcium-phosphate materials in the form of powders, granules, microparticles, plates, etc. and as composites with other materials can be prepared by various methods. Much information has been published about various applications of synthetic osteoplastic materials such as calcium sulfate, bioactive glass, cyanoacrylic acid esters, polyacrylamide gel, tricalcium phosphate, sodium alginate-based preparations, etc. Porous glass-ceramic bioimplants have been produced and studied; some are being subjected to clinical trials. The discovery of the ability of collagen to induce bone tissue regeneration in the 1970s gave an impetus to studies of biocomposite materials that contained both collagen and hydroxyapatite. Alveloform and Bigraft composites, containing purified fibrillary skin collagen (Collagen Corp., Palo Alto, U.S.) were developed for maxillofacial and dental surgery. These materials were successfully used to restore alveolar crest in surgery of periodontitis. Histological and ultrastructure investigations proved that the collagen/hydroxyapatite composition produced a favorable effect on bone regeneration, mainly serving as a scaffold, i.e., showing its osteoconductive property. A number of researchers think that biocomposites containing the skin collagen Ziderm and synthetic hydroxyapatite have some osteogenic potential. A series of such hybrid materials has been produced in Russia: Hydroxyapol, CollapAn® – composites of hydroxyapatite and collagen in the form of granules, plates, and gels, sometimes used as antibiotic carriers, designed by Intermedapatit, Rosdent, and Polystom; Bialgin® – a bioactive granulated osteoplastic material based on amorphous nano-disperse (5–10 nm), completely resorbable calcium hydroxyapatite incorporated in the polysaccharide matrix of sodium alginate. A great number of studies reported the use of hydroxyapatite/collagen composites. The Luga plant Belkozin produces two composites of collagen foam with hydroxyapatite, one of them incorporating chlorhexidine (an antiseptic) and thymmogen (an immunomodulator). Clinical trials showed that the biocomposite containing hydroxyapatite dramatically increased the rate of bone wound healing and practically eliminated the risk of postoperative complications and spontaneous fractures at the site where the cyst had been.

One of the recent approaches to improving mechanical properties of hydroxyapatite-based biomaterials (reducing rigidity, increasing elasticity), which has taken shape in the last 10–15 years, is preparation of hydroxyapatite-polymer composites. Although composites of hydroxyapatite and non-degradable polymeric materials (polyethylene, polysulphone) have good mechanical properties (Young's modulus, compressive strength, tensile strength, and bending strength), the presence of the polymers significantly reduces the biocompatibility of hydroxyapatite. In the opinion of specialists, the problem can be solved by constructing new composites based on hydroxyapatite and biopolymers. One of the solutions is to use polymeric materials that mimic interaction of the mineral part and the collagen matrix of bone tissue, which have been synthesized from natural polyesters (lactic, glycolic and other acids) capable of biodegradation and bioresorbtion. For instance, blending of PLA with hydroxyapatite dramatically (by a factor of 5–20) increases the mechanical strength of the composite compared with the hydroxyapatite/collagen systems. Reconstructive orthopedics still needs adequate materials to fill large bone cavities caused by osteoporosis or surgical removal of tumors. Common materials used in such cases are hydroxyapatite/collagen composites such as Collapan, but as this material is quickly hydrolyzed, it cannot be used to fill cavities larger than 4–6 cm^3. The use of polyurethane- and polymethyl methacrylate foam-based filling is still experimental

In order to improve bone healing, bone fragments are held by screws, nails and plates, which until recently were mainly made of metals or alloys. However, non-degradable internal fixation devices remain in the body after the fracture has healed, often causing discomfort and requiring removal. Surgical removal of internal fixation causes more pain and may induce infection. The use of polymer materials to construct fracture fixation devices has a number of advantages. First, they are lighter; second, they have better contact with tissues; and, third, they need not be removed. Both biostable and biodegradable polymer materials can be used. Devices fabricated from polymers by melt casting are suitable for fixing juxta-articular fractures. Good results were obtained with fixation devices coated with a thick layer (reaching 40% of the coating material) of porous polysulfone. Devices for external and internal fracture fixation fabricated from biodegradable polymers are even more attractive because they are completely replaced by the new bone tissue with time. Rather commonly used fixation devices are nails prepared from degradable polymers and reinforced with methyl methacrylate, nylon fibers, vinylpyrrolidone, etc. These constructs remain sufficiently strong for long periods, up to 6 months and even longer, allowing osteogenesis at the defect site to complete; they are fully resorbed *in vivo* in 1.5–2.0 years. Coatings for fracture fixation devices may contain antimicrobial agents to prevent infection. Not only polymer and composite materials but also adhesives (glues, cements) are used to join and hold together bone fragments. These substances, in addition to fixing the fragments and filling the fractures, also create an interface between the bone tissue and implant material. Acrylic cements, although insufficiently strong, are the most common adhesives, which have been used for a relatively long time to fix the shafts of the prostheses implanted into the channels of tubular bones.

Special consideration has been given to composites of biodegradable polymers of monocarboxylic acids (PLA, PGA) and hydroxyapatite. Filling of the polymers with hydroxyapatite enhances mechanical strength of the material and its affinity with bone tissue. These composites have properties similar to those of native bone tissue and high mechanical strength (compressive strength – over 100 MPa, tensile strength – 110–120 MPa, and impact strength – over 150 kJ/cm^2).

Thus, analysis of the market for new biomaterials and devices based on the available literature and databases show that the range of the medical-grade materials is rapidly expanding and their production outputs are growing. The focus has shifted to the use of high-molecular-weight synthetic and natural materials. Polymer materials that are biodegraded in biological media without releasing substances toxic to the organism have attracted special attention of researchers and manufacturers; their production outputs are growing continuously, and this is the tendency characterizing medical materials science in the 21st century. Degradable polymer materials have occupied a considerable market share; they are used to manufacture high-tech medical-grade products: functional suture material, various prostheses and implants, and biocompatible materials and composites for reconstruction of defects of the skin, soft tissues, and bones, and for fabrication of fillers and other materials.

KEYWORDS

- bioplastics
- clinical permission
- demands of medicine
- medical devices
- medical materials market
- novel biomaterials

REFERENCES

"Agenda 21" (Online): www.un.org/russian/conferen/wssd/agenda21/.

Biodegradable Materials Interest Community Association: *www.tcj.ru/2005/5/biorazl.*

Biosovmestimost (Biocompatibility). In: V. I. Sevastyanov. Information Center of Research Institute of Geosystems. Moscow, 1999, 368 p. (in Russian).

BioWorld (Online): http://www.cbio.ru/.

Chench, L. The challenge of orthopedic materials. Current Orthopaedics, 2000, 14, 7–15.

Ernst, Yung, L. L. P. American Biotechnology Report (Online): http://www.biovendor.com.

Fomin, V. A., & Guzeev, V. V. Biodegradable polymers, their current and potential uses. Plasticheskiye massy (Plastics), 2001, 2, 42–46 (in Russian).

German Biotech, C&EN, 12 June 2000 (Online): http://www.glycotope-bt.com/en/services/Biotechnological_Service.php.

Gerngross, T. U., & Martin, D. P. Enzyme-catalyzed synthesis of poly(R)-(3-hydroxybutyrate): formation of macroscopic granules *in vitro.* Proc. Natl. Acad. Sci. USA, 1995, 92, 6279–6283.

Glick, B., & Pasternak, J. Molecular Biotechnoology. Principles and Applications of Recombinant DNA. ASM Press. Washington, DC, 2010, 1000 p.

Hench, L., & Jones, J. Biomaterials, artificial organs and tissue engineering. Woodhead Publishing Limited: Cambridge, England, 2005, 284 p.

International Standard ISO 10993-1. Biological evaluation of medical devices. Reproduced by Global Engineering Documents with the Permission of ISO Under Royalty Agreement.

Jones, J. Artificial Organs. In: Biomaterials, Artificial Organs and Tissue Engineering. L. Hench, & Jones, J., Woodhead Publishing Limited. Cambridge, England, 2005, 142–152.

Lintex (Online): http://www.lintex.ru/index.php.

Ministry for Industry and Commerce of the RF (Online): http://minptomtorg.gov.ru.

Pickin, E. Biomaterials in Different Forms for Tissue Engineering: An Overview, Porous Materials for Tissue Engineering. In: D.-M. Liu, V. Dixit. Materials Science Forum, 1997, 250, 1–14.

Prudnikova, S. V., & Volova, T. G. Ekologicheskaya rol poligidroxialkanoatov: zakonomernosti biorazrusheniya v prirodnoy srede i vzaimodeistviya s mikroorganizmami (Ecological role of polyhydroxyalkanoates: biodegradation behavior in the natural environment and interactions with microorganisms). Krasnoyarskii Pisatel. Krasnoyarsk, 2012, 183 p. (in Russian).

Ratner, B. D. New ideas in biomaterials science – a path to engineered biomaterials. J. Biomed. Mater. Res., 1993, 27, 837–850.

Sevastyanov, V. I. Biomaterials for artificial organs. In: V. I. Shumakov. Iskusstvennyye organy (Artificial Organs). Meditsina. Moscow, 1990, 214–220 (in Russian).

Sevastyanov, V. I., & Kirpichnikov, M. P. Biosovmestimyye Materially (Biocompatible Materials) (course book). MIA Publishers. Moscow, 2011, 544 p. (in Russian).

Shishatsky, O. N., & Shishatskaya, E. I. Analiz rynka materialov i izdeliy meditsinskogo naznacheniya (Analysis of the Market for Materials and Devices for Medical Applications). Krasnoyarskii Pisate.l Krasnoyarsk, 2010, 144 p. (in Russian).

Shishatsky, O. N., Shishatskaya, E. I., & Volova, T. G. Razrushayemyye polimery: potrebnosti, proizvodstvo, primeneniye (Degradable biopolymers: need, production, applications). Novyye informatsionnyye tekhnologii (New Information Technologies). Krasnoyarsk, 2010, 156 p. (in Russian).

Shtilman, M. I. Polimery mediko-biologicheskogo naznacheniya (Polymers Intended for Biomedical Applications). Moscow: Akademkniga Publishers, 2006, 400 p. (in Russian).

Trademark "BIOPLASTOTANTM" Registration Certificate No. 315652 of the Federal Institute for Patent Examination for Application No. 2006703271/50, Priority of 15.02.2006.

Vasilov, R. G. Report: On the Strategy of Development of Biotechnology in Russia in 2010–2020. Cheboksary, 2009.

Vilensky, A. V., & Fedoseev, B. N. The market for medical devices and healthcare. Marketing v Rossii i za rubezhom (Marketing in Russia and Abroad), 2001, 3, 13–19 (in Russian).

VOLOT (Online): http://www.volot.ru.

Volova, T. G. Polyhydroxyalkanoates – Plastic Materials of the 21st Century: Production, Properties, and Application. Nova Science Pub., Inc. NY, USA, 2004, 282 p.

Volova, T. G., & Shishatskaya, E. I. Razrushayemyye biopolimery: polucheniye, svoistva, primeneniye (Degradable Biopolymers: Production, Properties, Applications). Krasnoyarskii Pisatel. Krasnoyarsk, 2011, 389 p. (in Russian).

Volova, T. G., Shishatskaya, E. I., & Sinskey, A. J. Degradable Polymers: Production, Properties and Applications. Nova Science Pub., Inc. NY, USA, 2013, 380 p.

WHO Oral Data Bank (Online): *www.medlinks.ru/sections.php/*1994.

CHAPTER 2

POLYHYDROXYALKANOATES: NATURAL DEGRADABLE BIOPOLYMERS

CONTENTS

2.1 INTRODUCTION

Polyhydroxyalkanoates (PHAs) – biopolymers of microbial origin – have very many attractive properties that make them promising materials for various applications, including biomedical ones. PHAs have significant advantages over other biomaterials, including polylactides:

- the high biocompatibility of PHAs, poly-3-hydroxybutyrate in particular, is accounted for by the fact that the monomers constituting this polymer – 3-hydroxybutyric acid – are natural metabolites of body cells and tissues;

- PHAs are not hydrolyzed in liquid media, as they undergo true biological degradation, which occurs via the cellular and the humoral pathways; the resulting monomers of hydroxybutyric acid do not cause abrupt acidification of tissues and, hence, do not give rise to any pronounced inflammatory reaction;

- PHAs bioresorption rates are much slower than those of polylactides and polyglycolides; PHA-based implants can function *in vivo* for between several months and 2–3 years, depending on their form and implantation site; moreover, PHA degradation can be controlled;
- PHAs are produced by direct fermentation; no multistage technology is needed (monomer synthesis, polymerization, addition of plasticizers and modifying components);
- PHAs can be synthesized from such feedstocks as saccharides, organic acids, alcohols, mixtures of CO_2 and H_2, products of plant biomass hydrolysis, industrial wastes of sugar and palm oil production, hydrogen-containing products of processing of brown coals and hydrolysis lignin;
- PHAs constitute a family of polymers of various chemical structures, consisting of monomers containing between 4 and 12 or more carbon units, including high-crystallinity thermoplastic materials and construction elastomers;
- PHAs properties (crystallinity, mechanical strength, temperature characteristics, and biodegradation rates) can be controlled by varying the composition of the culture medium and tailoring the chemical structure of the polymer; and
- PHAs can be processed from various phase states (powder, solution, gel, melt) using conventional techniques.

PHAs are very promising polymers as, being thermoplastic, like polypropylene and polyethylene, they also have antioxidant and optical properties as well as piezoelectricity. PHAs are highly biocompatible and can be biodegraded in biological media. In addition to poly-3-hydroxybutyrate (P3HB), there are various PHA copolymers, which, depending on their monomeric composition, have different basic properties (degree of crystallinity, melting point, ductility, mechanical strength, biodegradation rate, etc.).

2.2 CHARACTERIZATION OF PHAS: TYPES, SYNTHESIS, AND PROPERTIES

The first of the discovered polyhydroxyalkanoates (PHAs) was poly-3-hydroxybutyrate P(3HB). Until the end of 1973, interest in P(3HB) had been directed almost solely at its physiological significance in the functioning of microorganisms and at the influence of environmental factors on its synthesis and reutilization. The oil crisis in autumn 1973 and the subsequent increase in the price of oil, as a nonrenewable source of energy and feedstock, made OPEC members, which control the plastics market, realize the necessity of searching for alternative ways of producing plastic materials other than petrochemical synthesis of polyolefins.

In 1976, Imperial Chemical Industries (ICI) of England started the first commercial investigation of microbiological production of poly-3-hydroxybutyrate using sugar-containing substrates extracted from plant biomass (Senior, 1984). Not only could P(3HB) be synthesized from renewable sources, but some of its properties, such as thermoplasticity, resembled those of polypropylene (King, 1982). Other P3HB properties, such as biodegradability and biocompatibility, piezoelectric ability, and the possibility of using it as a source of optically active molecules (Howells, 1982), were recognized early on as additional assets and kept ICI's interest in P(3HB) alive after the oil crisis had begun to pass.

In succeeding years, interest in the process of biological synthesis of poly-3-hydroxybutyrate increased. It was found that P(3HB) could be synthesized by many prokaryotic microorganisms (more than 300 have been identified by now) with different efficiency, using various substrates. However, just a few species of microorganisms were chosen for commercial synthesis. They were the microorganisms that efficiently synthesized P(3HB) on a number of substrates: saccharides, methanol, hydrocarbons, and mixed hydrogen and carbon dioxide (the hydrogen-oxidizing bacteria such as *Alcaligenes eutrophus,* which is now known as *Ralstonia eutropha,* and *Alcaligenes latus,* the nitrogen-fixing bacterium *Azotobacter vinelandii,* the pseudomonade *Pseudomonas oleovorans,* the methylotrophs *Methylomonas* and *Methylobacterium organophilum* (Anderson and Dawes, 1990; Byron, 1987, 1994; Dawes, 1990; Braunegg et al., 1998).

However, pure poly-3-hydroxybutyrate is brittle and has a low extension to break. The lack of flexibility and thermal stability limits its range of applications. The discovery of the ability of microorganisms to synthesize PHA copolymers gave an impetus to extensive investigations of these biopolymers. It was found that the presence of 3-hydroxyvalerate

in PHA copolymers was a factor that significantly affected polymer properties, decreasing its melting point and crystallinity and making it more
ductile, elastic, and processable than poly-3-hydroxybutyrate (Luizier,
1992). Variations in monomer proportions of the PHA lead to considerable changes in its thermomechanical and fibrous properties. An extensive
search for microorganisms capable of synthesizing PHA copolymers was
conducted in many countries. Rather soon, it was found that some microorganisms under certain growth conditions were able to synthesize not
only homogenous poly-3-hydroxybutyrate but also various PHAs containing monomer units of 3-hydroxybutyrate and other hydroxy derivatives of
hydrocarbon acids. By now, over 150 various PHAs have been described,
but the PHAs that are practically produced and investigated are thebho-
mogenous poly-3-hydroxybutyrate (3HB) and copolymers of 3-hydroxy-
butyrate and 3-hydroxyvalerate, P(3HB/3HV), and 3-hydroxybutyrate and
3-hydroxyoctanoate, P(3HB/3HO). PHAs of different chemical composition differ in their structure and basic physicochemical properties. This
line of research has attracted much attention because even a slight variation in the proportions of monomer units in a PHA can lead to a fundamental change in its properties, including thermomechanical ones, which
is very important for practical purposes.

The main PHA structures are:

$$\left[O-\underset{\underset{R}{|}}{\overset{\overset{H}{|}}{C}}-(CH_2)_n-\underset{\underset{O}{\|}}{C}\right]_{100-30000}$$

n = 1   R = hydrogen-poly(3-hydroxypropionate),
        R = methyl-poly(3-hydroxybutyrate),
        R = ethyl-poly(3-hydroxyvalerate),
        R = propyl-poly(3-hydroxyhexanoate),
        R = pentyl-poly(3-hydroxyoctanoate),
        R = nonyl-poly(3-hydroxydodecanoate),
n = 2   R = hydrogen-poly(4-hydroxybutyrate),
n = 3   R = hydrogen-poly(5-hydroxyvalerate).

Most of the known hydroxyalkanoic acids have been detected as constituents of biosynthetic PHAs. Analysis of the diversity of polyhydroxyalkanoates showed that biosynthetic polymers contain HAs exhibiting the R-configuration but no HAs exhibiting the L-configuration. PHAs can be divided into three groups depending on the number of carbon atoms in the monomer units:

1. short-chain-length (SCL) PHAs, which consist of 3–5 carbon atoms;
2. medium-chain-length (MCL) PHAs, which consist of 6–14 carbon atoms; and
3. long-chain-length (LCL) PHAs, which consist of 17 and 18 carbon atoms.

The best-known representative of short-chain-length PHAs is poly-(R)-3-hydroxybutyrate, P3HB. Short-chain-length PHAs consist of poly-(R)-hydroxyalkanoates containing 3 to 5 carbon units. These polyesters are synthesized by different bacteria (*A. latus, Bacillus cereus, Pseudomonas pseudoflava – Hydrogenophaga pseudoflava, Pseudomonas cepacia, Micrococcus halodenitrificans, Azotobacter sp., Rhodospirillum rubrum, Ectothiorhodospira shaposhnikovii*, and *Cupriavidus necator*). *C. necator* (previously known as *Wautersia eutropha, Ralstonia eutropha* or *Alcaligenes eutrophus*) is a universally recognized producer of short-chain-length PHAs (Nakamura et al., 1991; Bear et al., 1997; Ballistreri et al., 1999; Babel et al., 2001; Pettinari et al., 2001; Zhang et al., 2004; Lenz and Marchessault, 2005; Stubbe et al., 2005; Hazer and Steinbüchel, 2007).

Medium-chain-length PHAs consist of aliphatic (R)-monomers of hydroxy acids, containing 6 to 14 carbon units. They are synthesized when the PHA producing strain is cultivated on the medium containing n-alkanoates or their precursors (Gross et al., 1989; Madison and Huisman, 1999; Witholt and Kessler, 1999).

Long-chain-length PHAs consist of more than 14 carbon atoms. This division of polymers into groups was based on the notion of high substrate specificity of PHA synthases, which were supposed to be able to accept only certain hydroxyalkanoic acids in the course of intracellular polymerization of monomers into PHAs (Anderson, Dawes, 1990).

PHAs that contain a single type of repeat unit are known as homopolymers, while PHAs containing a mixture of repeat units are known as copolymers (Madison and Huisman, 1999; Olivera et al., 2001a, b; Luengo et al., 2003). PHA characteristics, including physical and mechanical properties, are determined by the molar fractions of monomers in the polymer. The properties of PHA homopolymers are significantly different from those of PHA copolymers (Matsusaki et al., 2000; Chen et al., 2006; Laycock et al., 2013). Short-chain-length PHAs are thermoplastics that have a high degree of crystallinity; they are rigid and brittle crystalline materials that have low elongation-to-break. Medium-chain-length PHAs are elastomers with low crystallinity and a low melting point (Nakamura et al., 1991; De Koning, 1995; Kim et al., 2001; Zinn et al., 2001; Lenz and Marchessault, 2005).

Genetic construction of recombinant PHAs producing strains, with genes from PHAs produces introduced into these strains, has a great potential for synthesis of various polymers (Schmack et al., 1998; Steinbüchel and Hein, 2001; Lu et al., 2004; Chen et al., 2006; Wei et al., 2009).

As the diversity of PHAs is increasing, it has been proposed that they should be divided into two categories, based on the frequency of their occurrence: "usual" and "unusual" polyhydroxyalkanoates (Olivera et al., 2010).

Usual PHAs are PHAs intracellularly synthesized by microorganisms as storage macromolecules. This group includes polymers tailored from monomers (i.e., (R)-3-hydroxypropionate, (R)-3-hydroxybutyrate, (R)-3-hydroxyvalerate, (R)-3-hydroxyhexanoate, (R)-3-hydroxyoctanoate, (R)-3-hydroxydecanoate, and (R)-3-hydroxydodecanoate, or combinations thereof), which are obtained from different carbon sources (saccharides, alkanes, aliphatic fatty acids, triacylglycerols, etc.) through general pathways (usually fatty acid synthesis and fatty acid β-oxidation) involving the synthesis of (R)-3-hydroxyacyl-CoAs. In the final phase, these thioesters are polymerized to form poly-(R)-3-hydroxyalkanoates (Steinbüchel and Füchstenbusch, 1998; Rehm et al., 1998; Olivera et al., 2001a, b; Kessler and Witholt, 2001; Luengo et al., 2003, 2004).

"Unusual" PHAs are polyesters rarely found in nature that are constituted by:

- natural monomers with an unusual chemical structure (e.g., 4-hydroxyalkanoic acids, 5-hydroxyalkanoic acids, and 6-hydroxyalkanoic acids) that are synthesized by some microorganisms (Saito and Doi, 1994; Schmack et al., 1998; Choi et al., 1999a, b; Amirul et al., 2008); or

- unnatural monomers (generally obtained by chemical synthesis – xenobiotics) that can be taken up by the PHA producing microorganisms, activated to their CoA thioesteres, and used as substrates by the PHA polymerases.

"Unusual" PHAs include a group whose lateral chains contain double or triple bonds and/or different functional groups (Doi et al., 1987; Fritzsche et al., 1990a, b, c; Eggink et al., 1995; Kim et al., 1998; García et al., 1999; Lee et al., 1999, 2000; Imamura et al., 2001; Kim et al., 2001; Luengo et al., 2003). The term "unusual" PHAs (UnPHAs) unites many different PHAs, including semisynthetic and synthetic PHAs, PHA homopolymers and copolymers, and polymers produced by physical modification of natural polymers (blended polymers). UnPHAs include PHAs of microbial origin that have been synthesized either from natural monomers bearing different chemical functions, or from chemical derivatives of the natural ones and PHAs produced either by chemical synthesis or by physical modifications of naturally occurring polymers. Based on their chemical structure, UnPHAs can be grouped in four different classes (Olivera et al., 2010):

Class 1 includes PHAs whose lateral chains contain double or triple bounds or/and different functional groups (methyl, methoxy, ethoxy, acetoxy, hydroxyl, epoxy, carbonyl, cyano, phenyl, nitrophenyl, phenoxy, cyanophenoxy, benzoyl, halogen atoms, etc.).

Class 2 includes PHAs in which the length of the monomer participating in the oxoester linkage has been modified (the hydroxyl group to be esterified is not located at C-3).

Class 3 contains the polymers in which some oxoester linkages have been replaced by thioester functions (thioester-containing PHAs).

Class 4 includes those PHAs that have been manipulated chemically or physically.

Class 1 UnPHAs containing monomers with unsaturated bonds or functional groups have been obtained from different bacteria (*Pseudomonas aeruginosa*, *P. putida*, *P. oleovorans* and *P. resinovorans*) when cultured in media containing n-alkenes, n-alkynes, unsaturated fatty acids (undecenoic, undecynoic, oleic or linoleic acids), triglycerides and oils from different origins. Although saturated mcl-PHAs are mainly built from monomers whose carbon length ranges from C_6 to C_{14}, when unsaturated monomers are polymerized, the UnPHAs synthesized usually contain monomers with a longer carbon chain length (from C_6 to C_{16}). UnPHAs with elongated backbones (Class 2) contain as monomers hydroxyalkanoic acids different from (R)-3-hydroxyalkanoates (Saito and Doi, 1994; Schmack et al., 1998; Choi et al., 1999a, b; Amirul et al., 2008). These are representatives of scl-PHAs and mcl-PHAs, which have an elongated backbone. Examples of these UnPHAs are copolymers containing as repeating units (R)-3-hydroxybutyric acid and 4-hydroxybutyric acid – poly-(R)-3-hydroxybutyrate-co-4-hydroxybutyrate P(3HB/4HB).

Another way to optimize PHA processing and improve commercial properties of the product is to modify chemically the polymers that have been synthesized in biological processes and to prepare composites and blends of PHAs with natural and synthetic materials with different compositions. This research has been started quite recently, but preliminary results show that PHAs can be used to prepare blends and composites with enhanced physical and mechanical properties. The techniques employed to produce blended polymers are those based on either melt blending (direct mixing of component polymers in a molten state) or solution blending. The hydrophilicity, miscibility, crystallization behavior, thermal characteristics, mechanical properties, polymer morphology and biodegradation rates of PHA-containing blends are largely influenced by the nature of the different components (Verhoogt et al., 1994; Ha and Cho, 2002). In sum, blending procedures have allowed the production of many different PHA-containing mixtures with special characteristics and broad applications.

There is great activity in developing P(3HB)- and P(3HB-co-3HV)-based composites with a wide range of materials, such as polyethylene oxide (Avella et al., 1991), polyvinyl alcohol (Yoshi et al., 1992), synthetic atactic poly-3-hydroxybutyrate (Abe and Matsubara, 1995), polylactide, vinyl acetate (Yoon et al., 2000), polyethylene glycol (Hao and Deng, 2001), cellulose (Maekawa et al., 1999), polycaprolactone (Chum and Kim, 2000), etc.

The description and testing of new PHA producing strains, the design of novel methods for the detection of PHAs in mixed microbial populations, and recent advances in genetic, metagenomic, and metabolic engineering together with new approaches based on chemical synthesis and blending will expand the number of PHAs and their potential applications.

In sum, although PHAs were discovered relatively recently (Lemoigne 1926, 1927) and have been actively studied since the mid-1980s – the early 1990s, these compounds are among the most promising biomaterials of the 21st century.

Polyhydroxyalkanoates are storage macromolecules of the cell (energy and carbon reserves) synthesized by prokaryotic organisms under specific conditions of unbalanced growth, when synthesis of the major compounds (protein and nucleic acids) is reduced, but there is excess carbon in the medium. Very many microorganisms, including wild-type and genetically modified strains, are capable of storing PHAs (Braunegg et al., 1998; Madison and Huisman, 1999). The conditions under which constructive metabolism is directed to PHA synthesis and accumulation are determined by the redox state of the cytoplasm and intracellular concentrations of acetyl-CoA and free CoA (Oeding, Schlegel, 1973; Senior and Dawes, 1973). Under unbalanced growth, in a medium lacking one of the constructive elements (nitrogen, phosphates, etc.) or under oxygen deficiency, acetyl-CoA does not enter the tricarboxylic acid cycle and the level of free CoA is low. This is a favorable condition for activation of poly-3-hydroxybutyrate P(3HB) synthesis enzymes. Professor Chen's review (Chen, 2010a, b) describes various biochemical pathways of PHA synthesis by prokaryotes. The diversity of the pathways of PHA synthesis by different microorganisms is a basis for the production of a wide range of polymers belonging to this class.

PHAs are synthesized by microorganisms from various carbon sources. The cost of the carbon source contributes significantly to the overall production cost of a PHA. Thus, one of the main objectives of PHA investigations is to find inexpensive substrates for PHA production. This objective can be attained by finding new substrates for the already known PHA-producing organisms, discovering new PHA-producing strains, and engineering recombinant strains that can utilize various substrates, including new ones. In principle, PHAs can be produced on various substrates. Among the best known ones are individual compounds, such as carbon dioxide and hydrogen, saccharides, alcohols, and organic acids; byproducts of alcohol, sugar, and hydrolysis industries and of olive and palm oil production; and unusual substrates, including toxic ones. The choice of the substrate used for PHA production is determined by physiological-biochemical properties of PHA-producing microorganisms, economic efficiency of the production strategy, and the field of application of the ready produce. Since PHAs can be used in medicine, pharmacology, food industry, agriculture, and municipal engineering, the scale of their production may vary from dozens of kilograms (for medical and pharmacological applications) to several thousand tons a year. The quality and cost of the substrates for different production scales and applications may vary, too.

A virtually inexhaustible supply of feedstocks for large-scale PHA production can be provided by plant biomass generated in great quantities every year. Residues of agricultural crops contain polysaccharides of various structures and compositions; they can be hydrolyzed to yield a wide range of water-soluble sugars, including maize and soy hydrolysates, cotton plant vegetative biomass hydrolysates, etc. High cell concentrations and PHA yields can be obtained by growing various bacterial species on hydrolyzed hemicellulose containing xylose.

Cheese and sour milk production wastes, materials from sugar industry, and wastes from palm oil production can be used as substrates for PHA synthesis. A fundamental review addressing this issue was written by Dr. Koller and coauthors (Koller et al., 2010). In addition to their main components, complex waste streams can contain additional substances that make them advantageous in direct comparison with pure and expensive substrates. For example, some wastes provide the production strain in bioprocesses not only with carbohydrate lactose, but also with minor components such as minerals, protein residues, and vitamins that have positive impacts on the microbial cultivation (Purushothaman et al., 2001). On the other hand, such complex waste streams can also contain non-fermentable components and even compounds that have inhibiting effects on growth and production kinetics of microbes (Solaiman et al., 2006). For certain microbial strains, the composition of these substrates needs to be adjusted prior to the application of this substrate, e.g., for PHA production (Braunegg et al., 2007). Some waste materials contain all components necessary for PHA synthesis, and, thus, there is no need to supplement the culture medium with yeast or casein extracts (Neto, 2006).

Different fat-containing industrial wastes such as spent cooking oils and fats, which can be found in great quantities, residues of animal fats that can be processed into triacylglycerides, and meat and bone waste hydrolysates can be used as substrates for PHA synthesis. Whey from the dairy and cheese industries constitutes a waste and surplus material in many regions of the world, but it is rather costly to purify and use it (Audic et al., 2003). Lactose, the major carbohydrate in whey, can serve as a substrate for growth and product formation in numerous biotechnological processes, including PHA synthesis.

Lignocellulosic materials (about 60% of all plant biomass), consisting of lignin, cellulosic, and hemicellulosic fibers, are produced in great quantities during photosynthesis. These materials are a source of various saccharides (glycans, arabinose, xylose, galactose, alcohols, etc.) for biotechnological production. The composition of these waste products differs in terms of the proportions of lignin (10–25%), cellulose (30–60%) and hemicellulose (25–35%) (Kumar et al., 2008). Industrial branches generating the major shares of this waste are the agro-industry, the wood-processing industry and the paper industry. An estimation done by the FAO indicates that regarding the non-wood lignocellulosic biomass only, the amount produced annually is about 2.5×10^9 tons, and about 7×10^8 tons of different crop wastes are produced annually worldwide. As an example, the yearly cultivation of 6×10^8 tons of palm biomass generates a waste stream that contains about 90% of the entire palm plant (Kumar et al., 2008). Recently, increasing efforts have been dedicated worldwide to establishing biorefinery plants for the conversion of lignocellulosic and cellulosic waste to starting materials for the biotechnological production of bioethanol, biodiesel, and PHAs.

Coal and byproducts of coal processing have recently become an object of investigations for a number of researchers, as a substrate for some biotechnological processes (Catcheside and Ralph, 1999; Fakoussa and Hofirichter, 1999). This is a rather cheap carbon source available in large quantities and, therefore, it may be used as a substrate for the production of microbial biopolymers on a large scale. Füchtenbusch and Steibüchel (1999) investigated the production of medium-chain-length PHAs by several *Pseudomonas* cultures on media containing coal liquefaction product as a carbon source. *Ps.oleovorans* accumulated PHAs consisting of

3-hydroxyhexanoate, 3-hydroxtdecanoate, and 3-hydroxydodecanoate, while *Rhodobacter rubber* accumulated PHAs containing 3-hydroxybutyrate and 3-hydroxyvalerate. Coal liquefaction products obtained from *Trichoderma* were better substrates for *Ps.oleovorans* than products of coal chemical hydrolysis. In the future, two- or three-stage biotechnological processes can be organized: in the first stage, coal will be converted to water-soluble products by, e.g., fungi, and then these substrates will be used for PHA synthesis.

Among hydrogen bacteria there are organisms resistant to carbon monoxide (CO). This is the strain *Alcaligenes eutrophus* Z1 (currently known as *Ralstonia eutropha*) from Academician G.A. Zavarzin's collection (Institute of Microbiology RAS, Moscow) (Savelieva, Zhilina, 1968; Savelieva, 1979), as well as its fast-growing variant *A. eutrophus* B5786, isolated in the Institute of Biophysics SB RAS (Stasishina, Volova, 1996). It is important that CO does not inhibit hydrogenase of these organisms (Gruzinsky et al., 1977). Researchers of the Institute of Biophysics SB RAS were the first to grow these bacteria on hydrogen in the presence of CO (Volova et al., 1989, 1996), showing that CO did not affect kinetic and production characteristics of bacterial culture that produced PHA under nitrogen deficiency (Volova, 2004, 2009; Volova et al., 2002; Volova, Voinov, 2004). These results provided a scientific basis for PHA production on hydrogen-containing products derived from natural carboniferous raw materials, including mixtures of hydrogen, carbon monoxide, and carbon dioxide. These findings made it possible, for the first time in practical biotechnology, to synthesize high yields of polyhydroxyalkanoates from products of gasification of lignin and brown coals (RF Patent No. 2207375).

Thus, a great number of various compounds and wastes can be used for PHA synthesis and large-scale production. Feasibility studies show that PHAs can be produced commercially, using cost-efficient technologies and various feedstocks, including waste materials (Choi and Lee, 1997, 1999; Braunegg et al., 1998; Natano et al., 2001).

Polyhydroxyalkanoates can be used to fabricate specialized products by various methods (Amass et al., 1998; Sudesh et al., 2000). PHA properties vary depending on the structure of the lateral groups in the polymer chain and the distance between ester groups in a molecule. The linear structure of PHA molecules makes them thermoplastic. Furthermore, these polymers also have antioxidant and optical properties as well as piezoelectricity. PHAs are highly biocompatible and can be biodegraded in biological media. The combination of properties of PHAs makes them promising materials for various applications. This class of polymers can be used to produce materials with different physical and mechanical properties – from hard thermoplastics to construction elastomers.

One of the most important macroscopic parameters characterizing polymer properties is molecular weight, which determines the processability of the material. PHA molecular weight is a variable parameter, depending on a number of factors, including the source of carbon nutrition for bacteria, the length of cultivation, and the method of polymer recovery. The molecular weights of PHAs differ considerably; for instance, the weight average molecular weight may vary from several hundred to several million Da.

Thermal properties of P(3HB) and its ability to crystallize in the native state are its most important parameters because they determine the processability of the polymer. PHAs, like many other polymers, have a heat distortion temperature somewhat lower than the thermal degradation temperature. Thus, polymers cannot exist in the gaseous state, and the main type of phase equilibrium in them is a condensed state – crystalline, glassy, viscoelastic, and liquid. The ability of PHAs to crystallize is determined by the inner properties of its chains and is characterized by crystallization temperature, T_c. In a number of polymers, including PHAs, crystallization develops only partly for various reasons, and most of the polymers are semi-crystalline materials. The degree of crystallinity of PHAs with different structure can vary widely, between 10–20 and 70–80%, and this parameter can be controlled by changing the chemical composition of the polymers produced under different conditions of cultivation of PHA-producing bacterial strains and different carbon nutrition conditions (Volova, Shishatskaya, 2011; Volova, Shishatskaya, Sinskey, 2013; Laucock et al., 2013). In spite of the diversity of PHAs, few of them have been thoroughly studied and found suitable for commercial production (Table 2.1).

The database compiled by researchers of the Siberian Federal University and the Institute of Biophysics SB RAS [No. 2012620288] shows that

TABLE 2.1   Chemical Structures of the Best Studied PHAs, Their Producers, and Substrates for Their Synthesis

Producer, substrate	PHA type
Ralstonia eutropha CO_2, saccharides, organic acids, alcohols	$\{O-\overset{\displaystyle H}{\underset{\displaystyle CH_3}{C}}-CH_2-\overset{\displaystyle}{\underset{\displaystyle O}{C}}\}_x$ R-3-hydroxybutyrate
Ralstonia eutropha *Azotobacter* Saccharides, propionic acid	$\{O-\overset{\displaystyle H}{\underset{\displaystyle CH_3}{C}}-CH_2-\overset{\displaystyle}{\underset{\displaystyle O}{C}}\}_x\{O-CH_2-CH_2-CH_2-\overset{\displaystyle}{\underset{\displaystyle O}{C}}\}_y$ R-3-hydroxybutyrate 3-hydroxypropionate
Ralstonia eutropha *Alcaligenes latus* *Comamonas acidovorans* Saccharides, γ-butyric acid, γ-butyrolactone	$\{O-\overset{\displaystyle H}{\underset{\displaystyle CH_3}{C}}-CH_2-\overset{\displaystyle}{\underset{\displaystyle O}{C}}\}_x\{O-\overset{\displaystyle H}{\underset{\displaystyle C_7H_{11}}{C}}-CH_2-\overset{\displaystyle}{\underset{\displaystyle O}{C}}\}_y$ R-3-hydroxybutyrate 4-hydroxybutyrate
Ralstonia eutropha *Azotobacter* CO_2, saccharides, organic acids, valeric acid	$\{O-\overset{\displaystyle H}{\underset{\displaystyle CH_3}{C}}-CH_2-\overset{\displaystyle}{\underset{\displaystyle O}{C}}\}_x\{O-\overset{\displaystyle H}{\underset{\displaystyle C_3H_7}{C}}-CH_2-\overset{\displaystyle}{\underset{\displaystyle O}{C}}\}_y$ R-3-hydroxybutyrate 3-hydroxyvalerate
Ralstonia eutropha *Alcaligenes latus* *Aeromonas cavie* CO_2, saccharides, organic acids, hexanoic acid, vegetable oils, 1,5-pentanediol	$\{O-\overset{\displaystyle H}{\underset{\displaystyle CH_3}{C}}-CH_2-\overset{\displaystyle}{\underset{\displaystyle O}{C}}\}_x\{O-CH_2-CH_2-\overset{\displaystyle}{\underset{\displaystyle O}{C}}\}_y$ R-3-hydroxybutyrate 3-hydroxyhexanoate
Ralstonia eutropha *Pseudomonas sp.* Saccharides, vegetable oils, alkenes, 4-hydroxybutyric acid	$\{O-\overset{\displaystyle H}{\underset{\displaystyle CH_3}{C}}-CH_2-\overset{\displaystyle}{\underset{\displaystyle O}{C}}\}_x\{O-\overset{\displaystyle H}{\underset{\displaystyle C_2H_5}{C}}-CH_2-\overset{\displaystyle}{\underset{\displaystyle O}{C}}\}_y$ R-3-hydroxybutyrate 3-hydroxydecanoate

the properties of these polymers differ considerably depending on the type of monomer units and their proportions in the polymer chain.

The first among the isolated PHAs and the most fully characterized one is poly-3-hydroxybutyrate P(3HB). P3HB is a homopolymer of D(-)-3-β-hydroxybutyric acid, an isotactic polyester with regular $(C_4H_6O_2)$ units (Table 2.1). In contrast to complex synthetic polyesters, P3HB is a stereoregular optically active polymer, which forms helices in a solution and crystallizes in spherulites. P(3HB) is a colorless semi-crystalline hydrophobic substance. The density of the amorphous phase in P(3HB) is 1.177 g/cm^3 and the density of the crystalline phase is 1.23–1.26 g/cm^3. The melting temperature of P(3HB) varies between 160 and 185°C, and the T_m proper is 176–182°C. The thermal decomposition temperature of P(3HB) lies within the range between 275 and 280°C. After reheating, the melting temperature of P(3HB) decreases, while its crystallization temperature remains unchanged. The pronounced difference between the melting point and the temperature for the onset of degradation is an essential processing property of the polymer, enabling fabrication of films, filaments, hollow constructs, etc. by conventional methods of polymer processing (solution casting, extrusion, and injection molding). P3HB is isotactic and similar to the isotactic polypropylene, as both have pendant methyl groups attached to the main chain in a single conformation (Brandl et al., 1988). The P3HB chain has a 21 helix conformation, an orthorhombic unit cell, and space group P212121 with unit cell parameters a = 5.76 Å, b = 13.20 Å, and c = 5.96 Å (Yokouchi et al., 1974). The data obtained by X-ray structure analysis suggest a conclusion that the crystalline region is predominant in P(3HB). The degree of crystallinity of different P(3HB) samples is only slightly influenced by the conditions of polymer production, ranging between 70 and 80% (Lee, 1996; Madison and Huisman, 1999; Volova et al., 2000; Laucock et al., 2013). A disadvantage of P(3HB) is that it cannot be strain-crystallized, and, thus, it is very difficult to process. The resulting products show low impact strength and hardness and are prone to aging.

The physicochemical properties of PHA copolymers are more diverse, and, thus, they show greater promise. However, production of PHA copolymers is a very complicated technological task, and it cannot be achieved without fundamental knowledge of the structural-functional organization of the PHA cellular cycle and the relationship between the physicochemical properties of these polymers and their chemical composition.

The T_m and T_d of PHA copolymers are lower than those of P(3HB). The melting temperature of 3-hydroxybutyrate/3-hydroxyvalerate copolymers is lower than the melting peak of homogeneous P3HB, and this difference increases with an increase in the 3HV molar fraction. The melting and thermal decomposition temperatures of the copolymers consisting of 3-hydroxyhexanoate and 4-hydroxybutyrate monomer units are also lower than those of P(3HB). The low crystallization temperature of the homogeneous P(3HB) pose an obstacle to melt processing of this polymer. A possible way to increase the crystallization temperature of PHA is to synthesize 3HB copolymers tailored with 3HV and other monomers. The decrease in the melting temperatures and thermal decomposition temperatures of PHA copolymers does not reduce the difference between these parameters. Thus, different types of PHA copolymers retain their significant property – thermoplasticity.

The degrees of crystallinity of P(3HB/3HV)s are lower than that of P(3HB). The incorporation of 3HV monomers into the 3HB chain decreases the degree of crystallinity of P(3HB/3HV) copolymers. P(3HB/3HV) copolymers exhibit isodimorphism due to co-crystallization of the 3HB and 3HV units. Incorporation of 3HV into the 3HB polymer chain considerably influences the kinetics of polymer crystallization, including the rate of the process and the size of the spherulites. The degree of crystallinity of these copolymers is linearly related to the molar fraction of 3HV only in the specimens containing up to 25–35 mol.% 3HV (Volova, 2004; Volova et al., 2008). Specimens with higher 3HV molar fractions (over 35–40 mol.%) showed the degree of crystallinity of about 50%. This is similar to the C_x of the copolymers containing lower 3HV molar fractions. This may be ascribed to a change in the crystal lattice that occurs when the 3HB to 3HV ratio is changed. At less than 35–40 mol.% 3HV, 3HV units can crystallize in the 3HB lattice and at greater than 40 mol.% 3HV, 3HB units can crystallize in the 3HV lattice. In specimens with 3HV as the major fraction, 3HB and 3HV units co-crystallize in one lattice, and this does not seem to influence the degree of crystallinity of P(3HB/3HV), in spite of an increase in the 3HV molar fraction.

Similar changes in C_x are observed in 3-hydroxybutyrate/3-hydroxyhexanoate P(3HB/3HHx) copolymers. Incorporation of 3HHx, like 3HV in P(3HB/3HV), reduces the difference between the percentages of the amorphous and crystalline phases. As the 3HHx molar fraction is increased from 2.5% to 40–60%, the degree of crystallinity of the copolymer steadily decreases to 20–40%. Another PHA type – 3-hydroxybutyrate/4-hydroxybutyrate P(3HB/4HB) copolymers are highly elastic polymers, with elongation at break reaching 1000%, which is two orders of magnitude higher than the corresponding parameter of P3HB. The incorporation of 4HB influences considerably (to a greater extent than the incorporation of 3HV or 3HHx the ratio of crystalline to amorphous phase in the copolymer, significantly decreasing its crystallinity. P3HB/4HB copolymers with crystallinity degrees decreased to 10–12% have been described by some authors (Laucock et al., 2013; Volova et al., 2013, 2014a, b).

2.3 VOLUMES OF PRODUCTION AND APPLICATIONS OF PHAS

Many companies have been engaged in commercialization of PHA production technologies. Since the 1980s, they have been producing PHAs on a pilot scale or industrially (Table 2.2). The best-known companies and corporations that have been engaged in PHA activities are: Monsanto Co., Metabolix Inc., Procter & Gamble, Berlin Packaging Corp., Bioscience Ltd., BioVentures Alberta Inc., and Merk, which produce polymers with the trademarks Biopol®, Biopol™, TephaFLEX™, DegraPol/btc®, Nodax™, etc.

Imperial Chemical Industries (ICI) of England was the first industrial corporation to start commercial production of PHAs. Since 1992, Zeneka Seeds and Zeneka Bio Products (U.K.) began commercialization of a family of poly-3-hydroxybutyrate and P(3HB/3HV) copolymers with the tradename of Biopol®. The cost of Biopol®, which was produced at 10–15,000 tons per year, reached US$ 16,000/t. That was an order of magnitude higher than the global market value of polypropylene.

A world leader in PHA commercialization is Metabolix Inc. (U.S.), which was founded in Cambridge (MA) in 1992. The company has more than 500 owned and licensed patents and applications worldwide. The company is producing polymers using a recombinant strain – E.coli K12 – and sugars as substrate. The commercial names of Metabolix polymers are Biopol® and Biopol™. Their outputs reached 90 t in 2005 and 907 t in 2006. The company has many branches in different countries of the world. In 2004, Metabolix formed a strategic alliance with "Archer Daniels Midland Company" (ADM) to commercialize PHAs using the large fermentation capacity of ADM. In 2009, Metabolix marketed a new family of biodegradable plastics – Mirel. ADM has begun construction of the

TABLE 2.2   Companies Engaged in Scaling-Up Laboratory Processes and Establishing PHA Production Facilities

Company	Types of PHA	Production scale (tons/year)	Application
"ICI" (U.K.)	P(3HB/3HV)	300	Packaging
"Chemie Linz" (Austria)	P(3HB)	20–100	Packaging, drug delivery
"Biomers" (Germany)	P(3HB)	Unknown	Packaging, drug delivery
"BASF" (Germany)	P(3HB), P(3HB/3HV)	Pilot scale	Blending with Ecoflex
"Metabolix" (U.S.)	Several PHAs	Industrial scale	Packaging
"Tepha" (U.S.)	Several PHAs	Several hundred kg	Medical applications
"ADM" (U.S.) (with Metabolix)	Several PHAs	50,000	Feedstock
"P&G" (U.S.)	Several PHAs	Contract manufacture	Packaging
"Monsanto" (U.S.)	P(3HB), P(3HB/3HV)	Industrial production of PHAs	Feedstock
"Meredian" (U.S.)	Several PHAs	10,000	Feedstock
"Mitsubishi" (Japan)	P(3HB)	10	Packaging
"Biocycles" (Brasil)	P(3HB)	100	Feedstock
"Bio-On" (Italy)	PHA	10,000	Feedstock
"Zhejiang Tian An" (China)	P(3HB/3HV)	2,000	Feedstock
"Yikeman, Shandon" (China)	PHA	3,000	Feedstock
"Shantou Lianyl Biotech" (China)	Several PHAs	Pilot scale	Packaging and medical applications
"Tianjing Green Bio-Science" (China)	P(3HB/4HB)	10,000	Feedstock and packaging
"Bioplast" (Russia)	P(3HB), P(3HB/4HB), P(3HB/3HV)	Pilot scale	Feedstock, experimental devices

*Adapted from Chen (2009); Ienczak et al. (2013); Kaur and Roy (2015).

world's first Mirel biorefinery located in Clinton, Iowa. This new facility will produce 55,000 t of three Mirel varieties per year. Metabolix has signed a collaborative agreement with Australia's Cooperative Research Centre for Sugar Industry Innovation Through Biotechnology. Metabolix works in cooperation with British Petroleum to further develop direct production of bioplastics in switchgrass. Metabolix has also received government support for its technology from the U.S. Department of Agriculture and the U.S. Department of Commerce's Advanced Technology Program.

Tepha Inc. (U.S.), founded in 1998 as a sister company of Metabolix, is engaged in the medical applications of PHAs. The company has over 30 licenses for the production of PHA-based articles. Polymer production is performed using a patented process of fermentation of transgenic microorganisms. The product has been trademarked as TephaFLEX™.

Procter & Gamble, Chemicals (U.S.) has been developing and producing PHA heteropolymers consisting of monomer units that contain 4 to 12 carbon atoms. Unlike Metabolix, this company is engaged in the development of PHAs produced by fermentation of sugars and fatty acids; the trademark of the product is Nodax™. Structurally, Nodax polymers are similar to LDPE; they have branched chains, and, thus, the melt and glass-transition temperatures and crystallinity of Nodax plastics are lower than those of Biopol, which makes them easier to process.

Almost all industrially developed countries are to a greater or lesser extent engaged in PHA production, but large-scale production and application of PHAs is impossible without reducing their cost (Chen, 2009, 2010a, b). Austrian companies produce a homopolymer of 3-hydroxy-butyric acid using *Alcaligenes latus,* which can accumulate P(3HB) at a concentration of 90%. P(3HB) is produced from different substrates, including waste materials. The company produced P3HB in a quantity of 1,000 kg per week in a 15-m^3 fermentor. The P(3HB) production and the processing technology are now owned by Biomer, Germany. In 1995, the Brazilian sugar mill Copersucar assembled a pilot-scale P3HB production plant. The goal of this pilot plant was to produce enough P(3HB) to supply the market for tests and trials. Also, this pilot plant was intended as a training facility for future operators, and it is currently providing data for scale-up and economic evaluation of the process. Copersucar managed to produce 120–150 g/L CDW containing 60–65% P(3HB) with a productivity of 1.44 kg P3HB m^3 an hour and a P3HB yield of 3.1 kg sucrose per kilogram of P(3HB). In 2001, Copersucar began to produce PHA from sugar cane processing waste (Natano et al., 2001). In China, commercial production of PHAs is based on the use of efficient PHA producers such as naturally occurring and engineered strains of *Ralstonia eutropha* and other species. Chinese researchers have developed a process that can produce poly-3-hydroxybutyrate/3-hydroxyvalerate in high efficiency. Without supply of pure oxygen, *R. eutropha* grew to a density of 160 g/L CDW within 48 h in a 1,000-L fermentor (Chen, 2010a).

Rubber-like P(3HB/4HB) copolymers have recently attracted considerable attention. The wild-type *Ralstonia eutropha* and recombinant *Ralstonia* and *E. coli* strains are used by Chinese and U.S. companies to produce P(3HB/4HB). With the addition of 1,4-butanediol in different amounts, 4-hydroxybutyrate can be accumulated to 5–40 mol% in the copolymer, thus generating copolymers with various thermal and mechanical properties for various applications. Facilities with capacities of 10,000 and 50,000 tons of P(3HB/4HB) have been built in China and the U.S., respectively (Chen, 2009). P3HB/4HB may be the PHA available in the greatest quantity on the market. At the same time, companies in both countries are working to develop various bulk applications.

The least commercialized PHAs today are medium-chain-length copolymers consisting of monomer units containing 4 to 12 carbon atoms. The early development of this class of PHA copolymers was initiated by Procter & Gamble (USA) around the late 1980s. These PHAs consisted of various medium-chain-length monomer units, including copolymers without 3-hydroxybutyrate. A family of PHAs consisting of medium-chain-length monomer units with different chain lengths were synthesized and trademarked Nodax™ at Meredian (Bainbridge, GA, U.S.) (Poliakoff and Noda, 2004; Noda et al., 2005). In 2007, Meredian took over the Nodax™ technology from Procter & Gamble for the full commercialization of this class of bioplastics. A pilot facility was used to validate production and process design in 2009, prior to construction of its first full-scale PHA production facility, which was established in 2010. The planned annual output is about 300 million tons of PHA (Chen, 2009). Chinese and Korean researchers, in collaboration with Procter & Gamble, U.S., have been engaged in commercial production of this copolymer. The range of potential applications for this copolymer is getting wider, but its current production cost is still too high for real commercial applications.

In Russia, little research has been done on degradable polymers (Fomin and Guzeev, 2001; Shtilman, 2006). Biodegradable polymers have not been produced commercially yet. There are plans to establish polylactide production. Only a few research teams are engaged in PHA-related studies. These are the Institute of Microbiology RAS, the A.N. Bakh Institute of Biochemistry RAS, the Institute of Physiology and Biochemistry of Microorganisms RAS, and the Institute of Petrochemical Synthesis RAS. The Institute of Biophysics SB RAS was the first in Russia to establish pilot production of poly-3-hydroxybutyrate and 3-hydroxybutyrate/3-hydroxyvalerate copolymers in cooperation with the Biokhimmash company (within the framework of the ISTC project) in 2005 (Volova et al., 2006). Since that time, PHA synthesis processes have been considerably improved. During implementation of the mega projects supported by the Russian Government (Orders No. 219 and 220 of April 9, 2010), the team of researchers at the Siberian Federal University widened the range of 2- and 3-component PHAs with different chemical structures that contained major fractions of short- and medium-chain-length monomer units (Volova et al., 2008, 2013, 2014a, b) and established a new high-productivity pilot production facility equipped with an automatic fermentation system (Bioengineering, Switzerland) (Kiselev et al., 2014). The trademark "BIOPLASTOTAN" was registered for PHAs of different chemical compositions and PHA based products (Trademark "BIOPLASTOTAN").

As mentioned previously, polyhydroxyalkanoates have physicochemical properties similar to those of some synthetic polymers such as polypropylene, which are produced in large quantities and cannot be degraded in the environment. At the present time, the use of PHAs is limited by their rather high cost, but the range of their applications is wide. The growing environmental concern on the one hand, and the possibility of reducing the cost of biopolymers by increasing the production efficiency, on the other, make polyhydroxyalkanoates promising materials of the 21st century.

There are two possible ways to increase production and broaden applications of PHAs. One way is to develop large-scale PHA production, i.e., to increase PHA outputs and reduce their cost by manufacturing inexpensive items such as packaging materials, everyday articles, films and pots for agriculture, etc. Researchers in the U.S., Japan, EU countries, India, Malaysia, etc. are conducting extensive studies of PHAs for a variety of applications. They are mostly used to manufacture packaging items and garbage containers, food and cosmetic containers, and agricultural items (Plastics from Bacteria, 2010). These are extruded vials, jars, bottles, containers, and boxes for shampoos, lotions, etc. PHAs were initially used to make everyday articles such as shampoo bottles and packaging materials by Wella AG in Germany (Weiner, 1997). Short-chain-length PHAs are used to fabricate packaging films, shopping bags, containers and paper coatings, everyday use items such as housings for TV sets and computers, toys, sports equipment, disposable dishes, hygiene products, etc. by Biomers and Metabolix, and several other companies (Clarinval and Halleux, 2005; Noda, 2005). Some PHAs can be used to form gels and latexes, as a basis for producing glues and filling agents, including ones intended for stabilization of dyes. PHA laminates with paper and other polymers are successfully used as materials for producing garbage bags. PHAs can also be used to produce non-woven materials, various personal hygiene articles, etc. PHAs are used to produce dairy cream substitutes and flavor delivery agents in foods. There is a market for PHA depolymerization and hydrolysis products. These polymers can be converted to optically pure multifunctional hydroxy acids (Chen, 2009).

Potential nutritional qualities of PHAs have been discussed. Several research teams have evaluated monomers of (R)-3-hydroxybutanoic acid, as an alternative to the sodium salt of the monomer, for potential nutritional and therapeutic uses, such as treatment of metabolic acidosis. Use of these polymeric forms might provide controlled release systems for the monomer and, importantly, overcome the problems associated with administering large amounts of sodium ions *in vivo*. Tasaki et al. (1998) reported the results of infusing dimers and trimers of (R)-3-hydroxybutyrate into rats. PHAs have important applications for agriculture, such as those related to the production of packaging films for food and fertilizers, pots, nets, ropes, etc. A new and environmentally important PHA application may be delivery of agricultural chemicals, which are used to protect cultivated plants from pathogens and pests. Researchers of the Siberian Federal University were the first to prove that PHAs can be used as a degradable matrix enabling controlled release of pesticides and herbicides during the growing season of plants; pre-emergence formulations were developed, i.e., ones that can be buried in soil together with seeds (Prudnikova et al., 2013).

Another way is to establish small-scale facilities for the production of high-cost specialized items. PHAs show the greatest potential in medicine and pharmaceutics. The mild immune response to PHA implants and the sufficient duration of PHA degradation in biological media make these polymers attractive candidates for use as drug carriers in controlled-release drug delivery systems, implants and grafts for tissue and organ regeneration, materials for tissue engineering and designing of bioartificial organs (Volova et al., 2003; Volova and Shishatskaya, 2011; Amass et al., 1998; Pouton, 2001; Williams and Martin, 2002; Sudesh, 2004; Sudesh et al., 2000, 2007; Volova, 2004; Volova et al., 2013; Polyhydroxyalkanoates (PHAs): Biosynthesis, Industrial Production, and Applications in Medicine, 2014; Luef et al., 2015).

Press releases of such well-known companies as Tepha Inc., Metabolix, Procter & Gamble, and Monsanto suggest growing interest in PHAs and intensive research aimed at production, modification, and investigation of PHAs for cardiovascular surgery, dentistry, orthopedics, and pharmacology. Although PHAs are attracting more and more attention, there are rather few results of biomedical studies reported in the literature, as can be inferred from the analysis of biomedical studies of novel biomaterials, including PHAs, that are available in the major databases. Many aspects of PHA biotechnology and materials science remain unclear. Among them are production of high-purity PHAs and processes used to fabricate various PHA-based special biomedical items. Kinetics and mechanisms of PHA *in vivo* biodegradation need to be better understood. More studies should be performed to gain insight into mechanisms of *in vivo* interaction of PHA devices with cells and tissues and their medical and technical parameters in living organisms. The Food and Drug Administration (FDA) in the U.S. has approved the use of several products manufactured by Tepha Inc. in clinical trials. These are suture material, mesh grafts, and films for uroplastic surgery.

Analysis of available literature suggests that PHAs have been extensively studied as materials for cardiovascular surgery. These polymers are proposed as candidates for fabricating vascular grafts and coatings for grafts fabricated from synthetic materials (Marois et al., 1999a, b, 2000). Perhaps the most remarkable results with PHA polymers have been obtained in the development of cell-seeded tissue-engineered heart valves. Stock et al. (2000) successfully replaced the polyglycolide-polylactide valves in animals with poly-3-hydroxyoctanoate (P3HO) pulmonary conduits. Porous scaffolds prepared from elastic poly-3-hydroxyoctanoate were used to fabricate heart valves (Sodian et al., 1999, 2000). For 8 days, the cells proliferated and filled the pores; they also synthesized collagen and generated connective tissue between the outer and inner surfaces of the scaffold. Constructs of porous P(3HO/3HHx) were seeded with cells of vascular tissue and evaluated under pulsatile flow *in vitro* (Sodian et al., 1999).

One of the most advanced applications of PHA polymers in cardiovascular products has been the development of a regenerative PHA patch that can be used to close the pericardium after heart surgery, without formation of adhesions between the heart and sternum (Malm et al., 1992, 1994; Martin and Williams, 2003). Researchers of the Department of Thoracic and Cardiovascular Surgery at the University Hospital in Uppsala, Sweden, examined P(3HB) patches used as pericardial substitutes (Duvernoy et al., 1995). The patch was gradually degraded and replaced by native tissues strong enough to prevent the development of postoperative adhesions between the patch and the cardiac surface.

There are very few data on using PHAs to enhance biocompatibility of vascular stents. Unverdorben et al. (2002) tried to use poly-3-hydroxybutyrate stents in experiments on rabbits, but the experiment was not quite successful. Tepha (U.S.) researchers together with their German colleagues (Institute of Biomedical Engineering, Rostock) have tested polymer stents prepared from poly(tetrafluoroethane) and a blend of poly-lactide/poly-3-hydroxybutyrate (Grabow et al., 2007), comparing them with metallic stents. For the stents to withstand the pressure in vessels, they were rather bulky. This caused adverse response of the vessel wall exhibited as neointima growth and stronger inflammatory reaction than that caused by metallic stents.

PHA solutions can be used to prepare fibers. In one of the first studies of PHA fibers, Miller and Williams proved that P(3HB/3HV) mono-filaments are not biodegraded *in vitro* and *in vivo* (Miller and Williams,

1987). Tepha (U.S.) has developed suture material based on P(3HB/4HB). The filaments are melt-spun in a single-screw extruder by passing the material through 4 zones of the extruder, with temperatures 140, 190, 200, and 205°C, and multistage orientation (Martin et al., 2000). The extrusion process and subsequent orientation yield filaments with tensile strength over 126 MPa, which retain their properties for long periods of time. Having conducted all necessary tests, Tepha received the permission of FDA and marketed mono- and poly-filament fibers, meshes, and films under the trademark of TephaFLEX®.

PHAs are attractive materials for use in surgical reconstruction of bone tissues. There are data on preparation of mechanically strong PHA/HA composites, proving that incorporation of HA into a PHA enhances polymer strength (see Chapter 6 for details).

In recent years, much research effort has been devoted to the use of PHAs as matrices for drug delivery and as scaffolds for cell cultures intended to fabricate grafts for tissue engineering applications (see Chapter 4). In Russia, biomedical studies of polyhydroxyalkanoates and experimental prototypes of PHA products were initiated by the researchers of the Institute of Biophysics SB RAS and the Siberian Federal University united in Research-and-Education Center "Yenisei." In cooperation with the Institute of Transplant Surgery and Artificial Organs (currently the V.I. Shumakov Federal Research of Transplant Surgery and Artificial Organs Center), the research team studied films of poly-3-hydroxybutyrate and P(3HB/3HV) copolymers and proved that they did not produce any cytotoxic effect when contacting directly with the cultured cells and when implanted as sutures *in vivo*; high-purity specimens were suitable for contact with blood. Results of these studies were summarized in the two editions of the first Russian book on PHAs: "Polyhydroxyalkanoates – biodegradable polymers for medicine" (Volova et al., 2003). Since then, the research team has considerably widened the scope of its PHA studies by synthesizing PHAs with different chemical compositions and by designing and studying experimental films, barrier membranes, granules, filling materials, ultrafine fibers produced by electrospinning, solid and porous 3D implants for bone tissue defect repair, tubular biliary stents, mesh implants modified by PHA coating, microparticles for drug delivery, etc. In cooperation with the V.F. Voino-Yasenetsky Krasnoyarsk

Federal Medical University, the research team has conducted pioneering clinical trials. Results have been covered by several Russian patents, reported in papers published in peer-reviewed Russian and international journals, summarized in books and reviews (Shishatsky and Volova, 2008; Shishatsky and Shishatskaya, 2010; Volova et al., 2010; Volova and Shishatskaya, 2011; Volova, 2004; Volova et al., 2013).

Results of the experimental studies and clinical trials are reported in Chapters 3–8.

KEYWORDS

- polyhydroxyalkanoates
- natural degradable polymers
- structure
- properties
- production
- applications

REFERENCES

Abe, H., Marsubara, I., & Doi, Y. Physical properties and enzymatic degradability of polymer blends of bacterial poly((R)-3-hydroxybutyrate) stereoisomers. Macromol., 1995, 28, 844–853.

Amass, W., Amass, A., & Tighe, B. A Review of Biodegradale Polymers: Uses, Current Developments in the Synthesis and Characterization of Biodegradable Polyesters, Blends of Biodegradable Polymers and Recent Advances in Biodegradation Studies. Polymer Int., 1998, 47, 89–144.

Amirul, A. A., Yahya, A. R. M., Sudesh, K., Azizan, M. N. M., & Majid, M. I. A. Biosynthesis of poly(3-hydroxybutyrate-co-4-hydroxybutyrate) copolymer by a *Cupriavidus spp.* USMAA 1020 isolated from Lake Kulim, Malaysia. Bioresour. Technol., 2008, 99, 4903–4909.

Anderson, A. J., & Dawes, E. A. Occurrence, metabolism, metabolic role, and industrial uses of bacterial polyhydroxyalkanoates. Microbiol. Rev., 1990, 54, 450–472.

Audic, J. L., Chaufer, B., & Daufin, G. Non-food applications of milk components and dairy co-products: a review. Lait, 2003, 83, 417–438.

Avella, M., Martuscelli, E., & Greco, P. Crystallization behavior of poly(ethylenen oxide) from poly-D-(–)-(3-hydroxybutyrate)/poly(ethylenen oxide): phase structuring, morphology and thermal behavior. Polymer, 1991, 32, 1647–1653.

Babel, W., Ackerman, J. U., & Breuer, U., Physiology, regulation, and limits of the synthesis of poly(3HB). Adv. Biochem. Eng. Biotechnol., 2001, 71, 125–157.

Ballistreri, A., Giuffrida, M., Impallomeni, G. et al., Characterization by mass spectrometry of poly(3-hydroxyalkanoates) produced by *Rhodospirillum rubrum* from 3-hydroxyacids. Int. J. Biol. Macromol., 1999, 26, 201–211.

Bear, M. M., Leboucher-Durand, M. A., Langlois, V., Lenzb, R. W., Goodwinc, S., & Guérin, P. Bacterial poly-3-hydroxyalkenoates with epoxy groups in the side chains. React. Funct. Polym., 1997, 34, 65–77.

Bear, M. M., Renard, E., Randriamahefa, S., Langlois, V., & Guérin, P. Preparation of a bacterial polyester with carboxy groups in side chains. C. R. Acad. Sci. Chemistry, 2001, 4, 289–293.

Brandl, H., Gross, R. A., Lenz, R. W., & Fuller, C. W. *Pseudomonas oleovorans* as a source of Poly(β-hydroxyalkanoates) for potential application as biodegradable polyesters. Appl. Environ. Microbiol., 1988, 54, 1977–1982.

Braunegg, G., Koller, M., Hesse, P. J., Kutschera, C., Bona, R., Hermann, C., Horvat, P., Neto, J., & Dos Santos Pereira, L. Production of plastics from waste derived from agrofood industry. In: Graziani, M., Fornasiero, P. Renewable resources and renewable energy: a global challenge. CRC Press, Taylor and Francis Group. Boca Raton, 2007, 119–135.

Braunegg, G., Lefebvre, G., & Genzer, K. F. Polyhydroxyalkanoates, biopolyesters from renewable resources: Physiological and engineering aspects (Review article). J. Biotechnol., 1998, 65, 127–161.

Byron, D. Polyhydroxyalkanoates. In: Mobley, D. P. Plastics from microbes: microbial synthesis of polymers and polymer precursors. Hanser Munich, 1994, 5–33.

Byron, D. Polymer synthesis by microorganisms: technology and economics. Trends Biotechnol., 1987, 5, 246–250.

Byron, D. Production of poly-β-hydroxybutyrate: polyhydroxyvalerate copolymers. FEMS Microbiol. Rev., 1992, 103, 247–250.

Catcheside, D. E. A., & Ralph, J. P. Biological processing of coal. Appl. Microbiol. Biotechnol., 1999, 52, 16–24.

Chen, G. Q. A microbial polyhydroxyalkanoates (PHA) based bio- and materials industry. Chem Soc., 2009, 38, 2434–2446.

Chen, G. Q. Industrial production of PHA. In: Chen, G. Q., & Steinbüchel, A. Microbiol. Monogr. Plastics from bacteria. Natural functions and applications. Springer, 2010b, 14, 121–132.

Chen, G. Q. Plastics completely synthesized by bacteria: polyhydroxyalkanoates. In: G. Q. Chen, A. Steinbüchel. Microbiol. Monogr. Plastics from bacteria. Natural functions and applications. Springer, 2010a, 14, 17–38.

Chen, J. Y., Song, G., & Chen, G. Q. A lower specificity of PhaC2 synthase from *Pseudomonas stutzeri* catalyzes the production of copolyesters consisting of short-chain-length and medium-chain-length 3-hydroxyalkanoates. Antonie Van Leeuwenhoek, 2006, 89, 157–167.

Choi, J., & Lee, S. Y. Efficient and economical recovery of poly-(3-hydroxybutyrate) from recombinant *Escherichia coli* by simple digestion with chemicals. Biotechnol. Bioeng., 1999a, 62, 546–553.

Choi, M. H., Yoon, S. C., & Lenz, R. W. Production of poly(3-hydroxybutyric acid-co-4-hydroxybutyric acid) and poly(4-hydroxybutyric acid) without subsequent degradation by *Hydrogenophaga pseudoflava*. Appl. Environ. Microbiol., 1999b, 65, 1570–1577.

Chun, Y. S., & Kim, W. N. Thermal properties of poly(hydroxybutyrate-co-hydroxyvalerate) and poly(caprolactone) blends. Polymer, 2000, 41, 2305–2308.

Clarinval, A. M., & Halleux, J. Classification of biodegradable polymers. In: R. Smith. Biodegradable Polymers for Industrial Applications. Woodhead Publishing. Cambridge, 2005, 3–31.

Dawes, E. A. Novel biodegradable microbial polymers. Kluwer Academic, Dordrecht. The Netherlands, 1990, 287 p.

De Koning, G. J. M. Physical properties of bacterial poly((R)-3-hydroxyalkanoates). Can. J. Microbiol., 1995, 41, 303–309.

Doi, Y., Tamaki, A., Kunioka, M., & Soga, K. Biosynthesis of terpolyesters of 3-hydroxybutyrate, 3-hydroxyvalerate, and 5-hydroxyvalerate in *Alcaligenes eutrophus* from 5-chloropentanoic and pentanoic acids. Makromol. Chem. Rapid. Commun., 1987, 8, 631–635.

Duvernoy, O., Malm, T., Ramstrum, J., & Bowald, S. A biodegradable patch used as a pericardial substitute after cardiac surgery: 6- and 24-month evaluation with CT. Thorac. Cardiovasc. Surg., 1995, 43, 271–274.

Eggink, G., de Waard, P., & Huijberts, G. N. M. Formation of novel poly(hydroxyalkanoates) from long-chain fatty acids. Can. J. Microbiol., 1995, 41, 14–21.

Fakoussa, R. M., & Hofrichter, M. Biotechnology and microbiology on coal degradation. Appl. Microbiol. Biotechnol., 1999, 52, 25–40.

Fomin, V. A., & Guzeev, V. V. Biodegradable polymers, their current and potential uses. Plasticheskiye massy (Plastics), 2001, 2, 42–46 (in Russian).

Fritzsche, K., Lenz, R. W., & Fuller, R. C. Bacterial polyesters containing branched poly(b-hydroxyalkanoate) units. Int. J. Biol. Macromol., 1990b, 12, 92–101.

Fritzsche, K., Lenz, R. W., & Fuller, R. C. An unusual bacterial polyester with a phenyl pendant group. Makromol. Chem., 1990a, 191, 1957–1965.

Fritzsche, K., Lenz, R. W., & Fuller, R. C. Production of unsaturated polyesters by *Pseudomonas oleovorans*. Int. J. Biol. Macromol., 1990c, 12, 85–91.

Füchtenbuch, B., & Steinbüchel, A. Biosynthesis of polyhydroxyalkanoates from low-rank coal liquefaction products by *Pseudomonas oleovorans* and *Rhodococcus rubber*. Appl. Microbiol. Biotechnol., 1999, 52, 91–95.

García, B., Olivera, E. R., & Minambres, B. Novel biodegradable aromatic plastics from a bacterial source. Genetic and biochemical studies on a route of the phenylacetyl-CoA catabolon. J. Biol. Chem., 1999, 274, 29228–29241.

Grabow, N., Bünger, C. M., Schultze, C., Schmohl, K., Martin, D. P., Williams, S. F., Sternberg, K., & Schmitz, K. P. A biodegradable slotted tube stent based on poly(L-lactide) and poly(4-hydroxybutyrate) for rapid balloon-expansion. Ann. Biomed. Eng., 2007, 35, 2031–2038.

Gross, R. A., DeMello, C., Lenz, R. W. Brandl, H., & Fuller, R. C. Biosynthesis and characterization of poly(β-hydroxyalkanoates) produced by *Pseudomonas oleovorans*. Macromolecules, 1989, 22, 1106–1115.

Gruzinskii, I. V., Gogotov, I. N., Bechina, E. M., & Semyonov, Y. V. Hydrogenase activity of the hydrogen-oxidizing bacterium *Aicaligenes eutrophus*. Mikrobiologiya (Microbiology), 1977, 46, 625–630 (in Russian).

Ha, C. S., & Cho, W. J. Miscibility, properties and biodegradability of microbial polyester containing blends. Prog. Polym. Sci., 2002, 27, 759–809.

Hao, J., & Deng, X. Semi-interpenetrating networks of bacterial poly(3-hydroxybutyrate) with net-poly(ethylene glycol). Polymer, 2001, 42, 4091–4091.

Hazer, B., & Steinbüchel, A. Increased diversification of polyhydroxyalkanoates by modification reactions for industrial and medical applications. Appl. Microbiol. Biotechnol., 2007, 74, 1–12.

Howells, E. R. Opportunities for biotechnology for the chemical industry. Chem. Ind., 1982, 8, 508–511.

Ienczak, J. L., Schmidell, W., & de Aragao, G. M. F. High-cell-density culture strategies for polyhydroxyalkanoate production: review. J. Int. Microbiol. Biochem., 2013, 40, 275–286.

Imamura, T., Kenmoku, T., & Honma, T. Direct biosynthesis of poly(3-hydroxyalkanoates) bearing epoxide groups. Int. J. Biol. Macromol., 2001, 29, 295–301.

Kaur, G., & Roy, I. Strategies for Large-scale Production of Polyhydroxyalkanoates. Chem. Biochem. Eng., 2015, 29, 157–172.

Kessler, B., & Witholt, B. Factors involved in the regulatory network of polyhydroxyalkanoate metabolism. J. Biotechnol., 2001, 86, 97–104.

Kim, D. Y., Jung, S. B., & Choi, G. G. Biosynthesis of polyhydroxyalkanoate copolyester containing cyclohexyl groups by *Pseudomonas oleovorans*. Int. J. Biol. Macromol., 2001, 29, 145–150.

Kim, D. Y., Kim, Y. B., & Rhee, Y. H. Bacterial Poly(3-hydroxyalkanoates) Bearing Carbon-Carbon Triple Bonds. Macromol., 1998, 32, 4760–4763.

Kim, D. Y., Lütke-Eversloh, T., Elbanna, K., Thakor, N., & Steinbüchel, A. Poly(3-mercaptopropionate): a nonbiodegradable biopolymer. Biomacromolecules, 2005a, 6, 897–901.

Kim, H. W., Chung, C. W., & Rhee, Y. H. UV-induced graft copolymerization of monoacrylatepoly(ethylene glycol) onto poly(3-hydroxyoctanoate) to reduce protein absortion and platelet adhesion. Int. J. Biol. Macromol., 2005b, 35, 47–53.

King, P. P. Biotechnology: an industrial view. J. Chem. Technol. Biotechnol., 1982, 32, 2–8.

Kiselev, E. G., Demidenko, A. D., Baranovsky, S. V., & Volova, T. G. Scaling-up the process of the synthesis of biodegradable polyhydroxyalkanoates in a pilot facility. Journal of Siberian Federal University, Biology Series, 2014, 2, 134–147 (in Russian).

Koller, M., Atlić, A., Dias, M., Reiterer, A., & Braunegg, G. Microbial PHA production from waste raw materials. In: G. Q. Chen, A. Steinbüchel. Plastrics from bacteria – natural functions and applications. London, New York: Springer Heidelberg Dordrecht, 2010, 86–114.

Kumar, R., Singh, S., & Singh, O. V. Bioconversion of lignocellulosic biomass: biochemical and molecular perspectives. J. Ind. Microbiol. Biotechnol., 2008, 35, 377–391.

Laycock, B., Halley, P., Pratt, S., Werker, A., & Lant, P. The chemomechanical properties of microbial polyhydroxyalkanoate. Prog. Polym. Sci., 2013, 38, 536–583.

Lee, M. Y., Cha, S. Y., & Park, W. H. Crosslinking of microbial copolyester with pendant epoxyde groups by diamine. Polymer, 1999, 40, 3787–3793.

Lee, M. Y., Park, W. H., & Lenz, R. W. Hydrophylic bacterial polyesters modified with pendant hydroxyl groups. Polymer, 2000, 41, 1703–1709.

Lee, S. Y. Plastic bacteria. Progress and prospects for polyhydroxyalkanoate production in bacteria (Reviews). Tibtech., 1996, 14, 431–438.

Lemoingne, M. Etudes sur l'autolyse microbienne: origine de l'acide β-oxybutyrique forme par autolyse. Ann. Inst. Pasteur., 1927, 41, 148–165.

Lenz, R. W., & Marchessault, R. H. Bacterial polyesters: biosynthesis, biodegradable plastics and biotechnology. Biomacromolecules, 2005, 6, 1–7.

Lu, X. Y., Wu, Q., & Chen, G. Q. Production of poly(3-hydroxybutyrate-co-3-hydroxyhexanoate) with flexible 3-hydroxyhexanoate content in *Aeromonas hydrophila* CGMCC 0911. Appl. Microbiol. Biotechnol., 2004, 64, 41–45.

Luef, K. P., Stelzer, F., & Wiesbrock, F. Poly(hydroxy alkanoate)s in Madical Application. Chem. Biochem. Eng., 2015, 29, 287–2912.

Luengo, J. M., Arias, S., Sandoval, A., Arias-Barrau, E., Arcos, M., Naharro, G., & Olivera, E. R. From aromatic to bioplastic: the phenylacetyl-CoA catabolon as a model of catabolic convergence. In: Pandalai, S. G. Recent research developments in biophysics and biochemistry, Research Signpost, Kerala, 2004, 4, 257–292.

Luengo, J. M., García, B., & Sandoval, A. Bioplastics from microorganisms. Curr. Opin. Biotechnol., 2003, 6, 251–260.

Luizier, W. D. Materials derived from biomass/biodegradable materials. Proc. Natl. Acad. Sci. USA, 1992, 89, 839–842.

Madison, L. L., & Huisman, G. W. Metabolic engineering of poly(3-hydroxyalkanoates): From DNA to plastic. Microbiol. Mol. Biol. Rev., 1999, 63, 21–53.

Maekawa, M., Pearce, R., Marchessault, R. H., & Manley, R. S. J. Miscibility and tensile of poly(β-hydroxybutyrate)-cellulose propionate blends. Polymer, 1999, 40, 1501–1505.

Malm, T., Bowald, S., Bylock, A. et al. Enlargement of the right ventricular outflow tract and the pulmonary artery with a new biodegradazzble patch in transannular position. Eur. Surg. Res., 1994, 26, 298–308.

Malm, T., Bowald, S., Karacagil, S., Bylock, A., & Busch, C. A new biodegradable patch for closure of AF atrial septal defect. Scand. J. Thor. Cardiovasc. Surg., 1992, 26, 9–14.

Marois, Y., Zhang, Z., Vert, M., Beaulieu, L., Lenz, R. W., & Guidoin, R. In vivo biocompatibility and degradation studies of polyhydroxyoctanoate in the rat: A new sealant for the polyeatere arterial prosthesis. Tissue Eng., 1999b, 5, 369–386.

Marois, Y., Zhang, Z., Vert, M., Deng, X., Lenz, R. W., & Guidoin, R. Bacterial polyesters for biomedical applications: In vitro and in vivo assessment of sterilization, degradation rate and biocompatibility of poly(β-hydroxyoctanoate (PHO). In: Agrawal, C. M., Parr, J. E., & Lan, S. T. Synthesis Bioabsorbable Polymers for Implants. Scrawal: ASTM, 2000, 12–38.

Marois, Y., Zhang, Z., Vert, M., Deng, X., Lenz, R. W., & Guidoin, R. Effect of sterilization on the physical and structural characteristics of polyhydroxyoctanoate (PHO). J. Biomater. Sci. polymer. Edn., 1999a, 10, 469–482.

Martin, D., & Williams, S. Medical application of polyhydroxybutyrate: a strong flexible absorbable biomaterial. Biochem. Engiin. J., 2003, 16, 97–105.

Martin, D. P., Peoples, O. P., & Williams, S. F. Nutritional and therapeutic uses of 3-hydroxyalkanoate oligomers. PCT Patent Application No. WO 00/04895, 2000.

Matsusaki, H., Abe, H., & Doi, Y. Biosynthesis and properties of poly(3-hydroxybutyrate-co-3-hydroxyalkanoates) by recombinant strains of *Pseudomonas sp.* 6 1–3. Biomacromolecules, 2000, 1, 17–22.

Miller, N. D., & Williams, D. F. On the biodegradation of poly-β-hydroxybutyrate (PHB) homopolymer and poly-β-hydroxybutyrate-hydroxyvalerate copolymers. Biomaterials, 1987, 8, 129–137.

Nakamura, S., Kunioka, M., & Doi, Y. Biosynthesis and characterization of bacterial poly(3-hydroxybutyrate-co-3-hydroxypropionate). J. Macromol. Sci., 1991, 28, 15–24.

Natano, R. V., Mantelatto, P. E., & Rossell, C. E. Integrated production of biodegradable plastic, sugar and ethanol. Appl. Microbiol. Biotechnol., 2001, 57, 1–5.

Neto, J. New strategies in the production of polyhydroxyalkanoates from glycerol and meat and bone meal: PhD thesis. Graz University of Technology, 2006.

Noda, I., Green Ph., Satkowski, M., & Schechtman, L. A. Preparation and properties of novel class of polyhydroxyalkanoate copolymers. Biomacromol., 2005, 6, 580–586.

Oeding, V., & Schlegel, H. G. Beta-ketothiolase from *Hydrogenomonas eutropha* H16 and its significance in the regulation of poly-beta-hydroxybutyrate metabolism. Biochem. J., 1973, 134, 239–248.

Olivera, E. R., Arcos, M., Naharro, G., & Luengo, J. M. Unusual PHA Biosynthesis. In: G. Q. Chen, A. Steinbüchel. Plastics from Bacteria-Natural Functions and Applications. Springer Heidelberg: Dordrecht, London, New York, 2010, 18–34.

Olivera, E. R., Carnicero, D., García, B., Miñambres, B., Moreno, M. A., Cañedo, L., Dirusso, C. C., Naharro, G., & Luengo, J. M. Two different pathways are involved in the β-oxidation of n-alkanoic and n-phenylalkanoic acids in *Pseudomonas putida* U: genetic studies and biotechnological applications. Mol. Microbiol., 2001b, 39, 863–874.

Olivera, E. R., Carnicero, D., Jodrá, R., Miñambres, B., García, B., Abraham, G. A., Gallardo, A., Román, J. S., García, J. L., Naharro, G., & Luengo, J. M. Genetically engineered Pseudomonas: a factory of new bioplastics with broad applications. Environ. Microbiol., 2001, 3, 612–618.

Pettinari, M. J., Vázquez, G. J., Silberschmidt, D., Rehm, B., Steinbüchel, A., & Méndez, B. S. Poly(3-hydroxybutyrate) genes in *Azotobacter sp.* strain FA8. Appl. Environ. Microbiol., 2001, 67, 5331–5334.

Plastic from Bacteria. In: Chen, G. Q. Natural Functions Applications. Springer-Verlag. Berlin, 2010.

Polyhydroxyalkanoates (PHA): Biosynthesis, Industrial Production and Applications in Medicine. Nova Sciences Publ. Inc. NY, USA, 2014.

Pouton, C. W. Polymeric materials for advanced drug delivery. Adv. Drug. Deliv. Rev., 2001, 53, 1–3.

Prudnikova, S. V., Boyandin, A. N., Kalacheva, G. S., & Sinskey, A. J. Degradable polyhydroxyalkanoates as herbicide carriers. J. Polym. Environ., 2013, 21, 675–682.

Purushothaman, M., Anderson, R. K., Narayana, S., & Jayaraman, V. Industrial byproducts as cheaper medium components influencing the production of polyhydroxyalkanoates (PHA)-biodegradable plastic. Bioprocess. Biosystem. Engin., 2001, 24, 131–136.

Rehm, B. H. A., Krüger, N., & Steinbüchel, A. A new metabolic link between fatty acid de novo synthesis and polyhydroxyalkanoic acid synthesis. J. Biol. Chem., 1998, 273, 24044–24051.

Saito, Y., & Doi, Y. Microbial synthesis and properties of poly(3-hydroxybutyrate-co-4-hydroxybutyrate) in *Comamonas acidovorans*. Int. J. Biol. Macromol., 1994, 16, 99–104.

Savelieva, N. D. Hydrogen bacteria and carbon monoxide. Mikrobiologiya (Microbiology), 1979, 48, 360–362 (in Russian).

Savelieva, N. D., & Zhilina, T. N. On taxonomy of hydrogen bacteria. Mikrobiologiya (Microbiology), 1968, 309, 223–226 (in Russian).

Schmack, G., Gorenflo, V., & Steinbüchel, A. Biotechnological production and characterization of polyesters containing 4-hydroxyvaleric acid and medium-chain-length hydroxyalkanoic acids. Macromolecules, 1998, 31, 644–649.

Senior, P. J. Polyhydroxybutyrate, a speciality polymer of microbial origin. In: Dean, A., Ellwood, D., & Evans, C. Continuous Culture. Ellis Horwood. Chichester, UK, 1984, 8, 266–271.

Senior, P. J., & Dawes, E. A. The regulation of poly-β-hydroxybutyrate metabolism in *Azotobacter beijerinckii*. Biochem. J., 1973, 134, 225–238.

Shishatsky, O. N., & Shishatskaya, E. I. Analiz rynka materialov i izdeliy meditsinskogo naznacheniya (Analysis of the market for materials and devices for medical applications). Krasnoyarskii Pisatel. Krasnoyarsk, 2010 (in Russian).

Shishatsky, O. N., Shishatskaya, E. I., & Volova, T. G. Razrushayemyye polimery: potrebnosti, proizvodstvo, primeneniye (Degradable biopolymers: need, production, applications). Novyye informatsionnyye tekhnologii (New information technologies). Krasnoyarsk, 2010 (in Russian).

Shtilman, M. I. Polimery mediko-biologicheskogo naznacheniya (Polymers intended for biomedical applications). Akademkniga Publishers. Moscow, 2006 (in Russian).

Sodian, R., Hoerstrup, S. P., Sperling, J. S., Daebritz, S. H., Martin, D. P., Schoen, F. J., Vacanti, J. P., & Mayer, J. E. Jr. Tissue engineering of heart valves: *in vitro* experiences. The Annal. Thorac. Surg., 2000, 70, 140–144.

Sodian, R., Sperling, J. S., Martin, D. P., Stock, U., Mayer, J. E. Jr., & Vacanti, J. P. Tissue engineering of a trileaflet heart valve. Early in vivo experiences with a combined polymer. Tissue Eng., 1999, 5, 489–493.

Solaiman, D. K. Y., Ashby, R. D., Foglia, T. A., & Marmer, W. N. Conversion of agricultural feedstock and coproducts into poly(hydroxyalkanoates). Appl. Microbiol. Biotechnol., 2006, 71, 783–789.

Stasishina, G. N., & Volova, T. G. Shtamm bakterii *Alcaligenes eutrophus* – produtsent belkovoi biomassy (The strain of the bacterium *Alcaligenes eutrophus* – a producer of protein biomass). RF Patent No. 2053292. BI, 1996, No. 1 (in Russian).

Steinbüchel, A., & Füchstenbusch, B. Bacterial and other biological systems for polyesters production. Trends Biotechnol., 1998, 16, 419–427.

Steinbüchel, A., & Hein, S. Biochemical and molecular basis of microbial synthesis of polyhydroxyalkanoates in microorganisms. Adv. Biochem. Eng. Biotechnol., 2001, 71, 81–123.

Steinbüchel, A., & Valentin, H. E. Diversity of bacterial polyhydroxyalkanoic acids. FEMS Microbiol. Lett., 1995, 128, 219–228.

Stock, U., Nagashima, M., Khalil, P. N., Nollert, G. D., Herden, T., Sperling, J. S., Moran, A., Lien, J., Martin, D. P., Schoen, F. J., Vacanti, J. P., & Mayer, J. E. Jr. Tissue-engineered valved conduits in the pulmonary circulation. J. Thorac. Cardiovasc. Surg., 2000, 119, 732 –7 40.

Stubbe, J., Tian, J., Sinskey, A. J. Lawrence, A. G., & Liu, P. Nontemplate-dependent polymerization processes: polyhydroxyalkanoate synthases as a paradigm. Annu. Rev. Biochem., 2005, 74, 433–480.

Sudesh, K., Loo, C. Y., Goh, L. K., Iwata, T., & Maeda, M. The oil-absorbing property of polyhydroxyalkanoate films and its practical application: a refreshing new outlook for an old degrading material. Macromol. Biosci., 2007, 7, 1199–1205.

Sudesh, K. Microbial polyhydroxyalkanoates (PHAs): an emerging biomaterial for tissue engineering and therapeutic applications. Med. J. Malaysia, 2004, 59, 55–66.

Sudesh, K., Abe, & H., Doi, Y. Synthesis, structure and properties of polyhydroxyalkanoates: biological polyesters. Prog. Polym. Sci., 2000, 25, 1503–1555.

Tasaki, O., Hiraide, A., Shiozaki, T., Yamamura, H., Ninomiya, N., & Sugimoto, H. The dimer and trimer precursor of 3-hydroxybutyrate oligomers as of ketone bodies for nutritional care. Parent. Enteral. Nutr., 1998, 23, 321–325.

Trademark "BIOPLASTOTAN" Registration Certificate No. 315652 of the Federal Institute for Patent Examination for Application No. 2006703271/50, Priority of 15.02.2006.

Verhoogt, H., Ramsay, B. A., & Favis, B. D. Polymer blends containing poly(3-hydroxyalkanoate)s. Polymer, 1994, 35, 5155–5169.

Volova, T. G. Hydrogen-Based Biosynthesis. Nova Science Pub., Inc. NY, USA, 2009, 287 p.

Volova, T. G. Polyhydroxyalkanoates – Plastic Materials of the 21st Century: production, properties, application. Nova Science Pub., Inc. NY, USA, 2004, 282 p.

Volova, T. G., Fedorova, Y. V., & Kalacheva, G. S. The effect of growth limitation on polyhydroxybutyrate accumulation in hydrogen-oxidizing bacteria. Proceedings of the USSR Conference "Limitation and inhibition of growth of microorganisms," Pushchino, 1989, 16–24 (in Russian).

Volova, T. G., Goncharov, D. B., Sukovatyi, A. G., Shabanov, A., Nikolaeva, E. D., & Shishatskaya, E. I. Electrospinning of polyhydroxyalkanoate fibrous scaffolds: effect on electrospinning parameters on structure and properties. J. Biomater. Sci., Polym. Ed., 2014b, 25, 370–393.

Volova, T. G., & Kalacheva, G. S. Sposob polucheniya polimera 3-oksimaslyanoi kisloty (A biotechnological technique of producing a polymer of 3-hydroxybutyric acid). RF Patent No. 2051967, BI, 1996, No. 3 (in Russian).

Volova, T. G., Kalacheva, G. S., Gitelson, I. I., Kuznetsov, B. N., Shchipko, A. M., & Shabanov, V. F. Sposob polucheniya polimera β-oksimaslyanoi kisloty (A biotechnological technique of producing a polymer of β-hydroxybutyric acid. RF Patent No. 2207375, 2003 (in Russian).

Volova, T. G., Kalacheva, G. S., & Steinbüchel, A. Biosinthesis multi-component polyhydroxyalkanoates by the bacterium *Wautersia eutropha.* Journal of Siberian Federal University. Biology, 2008, 1, 91–101.

Volova, T. G., Kiselev, E. G., Vinogradova, O. N. Nikolaeva, E. D., Chistyakov, A. A., Sukovatyi, A. G., & Shishatskaya, E. I. A glucose-utilizing strain, Cupriavidus eutrophus B-10646: growth kinetics, characterization and synthesis of multicomponent PHAs. PloS One, 2014a, 9, 87551–87566.

Volova, T. G., Sevastyanov, V. I., & Shishatskaya, E. I. Polioxialkanoaty – biorazrushaemyye polimery dlya meditsiny (Polyhydroxyalkanoates – biodegradable polymers for medicine). In: V. I. Shumakov, ed. Novosibirsk: SB RAS Publishers, 2003, 350 p. (in Russian).

Volova, T. G., & Shishatskaya, E. I. Biorazrushayemyye polimery: sintez, svoistva, primeneniye (Biodegradable Polymers: Synthesis, Properties, Applications). In: A. J. Sinskey, ed. Krasnoyarskii Pisatel. Krasnoyarsk, 2011, 49 p. (in Russian).

Volova, T. G., Shishatskaya, E. I., & Sinskey, A. J. Degradable Polymers: Production, Properties and Applications. Nova Science Pub., Inc. NY, USA, 2013, 380 p.

Volova, T. G., Vasiliev, A. D., & Zeer, E. P. Investigation of molecular structure of poly(3-hydroxybutyrate), a thermoplastic and degradable polymer. Biofizika (Biophysics), 2000, 45, 33–439 (in Russian).

Volova, T. G., & Voinov, N. A. Characterization of the culture of *Ralstonia eutropha* synthesizing polyhydroxyalkanoates from coal processing products. Prikladnaya biokhimiya i mikrobiologiya (Applied biochemistry and microbiology), 2004, 40, 296–300 (in Russian).

Volova, T. G., Voinov, N. A., Muratov, V. S., Bubnov, N. V., Gurulev, K. V., Kalacheva, G. S., Gorbuova, N. V., Plotnikov, V. F., Zhila, N. O., Shishatskaya, E. I., & Belyaeva, O. G. Pilot production of degradable biopolymers. Biotekhnologiya (Biotechnology), 2006, 6, 28–34 (in Russian).

Volova, T. G., Zhila, N. O., Shishatskaya, E. I., & Sukovatyi, A. G. Database "Fiziko-khimicheskiye svoistva poligidroxialkanoatov razlichnoi struktury. Usloviya biosinteza i produtsenty" ("Physicochemical properties of polyhydroxyalkanoates with different structure. Biosynthesis conditions and producers"). Certificate of VNIIGPE No. 2012620288 of the state record of the database of 15 March 2012 (in Russian).

Wei, X., Hu, Y. J., Xie, W. P., Lin, R. L., & Chen, G. Q. Influence of poly(3-hydroxybutyrate-co-4-hydroxybutyrate-co-3-hydroxyhexanoate) on growth and osteogenic differentiation of human bone marrow-derived mesenchymal stem cells. J. Biomed. Mater. Res. A, 2009, 90, 894–905.

Weiner, R. M. Biopolymers from marine prokaryotes. Trends Biotechnol., 1997, 15, 390–394.

Williams, S. F., Martin, D. P., Horowitz, D. M., & Peoples, O. P. PHA applications: addressing the price performance issue. I. Tissue engineering. Int. J. Biol. Macromol., 1999, 25, 11–121.

Witholt, B., & Kessler, B. Perspectives of medium chain length poly (hydroxyalkanoates), a versatile set of bacterial bioplastics. Curr. Opin. Biotechnol., 1999, 10, 279–285.

Yokouchi, M., Chatani, Y., Tadokoro, H., & Tani, H. Structural studies of polyesters. VII. Molecular and crystal structures of racemic Poly(β-ethyl-β-*propiolactone*). Polym. J., 1974, 6, 267–272.

Yoon, J. S., Lee, W. S., Kim, K., Chin, I. J., Kim, M. N., & Kim, C. Effect of poly(ethylene glycol)-block-poly(L-lactide) on the poly((R)–3-hydroxybutyrate)-poly(L-lactid) blends. Eur. Polym. J., 2000, 36, 435–442.

Yoshie, N., Azuma, Y., & Sakurai, M. Crystallization and compatibility of poly(vinyl alcohol)/poly(3-hydroxybutyrate) blends: Influence of blend composition and tacticity of poly(vinyl alcohol). J. Appl. Polym. Sci., 1992, 56, 17–24.

Zhang, S., Kolvek, S., Goodwin, S., & Lenz, R. W. Poly(hydroxyalkanoic acid) biosynthesis in Ectothior-hodospira shaposhnikovii: characterization and reactivity of the type III Pha synthase. Bio-Macromolecules, 2004, 5, 40–48.

Zinn, M., Witholt, B., & Egli, T. Occurrence, synthesis and medical applications of bacterial polyhydroxyalkanoate. Adv. Drug. Rev., 2001, 53, 5–21.

PART II

MICROCARRIERS OF PHAs FOR CELL TECHNOLOGY AND DRUG DELIVERY

CHAPTER 3

POTENTIALS OF POLYHYDROXYALKANOATES AS MATERIALS FOR CONSTRUCTING CELL SCAFFOLDS IN TISSUE ENGINEERING

CONTENTS

3.1 INTRODUCTION

Tissue engineering is intended to construct, outside the organism, living functional components that can be used to regenerate damaged tissues and/ or organs. The task of tissue engineering is to provide constructions that would restore, maintain, or improve the function of damaged organs and tissues. Materials used in tissue engineering must have definite properties and make the tissue-engineered constructs (grafts) similar to living tissues. These constructs must be able to: (i) regenerate themselves; (ii) provide blood supply, and (iii) change their structure and properties in response to the effects of outer factors, including mechanical stress. The lifetime of implants is limited, and, thus, their main function is not only to replace the damaged tissue but also to facilitate its regeneration (Shtilman, 2006; Hench and Jones, 2007).

The approach to producing bioactive implants and bio-artificial organs consists of the following steps: (i) fabrication of scaffolds for culturing the patient's autologous cells or cells obtained from the cell bank; (ii) culturing of cells and *in vitro* tissue formation; and (iii) implantation of the resulting bioconstructs to the recipient organism. To induce tissue generation at the implantation site, initial cell concentration must be very high (10^7–10^8 cells). As introduction of cell suspension may be ineffective, appropriate carrier materials for the delivery of cells need to be found. The most important step in the use of pre-cultured cells is the transplantation of these cells to the defect site. The success of this step depends on the proportion of cells that reach the defect site, their adhesion to the carrier, and their functional activity. One of the most difficult problems is the choice of an appropriate cell carrier: to realize their potential, cells need to stay attached to the carrier for a certain period of time. This may be related to the histogenetic property of these cells, which exhibit their proliferative capacity when organized in complex 3D structures.

The challenge facing tissue engineering is to optimize cell harvesting, proliferation, and differentiation and construct scaffolds or delivery systems that would be able to support and guide the *ex vivo* growth of 3D tissues. Scaffolds must serve as templates and inducers of cell proliferation and differentiation into specialized cells, which generate specific new tissue. The tissue can be grown on scaffolds, which, after being implanted into the host organism, will be resorbed, and there will be only new tissue in the defect. Another approach is implantation of a "biocomposite" consisting of the scaffold with a partly-developed new tissue. Having been implanted to the defect site, the tissue-engineered construct must retain its structure and functions for the time period necessary to regenerate normally functioning tissue in the defect site and then integrate into the surrounding tissues.

The properties of the scaffold are determined by the properties of the material it is made of and the processing technology employed. Thus, successful creation of effective bioconstructs is based on the availability of an appropriate biodegradable and biocompatible material. Scaffolds are constructed from inorganic and organic biostable and biodegradable materials (metals/alloys, polymers, ceramics, hydroxyapatites, composite materials, corals, collagen, gelatin, elastin, fibronectin, alginate, chitosan, etc.). As the implanted carriers of functioning cells act as temporary scaffolds, facilitating mature tissue formation, they should be preferably made of biodegradable materials. The use of non-degradable scaffolds can cause complications due to a long-term presence of foreign material in the organism.

Scaffolds should be multifunctional, have sufficient mechanical strength and elasticity, exhibit biocompatibility at protein and cellular levels, favor cell attachment, facilitate cell proliferation and differentiation, and be capable of neovascularization. The use of such scaffolds loaded with drugs (antibiotics, hormones, vitamins, protein factors, etc.) is a revolutionary and very promising direction in surgical reconstruction and transplant surgery. The most significant requirement that biodegradable scaffolds must meet is controlled biodegradation rate. The scaffold must be degraded into biologically harmless compounds at a rate corresponding to the rate of growth of a new functional tissue and be completely replaced by the tissue at a defect site.

The most widely-used scaffolds today are membranes, films, and meshes. The flat surface of the films is an appropriate substrate for *in vitro* cell cultures. These systems are simple to prepare and use. Surface properties and porosity of scaffolds are important for cell attachment and proliferation. The pore size is determined by the type of the cells. Culturing of osteoblasts in scaffolds with small-sized pores can lead to hypoxia and osteochondral conditions during bone formation, while larger pore sizes facilitate normal osteogenesis and vascularization of the implant. The size of the pores in the scaffold determines the release of cell metabolites and the supply of nutrients to cells. One method to fabricate porous polymeric scaffolds is to prepare a solution of two-component mixtures and then to leach out the water-soluble component from the scaffold in the solution.

To enhance cell adhesion to the surface, improve the gas-dynamic properties of scaffolds, and increase their permeability to substrates and cell metabolites, the scaffolds are treated by using physical factors or chemical reagents. Chemical modification involves modification of functional groups on the surface of the polymer material, enhancing its hydrophilicity and facilitating cell adhesion and growth. Three-dimensional scaffolds are more complex systems in the form of sponges and gels. Hydrogels used as scaffolds in tissue engineering have to fulfill load-bearing and/or volume maintenance functions. Hydrogels can also contain pores that are large enough for cell migration; they can serve as a basis for dissolution and disintegration, leading to formation of pores, where cells can migrate.

Among promising approaches to preparing ultrafine fibers, membranes, and micro- and nano-particles as models of cell scaffolds are nanotechnological methods such as microencapsulation and electrospinning. Depending on the polymer molecular weight, the density of the solution, and fabrication conditions, the resulting fibers may be of different diameters and structures. Electrospinning can be used to prepare ultrafine fibers and porous constructs based on them from solutions and melts of variously structured polymers. The development of novel biomedical technologies of cellular and tissue engineering has attracted the attention of researchers to this method. Thus, construction of scaffolds for tissue engineering has received great attention lately, and perfect scaffolds may be designed in the nearest future.

There are several approaches to the application of cell technologies to tissue repair. One of them is to introduce cell suspension of the required phenotype of definite concentration grown *in vitro* into the damaged tissues or into blood. Another, more technologically complex, approach is to grow cells *in vitro*, on a scaffold, and then implant the bioengineered construct or the new tissue to the recipient organism. Successful implementation of this approach depends on the properties of cell scaffolds (Wang et al., 2003; Hench and Jones, 2007).

Polymers of hydroxyalkanoic acids – polyhydroxyalkanoates (PHAs) – hold great promise for their use in reconstructive medicine, including cell and tissue engineering (Volova et al., 2006; Shtilman, 2006; Volova and Shishatskaya, 2014). The main advantage of PHAs is that they can contain monomer units with different carbon chain lengths. PHA copolymers are more promising materials, as their properties can vary within a fairly broad range, depending upon the proportions of different monomer units contained in them (Sudesh et al., 2000; Volova, 2004; Laucock et al., 2013).

The structure and functional properties of cell scaffolds and other biomedical polymer devices are considerably influenced by the basic properties of the material they have been made of. Studies of PHAs as materials for cell scaffolds mainly address two PHA types – homogenous P3HB and copolymers of 3-hydroxybutyrate and 3-hydroxyvalerate. The literature data regarding other PHA types are scant and contradictory.

3.2   FILMS AS CELL SCAFFOLDS MADE FROM PHAS WITH DIFFERENT CHEMICAL COMPOSITIONS

In recent years, degradable PHAs have been increasingly studied as materials for applications in cell and tissue engineering. These polymers show great potential for regeneration of damaged skin, repair of defects of soft tissues and bones, fabrication of prostheses for implantation in blood vessels and heart valves, etc. (Volova and Shishatskaya, 2011; Williams and Martin, 2004; Chen and Wu 2005; Volova et al., 2013).

The best-known representative of the PHA family is a homopolymer of 3-hydroxybutyric acid (poly-3-hydroxybutyrate, P3HB). This is a thermoplastic polymer with a high degree of crystallinity (over 70%). P3HB is a highly biocompatible material because hydroxybutyric acid is a natural metabolite found in cells and tissues of humans and animals. The main disadvantage of P3HB is that it cannot crystallize into ordered structure and is not readily processable; products based on this material have low shock resistance and are rigid and prone to "physical aging." PHA copolymers are more promising materials as their properties can vary within a fairly broad range, depending upon the proportions of different monomer units contained in them (Sudesh et al., 2000; Lakshmi and Laurencin, 2007; Laucock et al., 2013). However, since PHAs contain monomers other than hydroxybutyric acid, the biocompatibility of PHA copolymers needs to be thoroughly tested.

The second best-studied PHA type is copolymers of 3-hydroxybutyrate and 3-hydroxyvalerate, which have a lower degree of crystallinity (50–60%) than P3HB. It took 10 years, however, to prove their biocompatibility (Shishatskaya, 2009; Gogolewski et al., 1993; Martin and Williams, 2003; Shishatskaya and Volova, 2004).

There are only fragmentary data on other PHA types. Researchers of Tepha Inc. (U.S.) study a rubber-like polymer of 3-hydroxyoctanoic acid (polyhydroxyoctanoate, PHO), which has a low melting point (40–60°C) (Williams and Martin, 1999; Köse et al., 2005); 3-hydroxybutyrate/3-hydroxyhexanoate [P(3HB/3HHx)] copolymers have also been investigated (Martin and Williams, 2003). A few studies have been reported on PHA terpolymers comprising monomers of butyric, valeric, and hexanoic acids (Ji et al., 2008, 2009; Chen, 2005). Among very promising but little-studied copolymers is a 3-hydroxybutyrate/4-hydroxybutyrate copolymer. This PHA is rapidly degraded *in vivo;* it is an elastomer and exhibits higher values of elongation at break and a higher tensile strength than most of the other well-known PHAs (Martin and Williams, 2003; Wang et al., 2008, 2010). However, the published data are not consistent, and thorough research is needed to decide which types of PHAs hold greater promise.

Researchers of the Institute of Biophysics SB RAS and the Siberian Federal University reported results of constructing and comparing films of 4 different PHA types used as cell scaffolds (Nikolaeva et al., 2011; Volova et al., 2014a, b). High-purity PHA specimens – a homopolymer of 3-hydroxybutyric acid (P3HB) and 3-hydroxybutyric/4-hydroxybutyric acid P(3HB/4HB), 3-hydroxybutyric/3-hydroxyvaleric acid P(3HB/3HV), and 3-hydroxybutyric/3-hydroxyhexanoic acid P(3HB/3HHx) copolymers were used to produce films Their chemical composition and physicochemical properties are given in Table 3.1.

Differences in basic physical properties of the polymers influenced the characteristics of the films prepared from them. Electron microscopy of the surface microstructure of scaffolds prepared from PHAs that differed in their chemical composition and basic physicochemical properties showed certain dissimilarities (Figure 3.1).

The surface of the films prepared from the P3HB homopolymer was the smoothest and there were no pores on it. On the surface of the films prepared from the P(3HB/4HB) copolymer there were numerous pores of diameter

TABLE 3.1 Composition and Properties of PHAs Used to Prepare Cell Scaffolds (Nikolaeva et al., 2011) (Reprinted with permission from the Human Stem Cells Institute.)

Polymer composition, mol.%	Polymer properties					
	M_n, kDa	M_w, kDa	Đ	C_x, %	T_m, °C	T_d, °C
Polylactide (PLA)100	50.590	100	1.78	amorphous	50	-
P(3HB) 100	722.890	1 200	1.66	76	179.7	273
P(3HB/4HB) 89.3/10.7	477.138	1 100	2.32	43	171.9	268
P(3HB/3HV) 87/13	332.836	1 115	3.35	50	162	266
P(3HB/3HV) 72.4/27.6	347.419	1 077	3.1	45	157	263
P(3HB/HHx) 93/7	253.500	507	2.0	32	158	240

about 1 µm. The surface of the films prepared from 3-hydroxybutyrate/3-hydroxyvalerate copolymers was smoother and more homogeneous. The films prepared from the 3-hydroxybutyrate/-hydroxyhexanoate copolymer had the most uneven surface, with numerous pores of diameter varying between 0.5 µm and 5.0 µm. The surface of the scaffolds prepared from polylactide (PLA) (the reference material) had no pores and consisted of spherical layers.

Hydrophilic/hydrophobic balance of the surface is a major parameter that indirectly characterizes biological compatibility and influences cell adhesion and viability. This balance is expressed as water contact angle. Measurements of water contact angles provide a basis for calculating such significant parameters of the surface as cohesive forces, surface tension, and interfacial free energy. The highest value of contact angle (71.8±4.8°) was recorded for PLA films followed by P3HB films (70.0±0.4°) (Figure 3.1, Table 3.2). Contact angle values of the films prepared from PHA copolymers of three types were much lower. The contact angles of the films prepared from 3-hydroxybutyrate/3-hydroxyvalerate and 3-hydroxybutyrate/3-hydroxyhexanoate copolymers were similar to each other (60–62.5°). This value is comparable to the contact angle of polystyrene plates (control). The contact angle value of the films prepared from

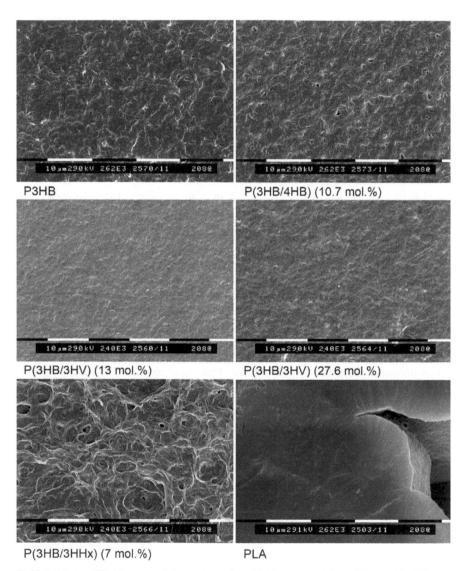

P3HB   P(3HB/4HB) (10.7 mol.%)

P(3HB/3HV) (13 mol.%)   P(3HB/3HV) (27.6 mol.%)

P(3HB/3HHx) (7 mol.%)   PLA

FIGURE 3.1   SEM images of the surface of scaffolds prepared from PHAs with different chemical compositions and polylactide (PLA) (Nikolaeva et al., 2011). (Reprinted with permission from the Human Stem Cells Institute.)

the 3-hydroxybutyrate/4-hydroxybutyrate copolymer was significantly lower – 57.4±0.6°. This is an intermediate position between hydrophobic and hydrophilic surfaces. The calculated parameters yielded the lowest hydrophilic/hydrophobic ratios for the copolymer specimens, suggesting that they were the best candidates for cell cultivation (Nikolaeva et al., 2011; Volova and Shishatskaya, 2011, 2014].

Surface energy is another major parameter that can influence the behavior of cells (Hallab et al., 2001; Kennedy et al., 2006). However, the hydrophilic/hydrophobic balance of the surface, which determines surface energy and other parameters, does not exert a universal effect on cells: in some cases, functions of cell structures are enhanced on hydrophilic surfaces, while in other cases – on hydrophobic ones. Results of the examination of cell scaffolds prepared from PHAs with different chemical compositions showed that films prepared from polylactide and poly-3-hydroxybutyrate, which were the least hydrophilic of the scaffolds examined, had the lowest values of surface tension and cohesive forces (about 31–32 and 95–97 erg/cm^2) (Table 3.2). For the copolymer scaffolds, the corresponding values were higher: 38.9–43.1 and 106.4–112.0 erg/cm^2 (Nikolaeva et al., 2011; Volova and Shishatskaya, 2014).

The nanometer surface roughness also affects cell adhesion and proliferation. However, some data suggest that cell adhesion is greater on rough surfaces than on polished ones, while other studies show that cells behave in a similar fashion on surfaces with different degrees of roughness. The study of the roughness of PHA film surfaces yielded the following results: the root mean squared roughness (Rq) of polylactide films was 241.629 nm – twice as high as the corresponding parameter of the films prepared

TABLE 3.2 Surface Properties of the Films Prepared From Different Types of PHAs (Nikolaeva et al., 2011) (Reprinted with permission from the Human Stem Cells Institute.)

PHA monomer composition, mol.%	Water contact angle, θ, °	Surface tension γ, erg/cm^2	Interfacial free energy, γ_{SL}, erg/cm^2	Cohesive forces, W_{SL}, erg/cm^2
Polylactide, 100	71.8±4.8	31.3	8.6	95.5
P(3HB), 100	70.0±0.4	32.8	7.9	97.7
P(3HB/4HB) 89.3/10.7	57.4±0.6	43.1	3.9	112.0
P(3HB/3HV) 87/13	60.3±2.8	40.7	4.6	108.9
P(3HB/3HV) 72.4/27.6	62.5±2.0	38.9	5.3	106.4
P(3HB/HHx) 93/7	60.9±1.6	42.4	4.1	111.1

from the PHA copolymers. The Rq values were similar for the P3HB homopolymer and for the copolymers of 3-hydroxybutyrate with 4HB, 3HV, and 3HHx, falling within the 109–113 nm range (Nikolaeva et al., 2011). In that study, biocompatibility of the films was evaluated *in vitro* in the culture of mouse fibroblast NIH 3T3 cells. Counts of cells stained using Romanowsky dyes attached to and growing on the PHA films that were performed at different time points of the experiment recorded significantly lower abundance of cells on the polylactide film than on the films prepared from any of the PHAs. None of the PHA films directly contacting fibroblast NIH 3T3 cells affected adversely their adhesion and growth. Cells attached to all films were well spread and most of them were star-shaped, which indicated their viability.

Cell counts obtained using the fluorescent DAPI DNA stain and FITC stain were similar to those obtained using Romanowsky stain (Figure 3.2).

In 24 h after fibroblast NIH 3T3 cells were seeded onto films, their counts on the control films (polystyrene) and PLA films were significantly lower than on PHA scaffolds; this difference was still observed at Day 4. By the end of the observation period, however, the number of cells on polystyrene membranes was comparable to that on PHA membranes, but cell counts on polylactide remained significantly lower, no more than 300±14.6 cells/field of view. On all PHA scaffolds, cell counts were similar, amounting to (cells/field of view) 440.0±27.4 on P3HB; 366.0±18.9 on P(3HB/4HB); 423.0±32.6 on P(3HB/3HV)-13; 452.0±13.8 on P(3HB/3HV) -27.6; and 402.0±19.4 on P(3HB/3HHx). As the differences in the number of cells on membranes prepared from different PHAs were insignificant, all PHAs tested in this study proved to be highly biocompatible and suitable for *in vitro* cell cultivation. This was also confirmed by examination of fibroblast morphology. In all phases of cultivation, the cells were viable and had a star-like shape.

Results of MTT assay provide another proof for high biocompatibility of the membranes prepared from all PHAs studied (Figure 3.3).

At 24 h after the cells were seeded on scaffolds, similar cell counts were obtained in the control (on polystyrene) and on all PHA films, but they were higher than cell counts on polylactide. After 4 days, the cell counts on all PHA films and in the control were comparable, but the number of cells on polylactide was almost twice lower. After 7 days, the same

Day 1   Day 4   Day 7

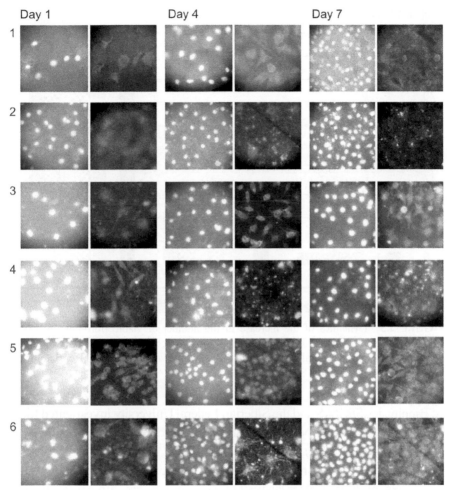

FIGURE 3.2 DAPI and FITC staining of NIH 3T3 fibroblast cells on PHA films: 1 – PLA, 2 – P3HB, 3 – P(3HB/4HB) (10.7 mol.%), 4 and 5 – P(3HB/3HV) (13 and 27.6 mol.%), 6 – P(3HB/3HHx) (7 mol.%) (Nikolaeva et al., 2011). (Reprinted with permission from the Human Stem Cells Institute.)

parameters and the same difference between cell counts on PHA films and on polylactide were observed. Thus, experiments with fibroblast NIH 3T3 cells showed that cells could be successfully cultivated on films prepared from any of the PHAs tested in this study.

Recent years have seen growing interest in PHA terpolymers, but their properties have not been sufficiently studied yet. Researchers of the Institute

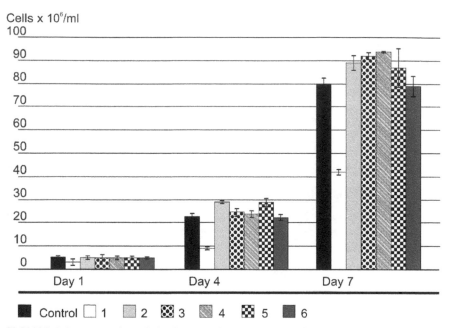

FIGURE 3.3 Dynamics of the increase in the number of viable fibroblast NIH 3T3 cells in MTT assay on scaffolds of different types: the control – polystyrene (numbers 1–6 denote the same materials as in Figure 3.2) (Nikolaeva et al., 2011). (Reprinted with permission from the Human Stem Cells Institute.

of Biophysics SB RAS and the Siberian Federal University investigated films prepared from PHA terpolymers with a decreased degree of crystallinity as cell scaffolds (Table 3.3).

Among the most important microscopic parameters characterizing properties of high-molecular-weight compounds are molecular weight characteristics, which determine processing parameters of polymers and their processability into devices intended for different applications. It is well-known that PHA molecular weight is a very variable parameter, depending on a number of factors, including the carbon source used, the length of cultivation, and the technique of polymer recovery employed. No clear relationship was found between PHA composition and values of M_w and M_n. The degree of crystallinity of polymers (C_x) is most significantly influenced by the composition and monomer fractions of PHA terpolymers. The ability of PHA to crystallize is determined by the inner properties of its chains and is characterized by crystallization temperature,

TABLE 3.3 Chemical Composition and Properties of PHAs Synthesized by Cupriavidus eutrophus B-10646 From Glucose Supplemented with Precursor Substrates: Propionate + γ-butyrolactone or valerate + hexanoate (Volova et al., 2014a)

PHA composition, mol. %				M_n,	M_w,		T_m,	T_d,	C_x,
3HB	3HV	4HB	3HHx	kDa	kDa	Đ	°C	°C	%
100	0	0	0	365	913	2.5	179	295	76
55.2	18.5	26.3	0	176	669	3.8	171	282	21
59.4	7.2	33.4	0	150	450	3.0	161	275	22
71.4	26.1	0	2.5	147	529	3.6	175	262	53
84.6	1.8	0	13.6	225	924	4.1	172	270	63

(From Volova T, Kiselev E, Vinogradova O, Nikolaeva E, Chistyakov A, Sukovatiy A, et al. (2014) A Glucose-Utilizing Strain, Cupriavidus euthrophus B-10646: Growth Kinetics, Characterization and Synthesis of Multicomponent PHAs. PLoS ONE 9(2): e87551. doi:10.1371/journal.pone.0087551. Reprinted via the Creative Commons License, https://creativecommons.org/licenses/by/4.0/.)

T_c. In many polymers, due to a number of reasons, crystallization develops only partly, and, thus, a large number of polymers are semi-crystalline materials. PHAs are also semi-crystalline polymers, and one of their important parameters is the degree of crystallinity, which characterizes the ratio of crystalline to amorphous phase and determines the processability of the material and the properties of the resulting products. The P(3HB/3HV/4HB) terpolymers showed the lowest degree of crystallinity, 9–20%. The difference between the C_x of P(3HB/3HV/3HHx) copolymers and P3HB, a highly crystalline polymer, was less pronounced: 53–63% and 76%, respectively. In all PHA terpolymers, the crystalline phase decreased while the amorphous, disordered regions increased, indicating better processability of these polymers.

Differences in physical properties of the PHA terpolymers influenced the properties of the films prepared from these PHAs. Electron microscopy of the surface structure of films prepared from PHAs that differed in their chemical composition and basic physicochemical properties showed certain dissimilarities (Figure 3.4). On the surface of the films prepared from P(3HB/3HV/4HB) (59.4/7.2/33.4), the pores were larger, reaching 5 μm, and more homogeneously sized. As the 4HB molar fraction increased, the film surface changed: the sample containing 3HV (17.7%) and 4HB (35.0%) had no pores, and the surface was covered by variously shaped protuberances and bubbles. The surface topography of the films prepared

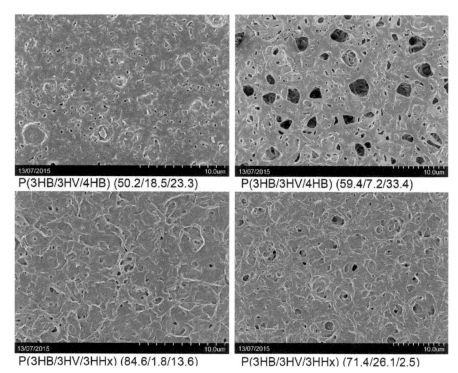

P(3HB/3HV/4HB) (50.2/18.5/23.3)   P(3HB/3HV/4HB) (59.4/7.2/33.4)

P(3HB/3HV/3HHx) (84.6/1.8/13.6)   P(3HB/3HV/3HHx) (71.4/26.1/2.5)

FIGURE 3.4 SEM images of the surfaces of the films prepared from PHA terpolymers with different compositions (Volova's data).

from the copolymer with 26.1 mol.% 3HV and 2.5 mol.% 3HHx was similar to the surface of P3HB films, but the surface of the P(3HB/3HV/3HHx) (84.6/1.8/13.6) sample had a rougher structure, with numerous pores reaching 4 µm.

Hydrophilic/hydrophobic balance of the surface is a major parameter that indirectly characterizes biological compatibility and influences cell adhesion and viability. This balance is expressed as water contact angle. Measurements of water contact angles provide a basis for calculating such significant parameters of the surface as cohesive forces, surface tension, and interfacial free energy. The lowest value of contact angle (70°00′) was recorded for P3HB films (Table 3.4). Films prepared from PHA terpolymers had larger water contact angles. The water contact angles of the samples that contained not only 3HB and 3HV but also 3HHx P(3HB/3HV/3HHx) were the greatest (above 90°). The terpolymers of the other type P(3HB/3HV/4HB), which contained 3HV and

TABLE 3.4 Surface Properties of Films From Solutions of PHAs With Different Chemical Compositions (Volova's data)

PHA composition, mol.%	Water contact angle, θ	Surface tension, γ, (erg/cm²)	Cohesive forces, W_{SL} (erg/cm²)	Interfacial free energy γ_{SL}, (erg/cm²)	Average roughness, R_a, nm	Root mean squared roughness, R_q, nm
P3HB (100)	70°00'	32.78	97.70	7.88	71.75	80.28
P(3HB/3HV/4HB) (55.2/18.5/26.3)	79°10'	25.73	86.56	11.97	113.74	153.39
P(3HB/3HV/4HB) (59.4/7.2/33.4	84°06'	22.16	80.33	14.63	43.32	56.24
P(3HB/3HV/3HHx) (71.4/26.1/2.5)	92°72'	16.51	69.34	19.97	19.54	22.89
P(3HB/3HV/3HHx) (84.6/1.8/13.6)	96°82'	14.13	64.15	22.78	208.29	243.62

4HB, showed similar water contact angles – 79°10'-89°20'. These values were somewhat higher than the water contact angle of polystyrene culture plates (67°12'). Surface energy is another major parameter that can influence the behavior of cells. However, the hydrophilic/hydrophobic balance of the surface, which determines surface energy and other parameters, does not exert a universal effect on cells: in some cases, functions of cell structures are enhanced on hydrophilic surfaces, while in other cases – on hydrophobic ones. All copolymer films had lower values of surface tension and cohesive forces than P3HB and higher values of interfacial free energy, 11.97 to 22.78 erg/cm².

It is well known that nanoscale surface roughness determines cell attachment, spreading, and motile activity and affects synthesis of specific proteins. However, some data suggest that rough surfaces favor cell attachment more effectively than polished ones, while other data indicate that changes in the surface roughness do not cause any changes in cell behavior. Results of investigating the surface roughness of the films prepared from PHA heteropolymers are given in Table 3.4. The values of root mean squared roughness (Rq) of the films prepared from PHA terpolymers differed depending on the monomer units constituting the PHA and their molar fractions. The highest Rq values (243 nm) were recorded in the P(3HB/3HV/3HHx) = 84.6/1.8/13.6 mol.% sample, and that was almost 4 times higher than the Rq value of the films prepared from the P3HB homopolymer (80 nm). The lowest Rq value was recorded for the sample containing different molar fractions of the same monomer units – P(3HB/3HV/3HHx) = 71.4/26.1/2.5 mol.%. Samples consisting of 3HB (50 mol.%), 4HB (35 mol.%), and 3HV (17–18 mol.%) monomer units had relatively high Rq values, 153 and 206 nm. At the same time, the roughness of PHA terpolymers containing different molar fractions of the same monomer units were significantly lower, 56 and 22 nm, respectively. These differences in the roughness are clearly demonstrated by results of AFM (Figure 3.5).

The physical-mechanical properties of a product should correspond to its intended use. Some devices need enhanced mechanical strength, determined by Young's modulus and tensile strength; in other cases, the product should

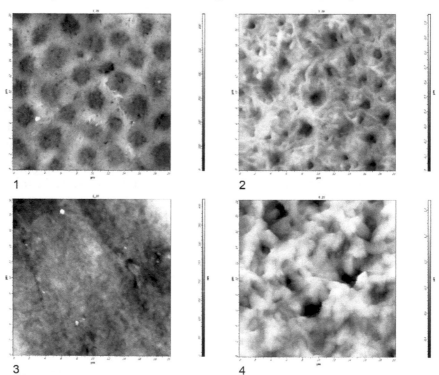

FIGURE 3.5   AFM images of the surfaces of films prepared from PHAs: 1 – P(3HB/3HV/4HB)=50.2/18.5/23.3; 2 – P(3HB/3HV/4HB)=59.4/7.2/33.4; 3 – P(3HB/3HV/3HHx)= 84.6/1.8/13.6; 4 – P(3HB/3HV/3HHx)= 71.4/26.1/2.5 (Volova's data).

be elastic, i.e., show high elongation at break. These parameters largely depend on the production technique employed and properties of the material used. Physical-mechanical properties of PHAs were investigated using films prepared from two types of copolymers, containing different monomer fractions: P(3HB/3HV/4HB) and P(3HB/3HV/3HHx). Table 3.5 compares physical-mechanical properties of PHA terpolymers with different composition and monomer fractions.

All copolymer films had significantly (1–2 orders of magnitude) higher values of elongation at break than films prepared from the highly crystalline poly-3-hydroxybutyrate. The presence on 3HHx monomer units in the polymer was the major factor influencing this parameter. Films with relatively low and relatively high 3HHx molar fractions (2.5 and 26.1%) showed relative elongation at break of 73.2%, while the films with 13.6 mol.% 3HHx and low 3HV (1.8 mol.%) had much higher ε – 390.5%. A similar effect was observed for the 4HB fraction: as the molar fraction of 4HB was increased, elongation at break of the films increased from 231.5 to 371.1%. Higher elasticity of PHA terpolymers was accompanied by lower mechanical strength of the films, whose Young's modulus was an order of magnitude lower than that of P3HB films. The tensile strength

TABLE 3.5 Physical-Mechanical Properties of Films From Solutions of PHAs With Different Chemical Composition (Volova et al., 2014a)

PHA composition, mol. %	Young's modulus, E, MPa	Tensile strength, σ, MPa	Elongation ε, %
P3HB (100)	2071.2	16.7	2.5
P(3HB/3HV/4HB) (55.2/18.5/26.3)	239.3	8.1	231.5
P(3HB/3HV/4HB) (59.4/7.2/33.4	37.5	10.1	371.1
P(3HB/3HV/3HHx) (71.4/26.1/2.5)	312.3	10.8	73.2
P(3HB/3HV/3HHx) (84.6/1.8/13.6)	257.5	9.4	390.5

(From Volova T, Kiselev E, Vinogradova O, Nikolaeva E, Chistyakov A, Sukovatiy A, et al. (2014) A Glucose-Utilizing Strain, Cupriavidus euthrophus B-10646: Growth Kinetics, Characterization and Synthesis of Multicomponent PHAs. PLoS ONE 9(2): e87551. doi:10.1371/journal.pone.0087551. Reprinted via the Creative Commons License, https://creativecommons.org/licenses/by/4.0/.)

of the terpolymer films was, however, only 1.5–2 times lower than that of P3HB films. Thus, mechanical properties of polymer films can be altered by varying the composition and fractions of monomer units in PHAs.

Biological properties of PHA films (adhesive properties and ability to facilitate cell proliferation) were studied in a culture of mouse fibroblast NIH 3T3 cells. MTT assay did not reveal any cytotoxic effect of PHA films as compared to the reference polystyrene films (Figure 3.6). The assay proved that all samples had high biological compatibility, favored cell attachment, and facilitated cell proliferation. In 24 h after seeding, the

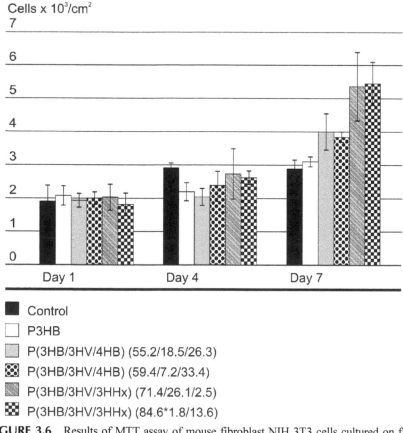

Cells x 10^3/cm^2

■ Control
□ P3HB
▨ P(3HB/3HV/4HB) (55.2/18.5/26.3)
▨ P(3HB/3HV/4HB) (59.4/7.2/33.4)
▨ P(3HB/3HV/3HHx) (71.4/26.1/2.5)
▨ P(3HB/3HV/3HHx) (84.6*1.8/13.6)

FIGURE 3.6 Results of MTT assay of mouse fibroblast NIH 3T3 cells cultured on films prepared from PHAs with different composition (Volova et al., 2014a). (From Volova T, Kiselev E, Vinogradova O, Nikolaeva E, Chistyakov A, Sukovatiy A, et al. (2014) A Glucose-Utilizing Strain, Cupriavidus euthrophus B-10646: Growth Kinetics, Characterization and Synthesis of Multicomponent PHAs. PLoS ONE 9(2): e87551. doi:10.1371/journal.pone.0087551. Reprinted via the Creative Commons License, https://creativecommons.org/licenses/by/4.0/.)

number of cells on PHA films was comparable with that on the reference (polystyrene) films. At Day 3, the number of viable cells on polystyrene and P(3HB/3HV/3HHx) (71.4/26.1/2.5) films was somewhat (5–30%) higher than on the other films. More significant differences were observed at Day 7 of the experiment, when on all terpolymer films the number of cells was greater than on the reference film. On P(3HB/3HV/4HB) (55.2/18.5/26.3 and 59.4/7.2/33.4) films, cell counts reached $4.00\pm0.55\times10^3$ and $3.84\pm0.17\times10^3$ cells/cm^2, respectively, while on polystyrene and P(3HB) films there were $2.91\pm0.28\times10^3$ and $3.13\pm0.12\times10^3$ cells/cm^2. The greatest number of viable cells was recorded on PHA films that contained not only 3-hydroxybutyrate and 3-hydroxyvalerate but also 3-hydroxyhexano-ate (2.5 and 13.6 mol.%) – $5.36\pm1.06\times10^3$ and $5.45\pm0.66\times10^3$ cells/cm^2, respectively; these counts were 1.8 times higher than on polystyrene and 1.7 times higher than on P3HB films. In spite of these differences, all PHA films showed high biological compatibility as cell scaffolds.

Results of investigations of fibroblast cells proliferating on films pre-pared from PHAs with different composition using the fluorescent DAPI DNA stain and FITC cytoplasm stain were generally similar to those obtained using MTT assay (Figure 3.7). On polystyrene and poly-3-hy-droxybutyrate films, cell counts were the lowest, and most of the cells were spherical.

On the films prepared from PHAs that contained not only 3HB and 3HV but also 3HHx, cell counts were the highest; the cells were spread well and had numerous pseudopodia, indicating good attachment to the surface and proliferation. On films prepared from PHAs containing 4HB as a third monomer, results were better than on the reference films, but cell counts were lower than on hexanoate-containing films; similarly to cell population on P3HB films, spherical and spindle-shaped cells were observed. Thus, the best adhesion and proliferation of mouse fibroblast NIH 3T3 cells were obtained on films prepared from PHA terpolymers containing 3HHx; at the same time, all PHA films outperformed polysty-rene and the P3HB homopolymer, which suggested their high biological compatibility with the cells cultured on them.

This study showed that PHA terpolymers had widely varying tem-perature parameters, degrees of crystallinity, and physical-mechanical properties; surface structure and properties of the films prepared from

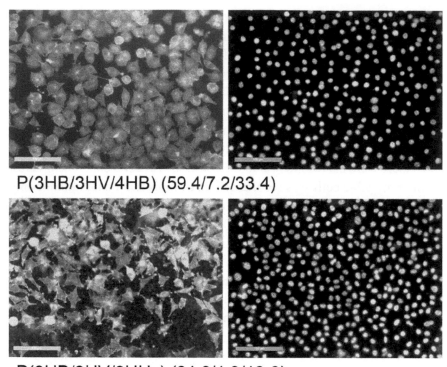

P(3HB/3HV/4HB) (59.4/7.2/33.4)

P(3HB/3HV/3HHx) (84.6/1.8/13.6)

FIGURE 3.7   Morphology of mouse fibroblast NIH 3T3 cells cultured on films prepared from PHAs with different composition at Day 7 of the culture: FITC and DAPI staining. Bar = 100 мм (Volova et al., 2014a).

them also differed significantly. All films prepared from PHA terpolymers with different composition facilitated attachment and proliferation of mouse fibroblast NIH 3T3 cells more effectively than polystyrene and the highly crystalline P3HB.

Cell scaffolds with smooth and dense surfaces, without holes or pores, have poor adhesive properties and may fail to ensure formation of new tissue in proper time. This hinders osteoblast migration from periosteum to the defect site and their attachment and proliferation. A new approach to modification of polymer products is to treat them by using physical methods or chemical reagents in order to enhance adhesive properties of the surface and facilitate attachment of the cultured cells, to improve gas-dynamic properties of the products, and to increase their permeability to substrates and metabolic products of cells and tissues. Laser treatment is a relatively new approach to modification of polymer products. Its main advantage over other treatments is that it modifies the surface selectively, without destroying the material or producing toxic substances. It has been assumed that treatment with laser radiation makes the surface more hydrophilic, thus, enhancing its adhesive properties (Jaleh et al., 2007; Ellis et al., 2011). However, laser treatment has scarcely been used to modify the surface of the products fabricated from P3HB or other PHAs.

The most powerful and commonly used continuous-light lasers are CO_2 electric-discharge gas lasers. CO_2 lasers use a mixture of carbon dioxide (CO_2), helium (He), nitrogen (N_2), sometimes hydrogen (H_2), water vapor, and/or xenon (Xe). In these lasers, light amplification occurs due to carbon dioxide molecules. Radiation is mainly generated at a wavelength of 10.6 μm. The efficiency of such lasers is higher than 10%, and they can generate high-quality radiation powers of several kilowatts. CO_2 lasers are commonly used for processing different materials – cutting, welding, and engraving – and in laser surgery.

In the Institute of Biophysics SB RAS, polymer films were prepared from the homopolymer P3HB of high purity. The films were about 40 μm thick. Laser treatment of the surface of flexible transparent polymer films was performed by moderate uniform irradiation of the surface, using CO_2 lasers LaserPro Explorer II (Coherent, U.S.) and LaserPro Spirit (Sunrad, U.S.). Pressed pellets intended for reconstructing bone tissue defects and perforated films were treated by using a LaserPro Spirit engraving system. CO_2 laser LaserPro Explorer II has the following characteristics: wavelength 10.6 μm, maximum power 30 W, and maximum speed 2 m/s. The laser is equipped with a standard SeZn lens, F=2'. The varied parameters were power and speed of processing and processing modes (focused and defocused (with no lens) modes). Power was varied between 1.5 and 16.5; speed between 0.8 and 2 m/s. CO_2 laser LaserPro Spirit has the following characteristics: wavelength 10.6 μm, maximum power 25 W, and maximum speed 1.5 m/s in the modes of raster and vector engraving, at a maximum resolution of 1000 dpi. The laser is equipped with a standard 2.0" lens.

In the first series of experiments, polymer films were treated by using a LaserPro Explorer II system, with the power varied between 1.5 and 16.5 W and the speed between 0.8 and 2 m/s. Under these conditions, the film surface is uniformly irradiated, and no considerable damage or perforation occurs. The treatment was performed in the focused and defocused modes. The defocusing of the laser beam on the film surface enlarged the irradiated area and, hence, decreased the radiative flux density, which enabled us to reduce surface deformation. Measurements of water contact angles on the film surface (Table 3.6) showed a decrease in this parameter on laser-treated films. In the focused mode, the decrease was more pronounced, and the angle was reduced to 67.4°, while in the defocused mode, it decreased to 79.4°, at the processing speed of 0.8 m/s and 1.8 m/s and power 9 W and 12 W, respectively. The use of the majority of irradiation modes caused a slight (10–16%) increase in the surface free energy of the film surfaces and a considerable increase (by a factor of 3–5) in its polar component, especially at higher power values. This may suggest that under high-energy impact, new polar functional groups may be generated on the surface and increase the water affinity of the polymer surface.

Changes in surface morphology influenced the adhesive properties of the surface and the number of the viable cells attached to it. Fluorescent and electron microscopy of the films with NIH 3T3 mouse fibroblast cells attached to the film surface (Figure 3.8) showed a great number of adherent viable cells. The most highly populated scaffolds

TABLE 3.6 Measurements of the Water Contact Angles and Surface Energy of Polymer Films Processed By Laser Radiation in Different Modes (Volova's data)

No samples	Process-ing mode	Speed (m/s)	Water con-tact angle, degrees	Surface energy, mN/m	Dispersive component, mN/m	Polar component, mN/m
1	Pristine P3HB		91.17±2.19	36.33±0.84	34.61±0.59	1.59±0.47
2	1.8	1.5	81.4±4.09	38.7±0.36	34.2±0.17	4.5±0.19
3	0.8	9.0	67.4±4.34	49.3±0.96	40.4±0.63	8.9±0.33
4	2.0	6.0	89.01±1.85	36.06±1.16	33.87±1.05	2.19±0.39

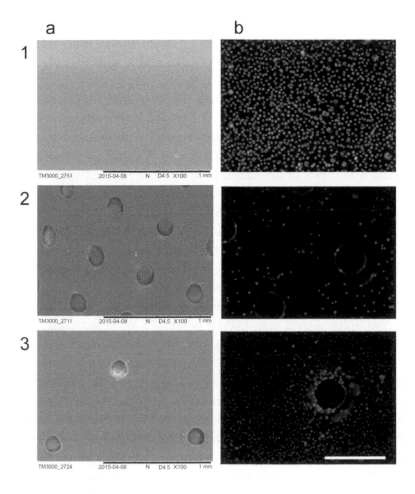

FIGURE 3.8 Fluorescent microscopy of NIH 3T3 mouse fibroblast cells attached to the films processed with laser radiation in different modes. DAPI (a) and FITC (b) staining, at 1000×; SEM images of fibroblast cells on laser-treated films (c); bar = 100 nm. Processing modes: 2 – 1.5 W/ 1.8 m/s; 3 – 9 W/0.8 m/s; 4– 6 W/2 m/s; 1 – control (untreated film) (Shishatskaya's data).

were the ones that had been treated at the power and speed of 1.5 W and 1.8 m/s, 3 W and 0.8 m/s, and 9 W and 2 m/s, respectively, in the focused mode and at 6 W and 2 m/s, respectively, in the defocused mode. Comparative counts of physiologically active and viable cells cultivated on laser-treated films in MTT assay showed that at Day 1, cell counts were similar on all films (about 0.28–0.30×10^5 cells/cm^2 (Figure 3.9a). At Day 4, the number of cells on laser-treated films was greater than in the control. The largest number of viable cells was observed on the films processed by laser radiation at 2 m/s (1.06–1.13×10^5 cell/cm^2). The other scaffolds, including the untreated film (control), had similar amounts of adherent cells: 0.72–0.84×10^5 cells/cm^2. At Day 7, the populations of cells on scaffolds treated with medium intensity of energy, i.e., 1.5 W/1.8 m/s; 3 W/0.8 m/s; 9 W/2 m/s of focused irradiation and 6 W/2 m/s of defocused irradiation, were similar to each other (8.1–8.8×10^5 cells/cm^2). These films showed a moderate decrease in the water contact angles (hydrophilicity increase). The number of cells on the films treated with more intensive laser radiation (9 W/0.8 m/s of focused irradiation and 12 W/1.8 m/s of defocused irradiation) was lower – 6.45 and 7.1×10^5 cells/cm^2. These results are consistent with the data obtained by microscopy of samples.

In the second series of experiments, films were treated by using a LaserPro Spirit system at its highest power, 25 W, in the modes that created uniform perforations in the films; the perforations were of preset diameter, 0.05 mm, with preset distances between them, 0.5 mm or 1 mm (Figure 3.10a). The pore diameter was 150 μm and the perforation density was 5 and 2 per 1 mm^2, respectively. At high magnification (1000× or higher), we could see melted edges around the pores and microcracks. The structure of the surface of the laser-treated samples was similar to that of the untreated sample. The wettability of the film surface (Table 3.6) decreased considerably after laser treatment. The greatest water contact angle reduction, to 65.7°, was observed on the films processed with h=0.5 mm. The polar component of the surface free energy increased considerably (to 5.5–13 mN/m, depending on the processing mode employed), suggesting a substantial enhancement of hydrophilicity of the laser-treated surface.

Thus, films prepared from PHAs with different composition, which were characterized in studies by Nikolaeva et al. (2011), Volova and Shishatskaya (2014), and Volova et al. (2014 a, b) have very good biological properties that make them suitable as experimental cell scaffolds.

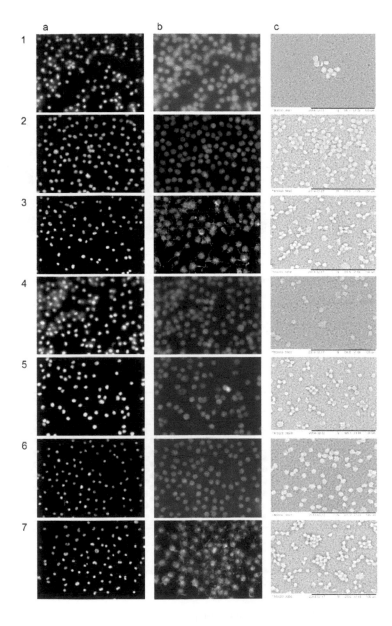

FIGURE 3.9 Counts of viable NIH 3T3 fibroblast cells on P3HB films processed with laser radiation in different modes (MTT assay). a) Non-perforated films. 1 – control (untreated film); 2 – 1.5 W/ 1.8 m/s; 3 – 9 W/0.8 m/s; 4 – 6 W/2 m/s. Samples 2–3 were processed in the focused mode, and sample 4 – in the defocused mode) (Shishatskaya's data).

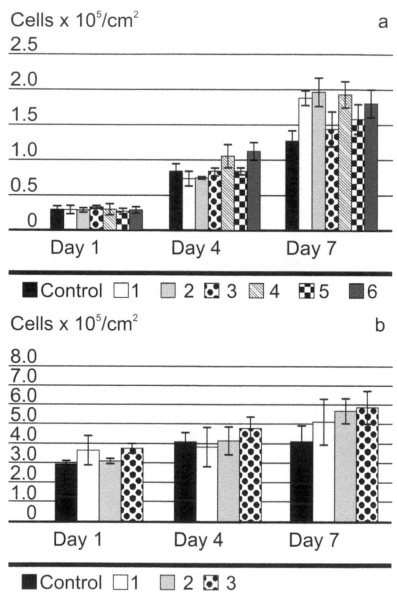

FIGURE 3.10 SEM images of film surfaces processed in different modes (LaserPro Spirit, power 25 W, speed 0.3 m/s), and morphology of cells: a) SEM images of films; b) DAPI staining of cells: 1 – h=0.5 mm; 2 – h=1 mm; 3 – control, untreated P3HB film. h – distance between points (Shishatskaya's data).

3.3 A STUDY OF NONWOVEN MEMBRANES COMPRISED OF ELECTROSPUN ULTRAFINE FIBERS AS CELL SCAFFOLDS

Electrospinning (electrostatic spinning) is a promising technique that can be used for fabricating micro- and ultrafine fibers and fibrous scaffolds (mats) and membranes. This technique was introduced in the 20th century to fabricate synthetic fibers. Several variants of production of fibers and devices using electrostatic force were patented in the 1930s–1940s. "Electrospinning" has become a widely used term quite recently (1994), but theoretical underpinning of electrospinning was produced by Taylor in the 1960s (Taylor, 1969). In the electrospinning process, ultrafine fibers are formed between two oppositely charged electrodes: one is placed in a polymer solution or melt and the counter electrode is a collecting metal screen. Electrospinning is used to produce ultra- and nano-fine fibers and porous structures based on them from solutions and melts of polymers with different structures. Electrospun products have been used in various applications such as filtration and protective apparel, tissue scaffolding and drug delivery, nanocomposites, and sensor applications. Fiber constructions prepared from various materials by electrospinning are promising candidates to be used as scaffolds for *in vitro* cell cultures and as medical devices for surgical reconstruction (wound dressings, barrier membranes for guided tissue regeneration in maxillofacial surgery, etc.). The process of electrospinning has been found to have great potential in cell and tissue engineering. Several types of electrospinning process have been described, which are used to produce fibers and membranes from both synthetic and natural polymers such as gelatin, carboxymethyl cellulose, polyethylene oxide, polylactide, polyurethanes, polyvinyl alcohol, polytrimethyl terephthalate, dimethylformamide, etc. A number of processing parameters (applied voltage, polymer solution feeding rate, solution density, working distance, capillary diameter, etc.) can greatly affect fiber formation and structure. In order to be able to tailor the properties of the resulting products, one needs to determine the processing parameters for each particular type of polymer material.

Electrospinning studies using PHAs have not been conducted until quite recently. The first papers reporting the employment of this technique to produce ultrafine fibers from poly-3-hydroxybutyrate and copolymers of

3-hydroxybutyrate with 3-hydroxyvalerate were published in 2006 (Volova et al., 2006; Tong and Wang, 2007). By now, the method has been tested to prepare electrospun products of PHA copolymers (Tong et al., 2012; Wang et al., 2012; Yu et al., 2012). A number of processing parameters (applied voltage, polymer solution feeding rate, solution density, working distance, capillary diameter, etc.) can greatly influence fiber formation and structure. In order to be able to tailor the properties of the resulting products, one needs to determine the processing parameters for each particular type of polymer material (Volova et al., 2006; Goncharov et al., 2012; Volova et al., 2014b). A number of studies described electrospinning of nonwoven PHA membranes and reported their structure, physical, mechanical, and biological properties as related to electrospinning parameters and polymer composition. Ultrafine fibers were electrospun from PHA solutions using a Nanon 01A automatic set-up (MECC Inc., Japan). Chloroform solutions with polymer concentration varied from 1 to 10 wt % were prepared from all types of PHAs. The polymer solution was poured into a plastic syringe (13 mm inside diameter). The syringe was fixed horizontally in the set-up, the solution feeding rate was varied from 4 to 8 mL/h, the applied voltage from 15 to 30 kV, and the working distance from 11 to 15 cm. Randomly oriented or aligned ultrafine fibers were collected on a flat steel plate or a rotating drum (at 1000 rpm), respectively; both collectors were covered with aluminum foil to collect ultrafine fibers more effectively.

The effect of the density of polymer solutions on fiber properties was studied using the homopolymer of 3-hydroxybutyric acid, in order to avoid the influence of the chemical composition of the PHA on the electrospinning process and properties of the products. P3HB chloroform solutions with polymer concentration varied from 1 to 10 wt. % were used. The process parameters were as follows: needle diameter – 1 mm, applied voltage – 30 kV, solution feeding rate – 5 mL/h, and working distance – 15 cm, a flat steel collection plate. Polymer concentration directly influences the quality of electrospun fibers. Polymer solutions of the density under 2 wt. % could not yield high-quality fibers, due to their low viscosity (below 100 cP). These solutions yielded a few very thin ultrafine fibers and fine spray, which consisted of microdrops rather uniformly distributed on the collector. As P3HB concentrations increased, the solution viscosities increased too (from 60 to 800 cP) because the entanglement of polymer molecular chains prevented the breakup of the electrically driven jet and allowed the electrostatic stresses to further elongate the jet.

Stable electrospinning of ultrafine fibers in the Nanon 01A set-up was attained from P3HB solutions with polymer concentrations between 2 and 8 wt. % (solution viscosity 200–800 cP). Polymer concentration significantly influenced the diameter of the ultrafine fibers, which varied from 0.45 to 3.14 μm. Within the study range of polymer concentrations, the diameter of the ultrafine fibers was linearly related to the solution density. The viscosity of the solutions with polymer concentrations above 8 wt.% was too high (about 1000 cP) to allow successful formation of ultrafine fibers. Figure 3.13 shows SEM images of electrospun P3HB ultrafine fibers prepared from the polymer/chloroform solutions with different polymer concentrations. Most of the fibers were cylindrical and smooth; their surface was virtually defect-free; there were spaces between the fibers. As polymer concentration of the solution was increased, the range of fiber diameters became wider (between 1 μm and 3 μm). As the fiber diameter increased, spaces between fibers in the fibrous mat widened from about 2 to about 10 μm and the thickness of the mats increased too (from 10 to 75 μm). P3HB concentrations of the solution ranging between 2 and 8 wt.% yielded good-quality fibers. However, some of the fibers produced by electrospinning from solutions with P3HB concentration of 8 wt.% fused together at several points (Figure 3.11d), probably because of the presence of residual solvent in larger-diameter fibers.

Polymer solution concentration considerably influenced the diameter of ultrafine fibers, which varied between 0.45 and 3.14 μm (Figure 3.12a). Within the study range of polymer concentration, the diameter of the fibers was linearly related to the solution density. As the fiber diameter increased, spaces between fibers in the nanofibrous mat increased from about 2 to about 10 μm and the thickness of the mats increased too (from 0.010 to 0.075 μm). The diameter of the fibers significantly influenced physical/mechanical properties of mats comprised of them (Figure 3.12b). As fiber diameter increased from 0.45 to 3.14 μm, the mats became more elastic, which was expressed as a considerable increase in elongation to break (from 4.5 to 10.6%), but less mechanically strong, with tensile strength decreasing from 23.16 to 6.65 MPa and Young's modulus from 1.16 to 0.3 GPa. Young's modulus of the specimens electrospun from P3HB

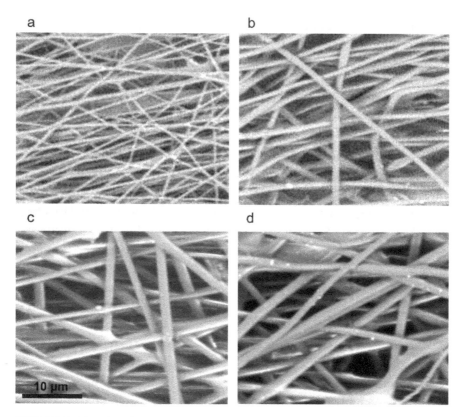

FIGURE 3.11 SEM images of nonwoven membranes comprised of ultrafine fibers prepared by electrospinning from P3HB solutions of different concentrations (2, 4, 6, and 8 wt.%) (Goncharov et al., 2012). (Reprinted with permission from the Journal of Siberian Federal University.)

(0.3–1.16 GPa) was comparable to those of polyethylene (0.2 GPa) and polypropylene (1.5–2.0 GPa). The mechanical properties of the nonwoven membranes were similar to the mechanical properties of human tendons (Young's modulus 0.06 GPa) and skin (Young's modulus 15–150 MPa, tension stress 5–30 MPa) (Ying et al., 2008). Thus, this is a promising material for tissue engineering applications.

The effect of the polymer solution-feeding rate on the properties of the fibers was studied by varying this parameter between 2 and 8 mL/h. At the solution-feeding rate below 2 mL/h, the jet was unstable; when the solution feeding rate reached 8 mL/h, polymer junctions were formed. Effects of changing applied voltages from 15 to 30 kV on fiber morphology and

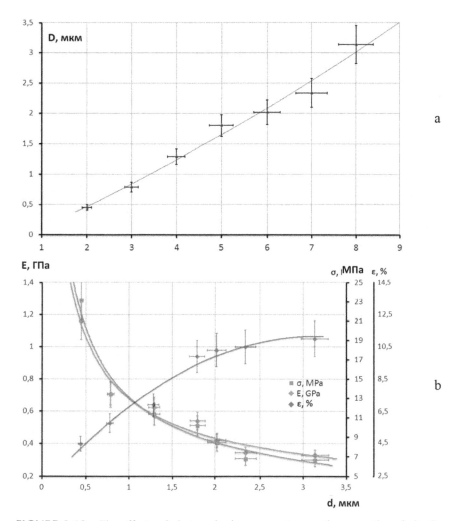

FIGURE 3.12 The effects of electrospinning parameters on the properties of ultrafine fibers: a – fiber diameter vs. polymer solution concentration; b – physical and mechanical properties of fibers with different diameters: Young's modulus (E, GPa) and tensile strength (ε, %) (Goncharov et al., 2012). (Reprinted with permission from the Journal of Siberian Federal University.)

diameter are shown in Figure 3.13. Other electrospinning parameters were maintained constant: needle diameter 1 mm, concentration of polymer solution 5 wt.%, solution feeding rate 5 mL/h, working distance 15 cm. Good fiber morphology was achieved at the voltage ranging between 15

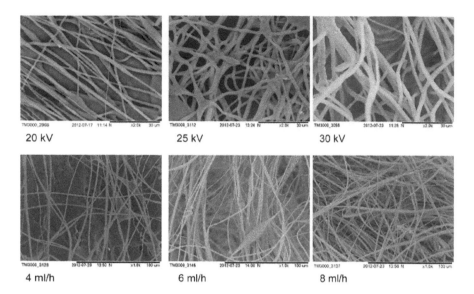

20 kV   25 kV   30 kV

4 ml/h   6 ml/h   8 ml/h

FIGURE 3.13   SEM images of nanofibers electrospun from P3HB solutions under varied applied voltage (20–30 kV) and solution feeding rate (4–8 mL/h) (Volova's data).

and 30 kV. Within this voltage range, the average fiber diameter changed from 1.25 to 2.5 µm. At voltages below 10–12 kV, no fiber formation occurred whatever the polymer concentration used.

Surface properties of nonwoven membranes comprised of ultrafine fibers of different diameters electrospun from P3HB solutions of different concentrations are listed in Table 3.7. Membranes consisting of the smallest-diameter fibers (0.45–1.29 µm) had the lowest values of the water contact angle (51.20–56.57 °).

The surface properties such as surface tension (γ_S) and interfacial free energy (γ_{SL}), calculated using de Gennes equations, had the lowest values at the highest cohesive force (W_{SL}). As fiber diameter increased above 1.8 µm, the water contact angle increased. The membranes became more porous as fiber diameter increased. The finest, 0.45-µm diameter, fibers showed the lowest porosity levels, about 43.5%. The porosity levels of the membranes formed by the fibers of diameters between 0.79 and 3.14 µm were similar to each other, but they were higher than those of the mats formed by the finest fibers (within the range of 62–65%).

Adhesive properties of the electrospun membranes comprised of ultrafine fibers of various diameters were studied in the culture of NIH 3T3

TABLE 3.7 Surface Properties of Nonwoven Membranes Comprised of Ultrafine Fibers with Different Diameters Electrospun From P(3HB) Solutions (Goncharov et al., 2012) (Reprinted with permission from the Journal of Siberian Federal University.)

Fiber diameter, μm	Water contact angle, θ °	Surface tension, γ, erg/cm²	Interfacial free energy, γ_{SL}, erg/cm²	Cohesive force, W_{SL}, erg/cm²
0.45±0.10	51.20±1.06	68.54	2.54	118.42
0.79±0.12	52.23±0.70	68.88	2.73	117.39
1.29±0.16	56.57±1.50	70.27	3.67	112.91
1.8±0.17	63.23±0.75	72.25	5.50	105.59
2.02±0.15	66.97±1.23	73.30	6.74	101.28
2.34±0.24	68.23±0.31	73.64	7.20	99.80
3.14±0.38	68.80±0.26	73.79	7.42	99.13

mouse fibroblast cells. All membranes, regardless of the fiber diameter, showed good properties as cell scaffolds. DAPI staining (Figure 3.14) did not show significant differences in the number of attached cells between the membranes consisting of the fibers prepared from 2 to 7% P3HB solutions and the reference membrane. On the membranes prepared from the densest polymer solution and comprised by the fibers of the largest diameter, cell proliferation was considerably slower than on other scaffolds; the membrane had defects in the form of junctions.

One of the indications of the functionality of cell scaffolds is the ability of cells to synthesize extracellular matrix proteins. SEM images of the cells obtained at Day 7 of the culture showed formation of intercellular substance on all types of membranes, and the greatest deposits were observed on the membranes prepared from the 2–5% solutions (with fiber diameter between 0.45 and 1.8 μm), which was indicative of the high metabolic activity of fibroblasts. At Day 7, the greatest amounts of intercellular substance were observed on the membranes consisting of the fibers between 0.45 and 1.8 μm in diameter. On the membranes prepared from denser P3HB solutions (6–7%), which consisted of the fibers between 2.02 and 3.14 μm in diameter, proteins were aligned along the fibers, and the deposits were smaller. On the membranes prepared from the densest polymer solutions, the number of cells was lower almost by a factor of 2; no extracellular matrix proteins were detected.

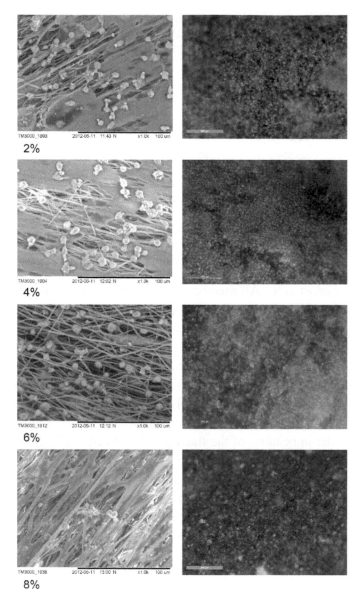

FIGURE 3.14   NIH 3T3 mouse fibroblast cells cultivated on membranes comprised of ultrafine fibers prepared from P3HB solutions of various concentrations; Day 7 of the culture: photographs to the left are SEM images, bar = 100 μm; photographs to the right are DAPI staining images, bar = 200 μm (Goncharov et al., 2012). (Reprinted with permission from the Journal of Siberian Federal University.)

Results of MTT assay of fibroblast cell viability are shown in Figure 3.15. The fiber diameter and the thickness of the membrane prepared from P3HB solutions of different concentrations (2–7%) did not significantly influence the physiological activity of fibroblasts. In 24 h after cell seeding, the number of cells on the membranes prepared from the densest polymer solution was comparable to the number of cells in the control (on polystyrene). On the other scaffolds, the number of cells was greater than on polystyrene almost by a factor of 2. The same results were observed in 3 days after seeding.

Thus, all membranes composed of ultrafine fibers prepared from P3HB solutions facilitated fibroblast cell proliferation. Membranes consisting of the finer fibers were better cell scaffolds, but they had poorer mechanical properties.

It is well-known that the structure and properties of polymer products are determined both by the technique employed to fabricate them and by the properties of the material. Volova et al. reported a study of the properties of membranes prepared by electrospinning from four PHA types, which differed in their physicochemical properties, by using different

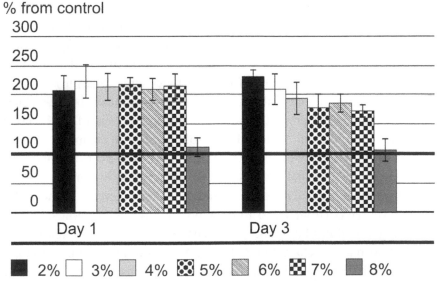

FIGURE 3.15 Viability of NIH 3T3 fibroblast cells on the membranes comprised of ultrafine fibers prepared by electrospinning from P3HB solutions of different concentrations (control is polystyrene) (Goncharov et al., 2012). (Reprinted with permission from the Journal of Siberian Federal University.)

types of collectors. Two types of collectors were used: a flat plate, used for the formation of randomly oriented fibers, and a rotating drum, for the formation of aligned fibers. The study addressed the effect of fiber alignment on the mechanical properties of the products, taking into account the chemical structure of PHAs used. Chloroform solutions were prepared from all types of PHAs (5% concentration); the following electrospinning parameters were used: solution feeding rate 5 mL/h, working distance 15 cm, applied voltage 30 kV, flat steel collection plate.

Figure 3.16(A) shows SEM images of the fibers prepared from chloroform solutions of different PHAs that contained similar molar fractions of their second monomers (3HV, 3HHx, 4HB) (Samples 1, 3, 6, and 9 in Table 3.8). The fibers were collected on a flat steel plate, i.e., randomly oriented fibers were produced. As Figure 3.16(A) shows, the fibers were good quality but different in diameter. The average diameter of the fibers prepared from P3HB was 2.9 μm, and the thickness of the fibrous scaffolds was 25 μm. Measurements of diameters of fibers prepared from copolymers with similar molar fractions of their second monomers showed that they had different average diameters and ranges of diameters, but the fibrous mats had similar thickness – about 20 μm. The average diameter of P3HB/10 mol.%-3HV fibers was 2.3 μm, while the average diameter of P(3HB/4HB) fibers was significantly smaller (1.7 μm). By contrast, the diameter of the fibers prepared from the solution of P3HB/10 mol.%-3HHx of the same concentration (5%) was twice larger (4.2 μm).

For randomly oriented fibers, no obvious relationship between their physical-mechanical properties and the molar fraction of the second monomer was observed. A more than 7-fold increase in the molar fraction of 3HV in P(3HB/3HV) (from 4.5 to 32.8%) caused a 1.6-fold increase in Young's modulus, but did not induce any change in tensile strength or elongation at break. However, a smaller (3-fold) increase in the molar fraction of 3HHx (from 4.9 to 13.6%) did not have any significant effect on Young's modulus but reduced the tensile strength by half and caused a 3-fold decrease in elongation at break. A 7.5-fold increase in the molar fraction of 4HB in P(3HB/4HB) did not affect mechanical strength of the samples but increased their elasticity considerably. The PHA samples with different chemical compositions had dissimilar degrees of crystallinity,

FIGURE 3.16 SEM images of the membranes comprised of randomly oriented ultrafine fibers electrospun from solutions of PHAs (A) and– FITC staining of NIH 3T3 fibroblast (7-day old culture) (B). PHA composition: a – P3HB; b – P3HB/10.5 mol.%-3HV; c – P3HB/10.0 mol.%-4HB; d – P3HB/10.0 mol.%-3HHx (Volova's data).

TABLE 3.8 The Physical-Mechanical Properties of the Fibrous Mats Composed of Aligned and Randomly Oriented Ultrafine Fibers (Volova's data)

No.	PHA composition	Young's modulus E, MPa	Tensile strength P, MPa	Elongation to break, %
Randomly oriented ultrafine fibers:				
1	P3HB (100 mol.%)	356.23±40.62	9.32±2.54	13.3±3.11
2	P(3HB-*co*-4.5 mol.%-3HV)	9.61±1.85	0.75±0.08	47.36±7.21
3	P(3HB-*co*-10.5 mol.%-3HV)	5.81±1.06	0.67±0.06	40.72±6.20
4	P(3HB-*co*-32.8 mol.%-3HV)	13.91±2.07	0.80±0.09	38.33±5.34
5	P(3HB-*co*-4.9 mol.%-3HHx)	22.58±3.24	0.94±0.12	63.25±9.85
6	P(3HB-*co*-10 mol.%-3HHx)	18.68±2.85	0.76±0.10	29.45±4.61
7	P(3HB-*co*-13.6 mol.%-HHx)	21.28±3.64	0.58±0.11	21.59±3.17
8	P(3HB-*co*-6.1 mol.%-4HB)	3.61±0.85	0.82±0.09	58.95±8.51
9	P(3HB-*co*-10 mol.%-4HB)	1.59±0.30	0.42±0.08	72.98±10.33
10	P(3HB-*co*-13.6 mol.%-4HB)	2.91±0.75	1.26±0.23	102.37±12.55
11	P(3HB-*co*-20.3 mol.%-4HB)	2.32±0.62	1.01±0.15	117.33±14.30
12	P(3HB-*co*-38 mol.%-4HB)	1.22±0.36	1.46±0.21	177.74±20.67
13	P(3HB-*co*-48.3 mol.%-4HB)	1.60±0.54	0.81±0.09	118.93±15.76
Aligned ultrafine fibers:				
14	P3HB (100 mol. %)	543.03±72.45	11.17±2.33	9,90±3.21
15	P(3HB-co-4.5 mol.%-3HV)	824.04±78.23	14.18±2.17	10.11±1.24
16	P(3HB-co-10.5 mol.%-3HV)	463.52±49.08	12,26±2.65	11.56±2.10
17	P(3HB-co-32.8 mol.%-3HV)	321.90±27.33	14.13±2.05	91.16±7.32
18	P(3HB-co-4.9 mol.%-3HHx)	563.61±61.56	16.27±1.91	14.83±2.13
19	P(3HB-co-10 mol.%-3HHx)	431.04±38.42	10.26±2.23	20.69±3.61
20	P(3HB-co-13.6 mol.%-3HHx)	479.95±52.62	12.71±2.32	50.00±4.66
21	P(3HB-co-6.1 mol.%-4HB)	817.50±76.81	15.33±3.71	13.10±2.00
22	P(3HB-co-10 mol.%-4HB)	193.70±22.30	5.70±0.82	52.46±6.32
23	P(3HB-co-13.6 mol.%-4HB)	113.54±9.85	9.30±1.24	180.33±20.71
24	P(3HB-co-20.3 mol.%4HB)	77.94±8.41	15.74±2.12	262.29± 32.41
25	P(3HB-co-38 mol.%-4HB)	5.50±4.22	11.91±1.56	245.90± 22.54
26	P(3HB-co-48.3 mol.%-4HB)	4.12±1.68	4.92±0.62	311.47± 44.37

which varied between 35 and 75%. These dissimilarities may be accounted for by differences in the micromolecular structure of polymer chains and crystallization processes of PHAs with different chemical compositions. Further work is required to establish this.

Aligned ultrafine fibers were prepared from PHAs of the same composition as those used in the previous study and under the same electrospinning conditions, but with a rotating drum (at 1000 rpm) used as a collector (Figure 3.17). The average diameter of aligned fibers prepared from P3HB was 2.1 μm. The diameter of copolymer-aligned fibers, in contrast to randomly oriented ones, was similar to the diameter of P3HB fibers. The diameter of aligned P(3HB/3HV) fibers was similar to that of the corresponding randomly oriented fibers (about 2.5 μm); the average diameter of P(3HB/3HHx) fibers was 2.2 μm and that of P(3HB/4HB) fibers 2.7 μm. The diameter range of the aligned fibers was about 2 μm, and that was close to the diameter range of randomly oriented fibers. The thickness of fibrous scaffolds composed of aligned fibers was 25–30 μm, i.e., comparable to the thickness of randomly oriented scaffolds.

The physical-mechanical properties of the fibrous mats composed of aligned and randomly oriented ultrafine fibers differed significantly (Table 3.8).

First, electrospun aligned fibrous scaffolds differed from randomly oriented ones in that they had much higher mechanical strength. Second, the effects of the second monomers of the copolymers used to prepare the fibers on the properties of the aligned fibrous mats were different from their effects on the properties of the randomly oriented fibrous mats. The aligned fibrous mats prepared from copolymers containing 3HV and 3HHx (Table 3.8) had similar values of tensile strength and Young's modulus, and they were not significantly lower than those of P3HB fibers, but their elasticity values differed by a factor of two. In P3HB/10 mol.%-4HB fibers, both parameters characterizing mechanical strength were lower than in P3HB ones, but this difference was not as significant as in randomly oriented fibers, while elasticity was more than 4 times higher.

The most important difference between randomly oriented and aligned copolymer fibrous scaffolds was that in the latter, increased molar fractions of the second monomers had pronounced effect on the properties of the scaffolds. As the molar fraction of 3HV in P(3HB/3HV) was

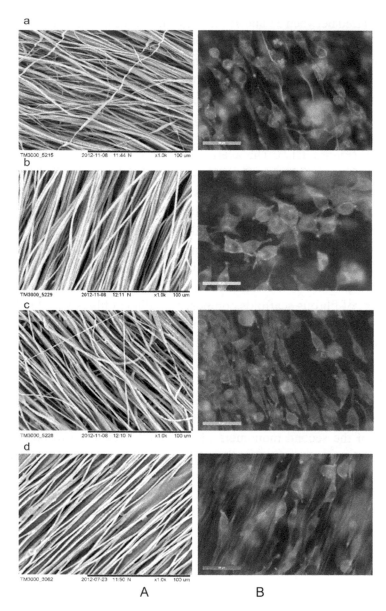

FIGURE 3.17 SEM images of the membranes comprised of aligned ultrafine fibers electrospun from solutions of PHAs. (A) and FITC staining of NIH 3T3 fibroblast (7-day old culture) (B). PHA composition: a – P3HB; b – P3HB/10.5 mol.%-3HV; c – P3HB/10.0 mol.%-4HB; d – P3HB/10.0 mol.%-3HHx (Volova's data).

increased from 4.5% to 32.8%, Young's modulus consistently decreased from 824.04±78.23 to 321.90±27.33 MPa, while tensile strength remained practically unchanged, amounting to 14 MPa, and elongation at break increased from 10.11±1.24% to 91.16±7.32%. For P(3HB/4HB), the increase in the molar fraction of 4HB from 6.1% to 48.3% also caused a gradual and more pronounced decrease in Young's modulus, from 817.50±76.81 to 4.12±1.68 MPa, and an insignificant decrease in tensile strength, with elongation at break rising from 13.10±2.00 to 311.47± 45.37%. In the fibers prepared from P(3HB/3HHx), the increase in the molar fraction of 3HHx from 4.9% to 13.6% did not cause any significant change in Young's modulus, but elongation at break gradually increased threefold.

Figure 3.18 clearly shows the effects of orientation and PHA composition on mechanical properties of electrospun fibrous scaffolds, demonstrating changes in tensile strength and elongation at break of randomly oriented and aligned fibrous scaffolds prepared from PHA copolymers that had different structures but contained the same molar fractions of their second monomers.

Scaffolds composed of aligned ultrafine fibers prepared from P3HB/10.5 mol.%-3HV had the highest values of elongation at break, while scaffolds composed of randomly oriented ultrafine fibers prepared from P3HB/10.0 mol.%-4HB showed the highest elasticity. The effects of the parameters investigated in this study on physical-mechanical properties of electrospun fibers can be ranked from strongest to weakest as follows: fiber orientation – PHA chemical composition – polymer solution density. The factor that has the strongest effect on fiber diameter and morphology is polymer solution density, while the effect of PHA chemical composition is less significant. Thus, by using different types of PHAs and different collectors, one can prepare electrospun products with tailored strength and elasticity.

Biological properties of fibrous scaffolds were studied in the culture of NIH 3T3 fibroblast cells. Considerable differences in cell culture parameters were revealed in experiments with fibrous scaffolds prepared from PHAs with different chemical compositions. Figure 3.19 shows results of investigating fibroblast cell morphology in experiments with cells stained with phalloidin conjugated to FITC and DAPI and SEM images showing the development of extracellular matrix. Results of the

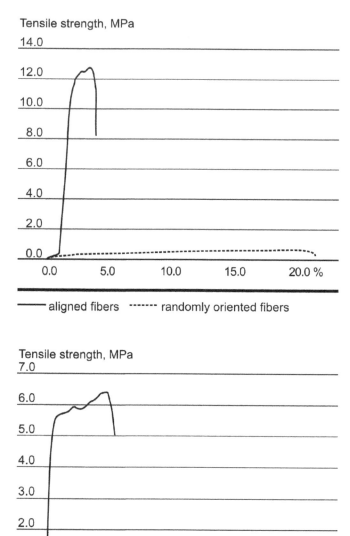

FIGURE 3.18   (Continued)

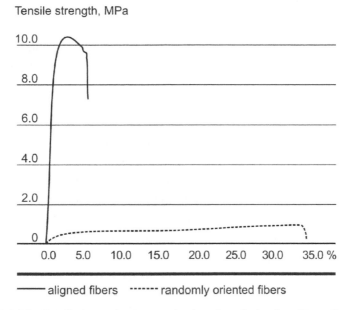

Tensile strength, MPa

aligned fibers ······· randomly oriented fibers

FIGURE 3.18 Tensile strength: A – randomly oriented ultrafine fibers, B – aligned ultrafine fibers produced by electrospinning from solutions of PHAs with different chemical compositions: a – P3HB/10.5 mol.%-3HV; b – P3HB/10.0 mol.%-4HB; c – P3HB/10.0 mol.%-3HHx (Volova's data).

counts of cells stained with fluorescent dyes are comparable with results of MTT assay. On aligned scaffolds, cells, mostly spindle-shaped ones, were arranged along the fibers. On randomly oriented scaffolds, cells were also arranged along the fibers; most cells were triangular or rectangular, with 3–4 appendages, used by cells to attach to the fibers. Larger cells were observed on the randomly oriented fibrous scaffolds. On the aligned P3HB scaffolds, the average cell size was 338.06±45.19 μm², while on randomly oriented scaffolds it was 177.05±45.20 μm². On randomly oriented scaffolds prepared from PHA copolymers, the average cell sizes were also greater than on copolymer aligned scaffolds: 381.73±74.97 μm² on P(3HB/3HV), 398.77±92.59 μm² on P(3HB/3HHx), and 429.93±158 μm² on P(3HB/4HB) 387.38±99.31 μm² on P(3HB/3HV), 370.82±91.87 μm² on P(3HB/3HHx), and 274.79±28.59 μm² on P(3HB/4HB). However, significant differences (P ≤ 0.05) were only obtained by comparing the results on aligned fibrous scaffolds prepared from P3HB and from copolymer PHAs. The differences between randomly oriented scaffolds were insignificant.

SEM images of cells proliferating on aligned and randomly oriented electrospun fibers are shown in Figures 3.16 and 3.17. Spherical cells prevailed on all types of scaffolds; cell infiltration between fibers was observed. Greater deposits of extracellular matrix protein were recorded on randomly oriented scaffolds. The number of cells on the reference scaffold (polystyrene) was lower than the number of cells on PHA scaffolds (half or even less than half of that number). At Day 7 after seeding, counts of viable cells using MTT assay showed that on randomly oriented fibrous scaffolds prepared from copolymers containing 3HV or 4HB as their second monomer the increase in the number of cells was about 1.4 times higher than on P3HB scaffolds or on the scaffolds prepared from the copolymer containing 3HHx. The number of cells on randomly oriented P3HB, P(3HB/3HV), P(3HB/4HB), and P(3HB/3HHx) scaffolds reached $1.81 \pm 0.15 \times 10^5$, $2.33 \pm 0.17 \times 10^5$, $2.41 \pm 0.20 \times 10^5$, and $1.803 \pm 0.21 \times 10^5$ cells/cm^2, respectively. No statistically significant ($P \leq 0.05$) differences were found between P(3HB/3HV) and P(3HB/4HB) scaffolds. Also, fiber orientation had a significant effect on the growth and viability of fibroblast cells. It is apparent from Figure 3.18 that the number of viable fibroblasts on all aligned scaffolds was significantly lower ($1.14 \pm 0.19 \times 10^5$, $1.56 \pm 0.25 \times 10^5$, $1.83 \pm 0.31 \times 10^5$ and $1.16 \pm 0.12 \times 10^5$ cells/cm^2 at Day 7) than on randomly oriented fibrous scaffolds. The effect of the PHA chemical composition on the number of cells on aligned scaffolds was similar to that on randomly oriented ones; scaffolds prepared from P(3HB/4HB) copolymers provided the highest increase in cell numbers. On the reference scaffolds (polystyrene plates), the number of proliferating cells was lower too.

Thus, MTT assay showed that all PHA fibrous scaffolds facilitated fibroblast cell growth better than the reference scaffolds. Results of MTT assay suggest that all types of fibrous scaffolds facilitate proliferation of fibroblast cells more effectively than the reference scaffolds and that randomly oriented scaffolds are more advantageous for growth and development of this kind of cells than aligned ones.

This study investigated the main parameters of electrospinning of fibers from solutions of PHAs with different compositions that influenced

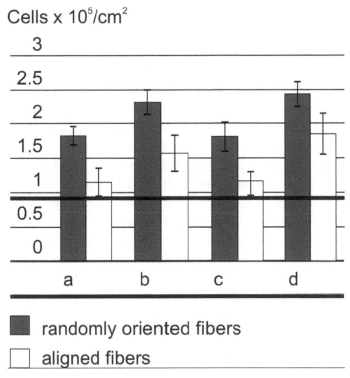

FIGURE 3.19 The results of MTT assay in the culture of NIH 3T3 fibroblast cells on fiber scaffolds at Day 7: A – randomly oriented scaffolds; B – aligned scaffolds, prepared from P3HB and copolymers with different compositions and similar molar fractions of their second monomers: a – P3HB; b – P(3HB/10.5 mol.%-3HV); c – P(3HB/10.0 mol.%-4HB); d – P(3HB/10.0 mol.%-3HHx) (the line shows the number of cells on polystyrene). * – P ≤ 0.05 for randomly oriented fibers relative to the aligned ones

fiber diameter and properties. The study revealed electrospinning parameters for the production of high-quality fibers from different types of PHA and determined which parameters should be varied to tailor the properties of the products (fiber diameter, surface morphology, physical-mechanical properties). This study was the first to compare biological and physical-mechanical parameters of PHAs with different chemical compositions as dependent upon the fractions of monomers constituting the polymers and fiber orientation. None of the fibrous scaffolds produced from PHAs by electrospinning had any adverse effects on attachment, growth, and viability of NIH 3T3 mouse fibroblast cells, and all of them were found to be suitable for tissue engineering applications.

3.4 EVALUATION OF PHA-BASED CELL SCAFFOLDS TO BE USED FOR PRECURSOR CELL DIFFERENTIATION

In modern clinical practice, there is very little potential left for improving surgical methods of defect repair to achieve better results of treating tissue diseases and traumas. Increasing attention has been paid recently to biotechnological methods, involving the use of cell engineering processes to repair defects (Deev, 2008). Since A.Ya. Fridenshtein discovered bone marrow mesenchymal stem cells (BM-MSCs), they have been regarded as the main material for posttraumatic tissue regeneration (Yarygin, 2008). When cultivated, BM-MSCs show high proliferative activity. The effectiveness of using BM-MSCs to repair bone defects was proved in experiments with animals, including rodents, dogs, sheep, and goats (Kruyt, 2004; Bruder, 2006). Results of clinical trials have also been reported (Schimming, 2004). BM-MSCs contribute considerably to bone repair as progenitor cells of osteoblasts and osteocytes; BM-MSCs were found to favor formation of fibrocartilage callus in the defect site (Yin, 2009; Yuan, 2010). During healing of the defect, BM-MSCs secrete anti-inflammatory cytokines, angiogenic factors, and other bioactive compounds, facilitating tissue reconstruction.

Bone marrow mesenchymal stem cells are capable of differentiating into osteoblastic and epidermal cells. Therefore, they are tested for tissue engineering and cell therapy applications. BM-MSCs can be administered to the defect site as suspension or loaded into a carrier (Deev et al., 2007; Hench, Jones, 2005; Uemura et al., 2003).

At the present time, the most widely used biodegradable polymers are polyesters of monocarboxylic acids, polylactides (PLA) and polyglycolides (PGA). The second most popular type of biodegradable polymers is polyhydroxyalkanoates (PHAs), which are more thermoplastic than PLA and have a less significant effect on tissue pH, and whose *in vivo* resorption time is longer (Shishatskaya and Volova, 2004; Rai et al., 2011). Sections 3.1 and 3.2 showed that smooth films and nonwoven membranes composed of ultrafine fibers were suitable for cultivating cells.

Shishatskaya reported a study of BM-MSCs growth and differentiation on scaffolds of resorbable PHAs prepared by various methods. Smooth films were prepared by casting polymer solution on degreased surface (polished metal); nonwoven membranes composed of ultrafine fibers were produced by electrospinning. Two processes were used to prepare porous 3D scaffolds. They were produced either by leaching or by freeze-drying. BM-MSCs were harvested from femurs of laboratory Wistar rats by conventional methods. To achieve BM-MSCs differentiation into keratinocytes or osteoblasts, they were cultured on the media that contained the corresponding differentiation factors. Two protein markers – keratin 14 and involucrin – were chosen to confirm BM-MSCs differentiation into keratinocytes and determine their degree of maturity. The markers that are most frequently used to identify keratinocytes are keratin 5 and 14, which are synthesized by cells of the epidermal basal layer, and filaggrin and involucrin, synthesized by the granular layer cells; keratin 1 and keratin 10, synthesized by the cells of the suprabasal layer, are used less frequently. To confirm BM-MSCs differentiation into osteoblasts, the authors detected alkaline phosphatase activity and osteopontin gene expression by real-time RT-PCR. Gene expression of protein markers of keratinocytes and osteoblasts was determined by real-time RT-PCR at Days 7, 14, 21, and 28. RT-PCR was performed in a Real-time CFX96 Touch amplifier (Bio-Rad LABORATORIES Inc., U.S.), using the corresponding primers, in accordance with the manufacturer's instructions. The housekeeping gene was β-actin. The relative expression of the marker genes was quantified by the threshold cycle method ($2^{-\Delta\Delta C}t$). Synthesis of protein markers was detected by immunocytochemical staining.

At Days 3, 7, and 14 of cell culture, the authors performed MTT assay to determine the number of viable cells. At Days 7, 14, 21, and 28, real-time RT-PCR was performed to determine the expression of mRNA of the markers indicating BM-MSCs differentiation into keratinocytes: keratin 14 and involucrin. The primers for keratin 14 (K14) were forward 5'-GGACGCCCACCTTTCATCTTC-3' and reverse 5'-ATCTG-GCGGTTGGTGGAGG-3'; the primers for involucrin (INV) were forward 5'-CTCTGCCTCAGCCTTACT-3' and reverse 5'-GCTGCT-GATCCCTTTGTG-3'.

At Day 7, an insignificant amount of K14 mRNA was detected (Figure 3.20). Gene expression of K14 reached its maximum at Day 21; INV gene expression was revealed at that time too, suggesting differentiation of basal keratinocytes into mature cells.

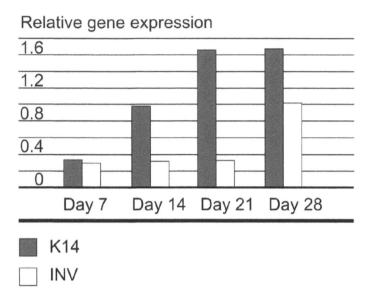

FIGURE 3.20 Dynamics of expression of mRNA of keratinocyte protein markers. Results were normalized relative to the housekeeping gene – β-actin (Shishatskaya's data).

Results of MTT assay showed that the number of cells was 9–11% greater on the collagen type I-coated membranes prepared by electrospinning from PHAs with different chemical composition than on membranes prepared from other PHA types [P3HB, P(3HB/3HV), and P(3HB/3HHx)], but these differences were within the range of statistical error (Figure 3.20). On the fibrin-coated scaffolds, the number of cells was 10–14.5% lower than on the collagen-coated ones. For *in vivo* studies, the authors chose scaffolds prepared from P(3HB/4HB) modified with collagen type I. In three weeks after the cells were seeded on the nano-scaffolds, synthesis of the protein marker of basal keratinocytes (K14) was confirmed by immunocytochemical (ICC) staining. ICC staining showed the presence of differentiated keratinocytes on the surface of scaffolds (Figure 3.21).

Thus, the study showed that keratinocytes could be differentiated from BM-MSCs cultivated on the specialized medium on cell scaffolds fabricated from PHAs by electrospinning.

To differentiate BM-MSCs into osteoblasts, the cells were cultured on the DMEM medium with 10% fetal bovine serum, a solution of antibiotics, 0.15 mM ascorbic acid, 10 nM dexamethasone, and 10 mM β-glycerophosphate.

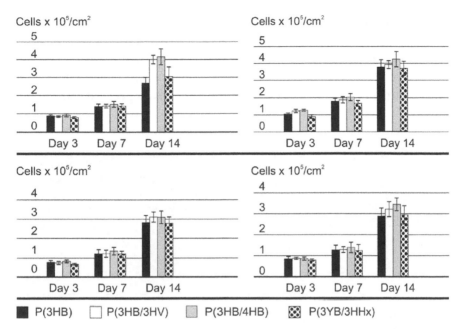

P(3HB)   P(3HB/3HV)   P(3HB/4HB)   P(3YB/3HHx)

FIGURE 3.21 The number of viable BM-MSCs on the membranes prepared from PHAs with different compositions, cultivated on the medium containing factors of BM-MSCs differentiation into keratinocytes (Shishatskaya's data).

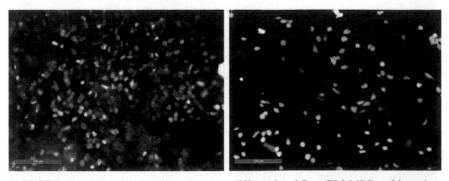

FIGURE 3.22 Photographs of keratinocytes differentiated from BM-MSCs cultivated on the surface of the membranes prepared by electrospinning from P(3HB/4HB) and modified by collagen type I: a – randomly oriented fibers; b – aligned fibers (Shishatskaya's data).

Cultivation was conducted for 21 days, with the medium refreshed every three days. One of the biochemical markers of bone tissue formation is activity of alkaline phosphatase, which is released by osteoblasts and which may be involved in mineralization of osteoids. Phosphatase was detected by using an Alkaline Phosphatase Detection Kit (Sigma) in accordance with the manufacturer's instructions. A wider range of polymer scaffolds were used to differentiate BM-MSCs into osteoblast-like cells (Figure 3.23).

The scaffolds had different porosities (Figure 3.23). The interlinked porous structure is a necessary condition for cells to penetrate inside the

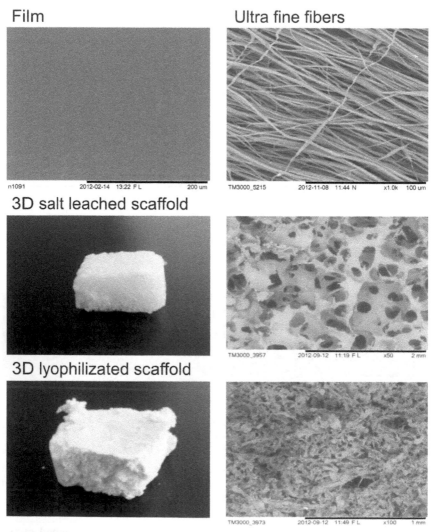

FIGURE 3.23 The types of scaffolds fabricated for differentiating osteoblast-like cells from BM-MSCs: left, photographs, bar = 1 cm; right, SEM images (Shishatskaya et al., 2013). (Reprinted with permission from the Human Stem Cells Institute.)

scaffolds. Leaching produced 3D scaffolds of high porosity (82–88%), with pore size of 100–200 μm. The size of the pores of freeze-dried scaffolds was between 50 and 60 μm.

BM-MSCs were seeded onto all polymer scaffolds, which contained the factors of osteogenic differentiation. MTT assay performed at Days 3, 7, and 14 showed that electrospun membranes and 3D scaffolds prepared by leaching carried the greatest number of cells (Figure 3.24). The lowest abundance of cells was observed on scaffolds prepared by freeze-drying. There are literature data suggesting that osteoblasts proliferate better in round 100–300-μm pores (Karageorgiou, Kaplan, 2005). The scaffolds

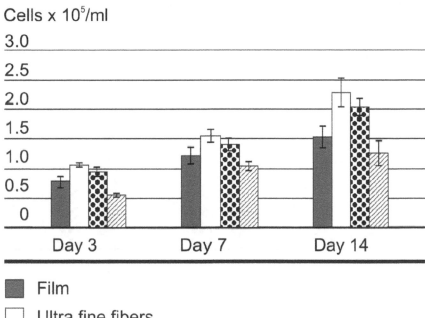

Cells x 10^5/ml

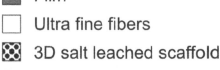

■ Film

□ Ultra fine fibers

▨ 3D salt leached scaffold

▨ 3D lyophilizated scaffold

FIGURE 3.24 The number of viable BM-MSCs on scaffolds fabricated from PHAs with various compositions at different time points of cultivation on the medium containing factors of BM-MSCs differentiation into osteoblasts; 1 – films fabricated by solution casting; 2 – membranes fabricated by electrospinning; 3 – 3D scaffolds fabricated by leaching; 4 – 3D scaffolds fabricated by freeze-drying (Shishatskaya et al., 2013). (Reprinted with permission from the Human Stem Cells Institute.)

fabricated by freeze-drying had smaller pores of irregular shape, which must have affected cell viability.

Alkaline phosphatase showed the highest activity in the cell culture on membranes fabricated by electrospinning and 3D scaffolds prepared by leaching: 4.023 and 3.908 mmol/min × 10^5 cells, respectively, at Day 21. That was indicative of higher metabolic activity of the cells cultured on solution-cast polymer films and 3D scaffolds fabricated by freeze-drying. These results were in good agreement with the results of MTT assay (Figure 3.24–3.25).

Differentiation of BM-MSCs into osteoblasts was confirmed by detecting osteopontin gene expression by real-time RT-PCR. Primers for

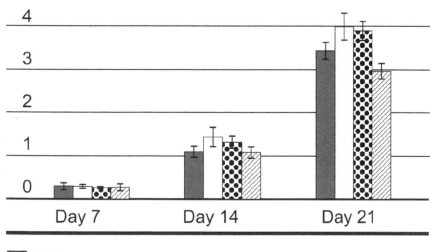

FIGURE 3.25 Alkaline phosphatase activity of osteoblasts derived from BM-MSCs cultured on scaffolds of various PHAs: 1 – films fabricated by solution casting; 2 – membranes fabricated by electrospinning; 3 – 3D scaffolds fabricated by leaching; 4 – 3D scaffolds fabricated by freeze-drying (Shishatskaya et al., 2013). (Reprinted with permission from the Human Stem Cells Institute.)

osteopontin – the major protein of bone tissue – were forward 5'-AAGGC-GCATTACAGCAAACACTCA-3' and reverse 5'-TCATCGGACTCCTG-GCTCTTCAT-3'. Between Days 7 and 21, the amount of osteopontin mRNA increased (Figure 3.26). Similar results were obtained for cells on 3D porous nano-scaffolds. The lowest amount of mRNA was recorded on the scaffolds prepared by freeze-drying, which must have been caused by the previously observed low proliferative activity of cells on these scaffolds.

Staining to detect osteopontin at Day 21 confirmed protein synthesis on the scaffolds (Figure 3.27). On the films and 3D scaffolds fabricated by

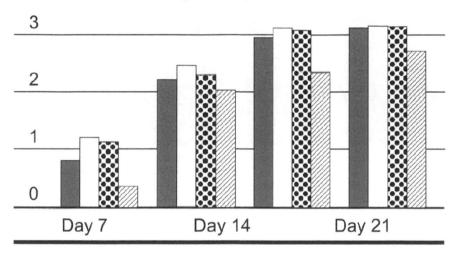

Relative osteopontin gene expression

■ Film

□ Ultra fine fibers

▨ 3D salt leached scaffold

▨ 3D lyophilizated scaffold

FIGURE 3.26 Expression of osteopontin mRNA at different time points of BM-MSCs cultivation on the medium containing factors of cell differentiation into osteoblasts on different types of scaffolds: 1 – films fabricated by solution casting; 2 – membranes fabricated by electrospinning; 3 – 3D scaffolds fabricated by leaching; 4 – 3D scaffolds fabricated by freeze-drying. Results are normalized relative to the housekeeping gene, β-actin (Shishatskaya's data).

Film   Ultra fine fibers

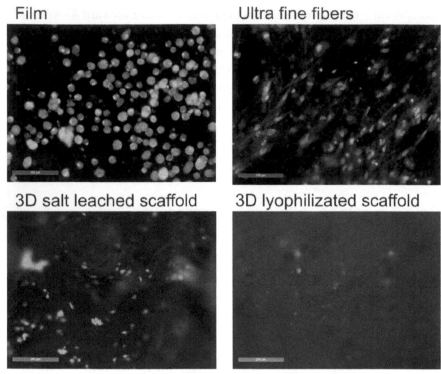

3D salt leached scaffold   3D lyophilizated scaffold

FIGURE 3.27   Photographs of osteoblasts cultured on scaffolds of PHAs. Staining with Anti-Osteopontin antibodies. Bar = 100 μm (Shishatskaya et al., 2013). (Reprinted with permission from the Human Stem Cells Institute.)

leaching, cells had a round shape; on the membranes, cells were aligned along the fibers. The lowest abundance of cells was observed on 3D scaffolds prepared by freeze-drying.

SEM images show that cells were of round shape and were uniformly distributed over the surface of the scaffolds (Figure 3.28). Increased amounts of mineral crystals (calcium phosphates) were recorded on all scaffolds except 3D constructs prepared by freeze-drying. Cells cultivated on 3D scaffolds fabricated by leaching were located not only on the surface but also inside the scaffolds.

Spectral analysis showed the presence of calcium-phosphate deposits on all scaffolds fabricated from PHAs. The highest amounts of Ca and P were detected on the membranes and 3D scaffolds with large pores, which had been fabricated by leaching –2.0–2.5 times higher than on the films;

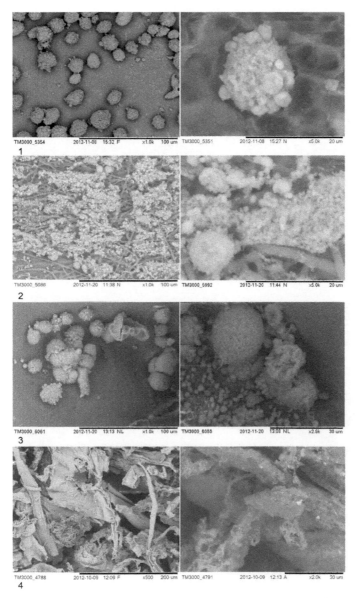

FIGURE 3.28   SEM images of osteoblasts differentiated from the BM-MSCs cultivated on different types of scaffolds: film (1), membrane (2), 3D scaffold fabricated by leaching (3), and 3D scaffold fabricated by freeze-drying (4) (Shishatskaya et al., 2013). (Reprinted with permission from the Human Stem Cells Institute.)

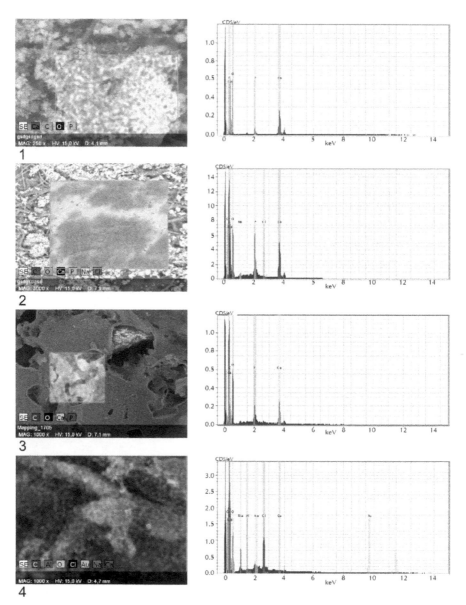

FIGURE 3.29 The elemental composition of osteoblasts differentiated from the BM-MSCs cultivated on different types of scaffolds: film (1), membrane (2), 3D scaffold fabricated by leaching (3), and 3D scaffold fabricated by freeze-drying (4) (Shishatskaya et al., 2013). (Reprinted with permission from the Human Stem Cells Institute.)

the lowest amounts of Ca and P were detected on the 3D scaffolds fabricated by freeze-drying (Table 3.9). The Ca/P ratio was 1.88 on the 3D scaffold prepared by freeze-drying, 2.46 on the membrane, 2.78 on the films, and 3.12 on the 3D scaffolds with large pores, fabricated by leaching.

3D scaffolds were also subjected to laser treatment to test whether it would influence adhesive properties of the scaffolds and their ability to facilitate BM-MSCs proliferation and differentiation into osteoblasts. A preliminary study of 3D scaffolds prepared by cold pressing showed poor cell adhesion to the smooth and pore-free surface of these scaffolds. That adversely affected cell viability, growth, and differentiation into osteoblasts. This type of 3D polymer constructs was proved to be unsuitable for using as cell scaffold (Volova's unpublished data).

3D constructs prepared from P3HB were subjected to different modes of laser treatment as a way to improve the properties of polymer devices intended to be used for reconstructive osteogenesis. P3HB plates produced by direct cold pressing (weighing 500 and 100 g and 40 and 13 mm in diameter) were treated with CO_2 lasers, LaserPro Explorer II (Coherent, U.S.) and LaserPro Spirit, by raster and vector engraving techniques, at a power of v=4%, speed of P = 100%, and resolution of dpi=1000. The beam parameters were varied as follows: diameter (D) between 0.05 and 0.1 mm and the height (h) – between 0.25 and 1.0 mm (Figure 3.30).

SEM images of the laser-treated 3D constructs were used to estimate the effect of the processing mode employed on the density and size of perforations (Figure 3.31). Laser treatment produced perforations running through the polymer constructs, which covered their entire surface. The

TABLE 3.9 Spectral Analysis of Mineral Precipitates Produced by Osteoblasts Differentiated From BM-MSCs Cultivated on Scaffolds Prepared From Different PHAs (Shishatskaya et al., 2013) (Reprinted with permission from the Human Stem Cells Institute.)

Type of scaffold	Elements, wt.%			
	C	O	Ca	P
Film	50.5	37.4	8.9	3.2
Membrane (electrospinning)	34.1	37.4	20.2	8.2
3D (leaching)	50	36.46	10.3	3.3
3D (freeze-drying)	55.7	39.4	3.2	1.7

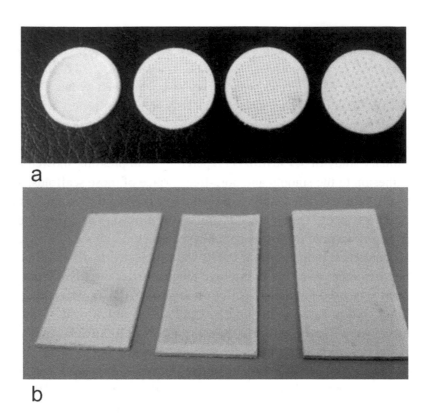

FIGURE 3.30 Photographs of pressed 3D constructs of P3HB treated with laser radiation: A – pellets 13 mm in diameter; B – 20×40-mm plates (Volova's data).

diameter and height of the laser beam influenced the diameter of the pores and the distance between them. At D=0.05 mm and h=1.0 mm, the distance between pores was about 600 µm, but at D=0.05 mm and h=0.25 mm, it was considerably shorter (100 µm). The smallest pores (100 µm) were produced under treatment with the beam of D=0.05 mm and h=0.25 mm, and the largest (300 µm) at D=0.05 mm and h=1.0 mm. Measurements of water contact angles on the surface of porous and rough products are often accompanied by trapping of air bubbles. Moreover, the contact angle hysteresis is influenced by the porosity of the products and the increase in the contact time between the two phases, which was clearly demonstrated by measurements on the polymer constructs tested in this study. The maximum decrease in the angle, to 53.2±18.4 degrees, was recorded on laser-treated plates with the smallest pores (100 µm) after a 15-min contact, while the angle on the plates with the largest pores (250 μm) was 70.1±4.34 degrees. Surface free energy increased on all samples treated with a CO_2 laser. Samples treated with laser radiation at D=0.05 mm and h=0.25 mm showed the highest Young's modulus (2.143 GPa) and tensile strength (11.655 MPa), which might be associated with the shortest distance between pores and fused edges. As the distance between pores and their diameter were further increased, the plate strength decreased. A possible reason for this fact may be internal damage done to samples by laser processing: microscopy at higher magnification revealed narrow cracks emanating from the pores.

Adhesive properties of the surface of laser-treated 3D plates were evaluated in the culture of BM-MSCs of Wistar rats. MTT assay showed that after 7 days of cultivation of BM-MSCs, the highest counts of viable cells – 1.2 ×10^5 cells/mL – were on the plates treated with a CO_2 laser at the beam parameters of D=0.05 mm and h=0.25 mm (Figure 3.31). The lowest counts – 0.37 ×10^5 cells/mL and 0.28 ×10^5 cells/mL – were on the plates treated at D=0.05 mm and h=0.5 mm and D=0.05 mm and h=1 mm, respectively. Results of MTT assay were supported by results of staining with fluorescent dyes DAPI. On the plates with small-diameter perforations (100 μm) and high pore density (72/mm²), cells were located both on the surface of the junction between pores and within the pores (Figure 3.31B). At a lower pore density (5/mm²) and larger pore diameter (300 μm), cells were mainly located within the pores close to microcracks and bumps – microdefects caused by laser treatment. A possible reason for the low counts of the cultured cells on the plates with the largest pores (250 μm) and the longest distance between them (600 μm) may be that they were washed off of the plate surface and proliferated within the pore or on the bottom of the culture plate.

In this study, polymer products were processed with laser radiation in different modes and tested to find their most promising biomedical uses. Processing of 3D constructs with laser radiation produced perforated scaffolds that either retained their mechanical properties or had better ones and were highly biocompatible with bone marrow mesenchymal stem cells, which suggested their good potential as devices for reconstructing bone tissue defects. Thus, targeted laser modification of polyhydroxyalkanoate 3D constructs improves their biomedically important properties.

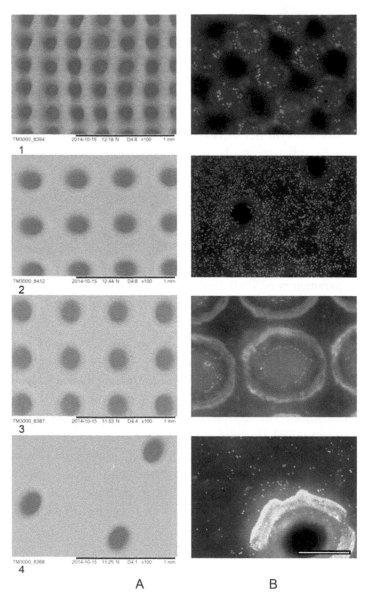

FIGURE 3.31 SEM images of P3HB 3D constructs treated with laser radiation in different modes, with varied diameter (D) and height (h) of the beam (b) (A): 1- D=0.05 mm, h=0.25 mm; 2 – D=0.1 mm, h=0.5 mm; 3 – D=0.05 mm, h=0.5 mm; 4 – D=0.05 mm, h=1.0 mm, where D is preset size of the perforation, h – distance between points. Bar = 1 mm, and DAPI staining of NIH 3T3 fibroblast (7-day old culture) (B) (Volova's data).

Thus, all types of scaffolds fabricated from PHAs with different chemical compositions, employing various processes, have been proved to be suitable for cultivation of BM-MSCs, enabling BM-MSC differentiation into osteoblastic and epidermal cells.

In spite of numerous advantages of using BM-MSCs, the processes involving them have their limitations. Recent findings showed that as the duration of the cultivation increases, BM-MSC populations proliferate at lower rates, populations are very heterogeneous, and there are no specific CD markers (Vakhrushev et al., 2011). Moreover, BM-MSCs are prone to genetic disorders caused by errors in DNA replication and external factors (UV radiation, toxic effects, etc.). Extraction of BM-MSCs by lumbar puncture is painful and may be associated with infection or other risks.

Mesenchymal stem cells can be isolated from various sources such as adipose tissue, muscles, periosteum, synovial membrane, umbilical cord blood, etc. (Kestendjieva et al., 2008). Adipose tissue is an alternative source that can be obtained by a less invasive method and in larger quantities than bone marrow. Adipose tissue mesenchymal stem cells (AT-MSCs) have multilineage differentiation capacity that is not age-related. The cell phenotype of AT-MSCs is similar to that of BM-MSCs (Kern, 2006) and they have comparable osteogenic capacities (Seo, 2004). Osteogenic differentiated cells from AT-MSCs can be implanted on the scaffold; they do not need to be pre-cultivated to form a layer of bone matrix. The quantity of stem cells isolated from the adipose tissue by the conventional technique is greater than the quantity of cells derived from bone marrow, speeding up the transition from cultivation and differentiation to implantation of the seeded graft to the defect site. Cedola (2006) reported that the rate of new bone formation is faster on porous hydroxyapatite scaffolds seeded with BM-MSCs and AT-MSCs. The defect was completely repaired in 12 weeks after implantation owing to the high osteogenic potential (Cowan et al., 2004).

Shishatskaya et al. (2000) studied osteogenic potential of adipose tissue stromal cells and compared grafts seeded with osteoblasts differentiated from BM-MSCs and AT-MSCs. Cell scaffolds (5×5-mm P3HB 3D porous constructs with 88% overall porosity) were sterilized and placed

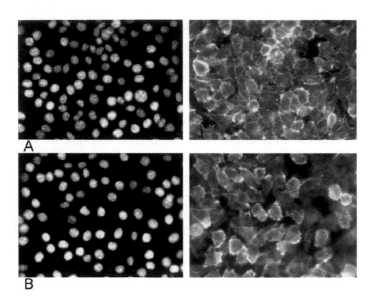

FIGURE 3.32 Photographs of osteoblasts differentiated from AT-MSCs (A) and BM-MSCs (B); light microscopy of staining with fluorescent dyes DAPI and FITC (Shishatskaya's data).

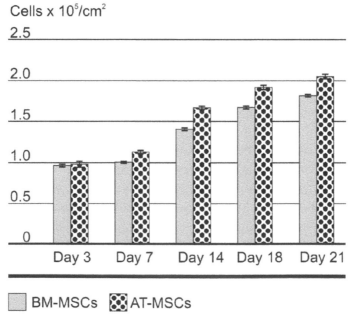

FIGURE 3.33 Results of MTT assay in osteoblast populations cultivated on P3HB scaffolds, differentiated from BM-MSCs and AT-MSCs (Shishatskaya's data).

into 24-well culture plates. After 3 passages, cells were seeded onto sterile scaffolds: 100 000 cells/mL. Cells were cultured on the DMEM medium with 10% fetal bovine serum, a solution of antibiotics, 0.15 mM ascorbic acid, 10 nM dexamethasone, and 10 mM β-glycerophosphate for 21 days, with the medium refreshed. Figure 3.32 shows photographs of osteoblasts differentiated from AT-MSCs and BM-MSCs.

Results of the comparative study of the number and activity of viable osteogenic cells differentiated from two sources (bone marrow and adipose tissue) are shown in Figures 3.33 and 3.34.

From Day 14 onward, the number of cells differentiated into osteoblasts from AT-MSCs was significantly higher than the number of osteoblasts differentiated from BM-MSCs. The difference between the activities of alkaline phosphatase (AP) (the marker of osteoblast differentiation) was even greater: at Day 21, the AP activity of osteoblasts differentiated from AT-MSCs was higher than that of the cells differentiated from BM-MSCs by a factor of 1.5 (Figure 3.34).

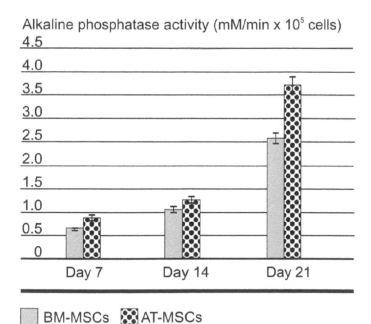

FIGURE 3.34 Alkaline phosphatase activity of osteoblasts differentiated from BM-MSCs and AT-MSCs during cultivation of cells on P3HB scaffolds (Shishatskaya's data).

Figure 3.35 shows SEM images of 3D cell scaffolds of P3HB seeded with the primary osteoblasts differentiated from BM-MSCs and AT-MSCs during their development.

Micrographs of the scaffold surfaces show that at Day 7 after seeding, cells differentiated from AT-MSCs formed colonies. At Day 14, numerous

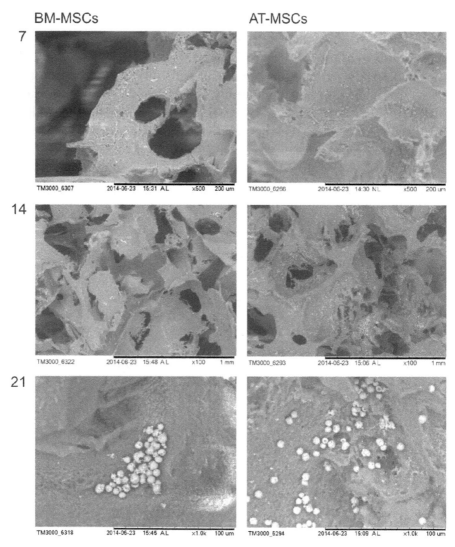

FIGURE 3.35 SEM images of 3D cell scaffolds of P3HB seeded with the primary osteoblasts differentiated from BM-MSCs and AT-MSCs during their development (Shishatskaya's data).

colonies and groups of osteoblasts could be seen not only on the surface of the scaffold but also in its pores. On the scaffolds seeded with osteoblasts differentiated from BM-MSCs, fewer colonies were observed, but cells were located both on the surface and in the pores of the scaffold. Micrographs taken at Day 21 show osteoblasts differentiated from BM-MSCs and AT-MSCs and small crystals of salt precipitates secreted by osteoblasts. Osteoblasts differentiated from BM-MSCs and AT-MSCs were of round shape, retained high activity of alkaline phosphatase, and produced salt precipitates (Figure 3.36).

The comparative study of osteoblasts differentiated from different sources and proliferating on P3HB 3D scaffolds suggested that osteoblast differentiation from AT-MSCs was more effective, although the osteoblasts differentiated from both sources showed comparable activities of alkaline phosphatase and production rates of calcium and phosphate precipitates.

Results of the studies described in this chapter show that designing cell scaffolds of PHAs is a promising line of PHA biotechnology.

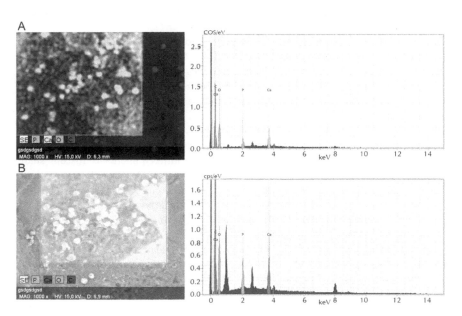

FIGURE 3.36 The elemental composition of osteoblasts differentiated from AT-MSCs (a) and BM-MSCs (b) cultivated on P3HB 3D scaffold (Shishatskaya's data).

KEYWORDS

- attachment
- cell differentiation
- films
- MTT assay
- nonwoven membranes
- proliferation
- scaffolds
- surface structure

REFERENCES

Cedola, A., Mastrogiacomo, M., Burghammer, M., Komlev, V., Giannoni, P., Favia, A., Cancedda, R., Rustichelli, F., & Lagomarsino, S. Engineered bone from bone marrow stromal cells: a structural study by an advanced x-ray microdiffraction technique. Phys. Med. Biol., 2006, 51, 109–116.

Chen, G. Q., & Wu, Q. The application of polyhydroxyalkanoates as tissue engineering materials. Biomaterials, 2005, 26, 6565–6578.

Deev, R. V., Isaev, A. A., Kochish, A. Y., & Tihilov, R. M. Cellular technologies in trauma surgery and orthopedics: ways of development. Kletochnaya transplantologiya i tkanevaya inzheneriya (Cell Transplantation and Tissue Engineering), 2007, 2(4), 18–30 (in Russian).

Ellis, G., Cano, P., Jadraque, M., Martín, M., López, L., Núñez, T., de la Peña, E., Marco, C., & Garrido, L. Laser microperforated biodegradable microbial polyhydroxyalkanoate substrates for tissue repair strategies: an infrared microspectroscopy study. Anal Bioanal Chem., 2011, 399, 2379–238.

Gogolewski, S., Javanovic, M., Perren, S. M., Dillon, J. G., & Hudges, M. K. Tissue response and in vivo degradation of selected polyhydroxyacids: Polylactides PLA), poly(3-hydroxybutyrate) (PHB), and poly(3-hydroxybutyrate-co-3-hydroxyvalerates (PHB/PHV). J. Biomed. Mater. Res., 1993, 27, 1135–1148.

Goncharov, D. B., Nikolaeva, E. D., Sukovatyi, A. G., Shabanov, A. V., Shishatskaya, E. I., & Markelova, N. M. Characterization of ultrafine fibers prepared by electrospinning from poly-3-hydroxybutyrate. Journal of Siberian Federal University. Biology, 2012, 5 (4), 417–426 (in Russian).

Hallab, N. J., Bundy, K. J., O'Connor, K. J., Moses, R. L., & Jacobs, J. J. Evaluation of metallic and polymeric surface energy and surface roughness characteristics for directed cell adhesion. Tissue Eng., 2001, 7, 55–71.

Hench, L., & Jones, J. Biomaterials, artificial organs and tissue engineering. Woodhead Publishing Limited. Cambridge, England, 2005, 284 p.

Jaleh, B., Parvin, P., Sheikh, N., Zamanopour, Z., & Sajad, B. Hydrophilicity and morphological investigation of polycarbonate irradiated by ArF excimer laser. Nucl. Instrum. Meth. B., 2007, 265, 330–333.

Ji, G. Z., Wei, X., & Chen, G. Q. Growth of human umbilical cord Wharton's jelly-derived mesenchymal stem cells on the terpolyester poly(3-hydroxybutyrate-co-3-hydroxyvalerate-co-3-hydroxyhexanoate). J. Biomater. Sci. Polym. Ed., 2009, 20, 325–339.

Ji, Y., Li, X. T., & Chen, G. Q. Interactions between a poly(3-hydroxybutyrate-co-3-hydroxyvalerate-co-3-hydroxyhexanoate) terpolyester and human keratinocytes. Biomaterials, 2008, 29, 3807–3814.

Kennedy, S. B., Washburn, N. R., Simon, C. G., & Amis, E. J. Combinatorial screen of effect of surface energy on fibronectin mediated osteoblast adhesion, spreading and proliferation. Biomaterials, 2006, 27, 3817–3824.

Kern, S., Eichler, H., Stoeve, J., Klüter, H., & Bieback, K. Comparative analysis of mesenchymal stem cells from bone marrow, umbilical cord blood, or adipose tissue. Stem Cells, 2006, 24, 1294–1301.

Kestendjieva, S., Kyurkchiev, D., Tsvetkova, G., Mehandjiev, T., Dimitrov, A., Nikolov, A., & Kyurkchiev, S. Characterization of mesenchymal stem cells isolated from the human umbilical cord. Cell Biol. Int., 2008, 32, 724–732.

Kruyt, M. C., Dhert, W. J., Yuan, H., Wilson, C. E., van Blitterswijk, C. A., Verbout, A. J., & de Bruijn, J. D. Bone tissue engineering in a critical size defect compared to ectopic implantations in the goat. J. Orthop. Res., 2004, 22, 544–551.

Köse, G. T., Kokusuz, F., Özkul, A., Soysal, Y., Ozdemir, T., Yildiz, C., & Hasirci, V. Tissue engineered cartilage on collagen and PHBV matrices. Biomaterials, 2005, 26, 5187–5197.

Lakshmi, S., & Laurencin, C. Biodegradable polymers as biomaterials. Prog. Polym. Sci., 2007, 32, 762–798.

Laycock, B., Halley, P., Pratt, S., Werker, A., & Lant, P. The chemomechanical properties of microbial polyhydroxyalkanoate. Prog Polym Sci., 2013, 38, 536–583.

Martin, D. P., & Williams, S. Medical applications of poly-4-hydroxybutyrate: a strong flexible absorbable biomaterial. Biochem. Eng. J., 2003, 16, 97–105.

Nikolaeva, E. D., Shishatskaya, E. I., Mochalov, K. E., Volova, T. G., & Sinskey, A. J. A comparative study of cell scaffolds prepared from resorbable polyhydroxyalkanoates with different chemical compositions. Kletochnaya transplantologiya i tkanevaya inzheneriya (Cell Transplantation and Tissue Engineering), 2011, 6(4), 63–67 (in Russian).

Rai, R., Keshavarz, T., Roether, J. A., Boccaccini, A. R., & Roy, I. Medium chain length polyhydroxyalkanoates, promising new biomedical materials for the future. Mat. Sci. Eng. R. Rep., 2011, 72, 29–47.

Seo, B. M., Miura, M., Gronthos, S., Bartold, P. M., Batouli, S., Brahim, J., Young, M., Robey, P. G., Wang, C. Y., & Shi, S. Investigation of multipotent postnatal stem cells from human periodontal ligament. Lancet, 2004, 364, 149–155.

Shishatskaya, E. I. Biotekhnologiya poligidroxialkanoatov: nauchnyye osnovy mediko-biologicheskogo primeneniya (Biotechnology of polyhydroxyalkanoates: scientific foundations of biomedical applications): Summary of Doctorate Dissertation. Krasnoyarsk, 2009, 47 p. (in Russian).

Shishatskaya, E. I., & Volova, T. G. A comparative investigation of biodegradable polyhydroxyalkanoate films as matrices for in vitro cell cultures. J. Mater. Sci.: Mater. Med., 2004, 15, 915–923.

Shishatskaya, E. I, Eremeev, A. V., Gitelson, I. I., Setkov, N. A., & Volova, T. G. A study on cytotoxicity of polyhydroxyalkanoates in cultures of animal cells. Doklady RAN (Proceedings of the Russian Academy of Sciences), 2000, 374(4), 561–564 (in Russian).

Shishatskaya, E. I., Nikolaeva, E. D., Shumilova, A. A., Shabanov, A. V., & Volova, T. G. Cultivation and differentiation of bone marrow multipotent mesenchymal cells on scaffolds of the resorbable BIOPLASTOTAN. Kletochnaya transplantologiya i tkanevaya inzheneriya (Cell Transplantation and Tissue Engineering), 2013, 8(14), 57–65 (in Russian).

Shtilman, M. I. Polimery mediko-biologicheskogo naznacheniya (Polymers intended for biomedical applications). Akademkniga Publishers. Moscow, 2006, 399 p. (in Russian).

Shumilova, A. A. Potentsial biorazrushaemykh poligidroxialkanoatov v kachestve kostnoplsticheskikh materialov (Potential of biodegradable polyhydroxyalkanoates as bone replacement materials). PhD thesis in biology. Specialty 03.01.06: Biotechnology. Krasnoyarsk, Institute of Biophysics SB RAS, 2015, 145 p. (in Russian).

Sudesh, K., Abe, H., & Doi, Y. Synthesis, structure and properties of polyhydroxyalkanoates: biological polyesters. Prog. Polym. Sci., 2000, 25, 1503–1555.

Taylor, G. Electrically driven jets. Proc. Natl. Acad. Sci. London, 1969, A313(1515), 453–475.

Tong, H. W., & Wang, M. Electrospinning of aligned biodegradable polymer fiber and composite fibers for tissue engineering applications. J. Nanosci. Nanotechnol., 2007, 7, 3834–3841.

Tong, H. W., Wang, M., & Lu, W. L. Electrospinning of Poly(Hydroxybuturate-co-hydroxyvalerate) Fibrous Membranes Consisting of Paralleled-Aligned Fibers or Cross-Aligned Fibers: Characterization and Biological Evaluation. J. Biomater. Sci., 2012, 22, 2475–2497.

Uemura, T., Dong, Y., Wang, Y., Kojima, H., Saito, T., Iejima, D., Kikuchi, M., Tanaka, J., & Tateishi, T. Transplantation of cultured bone calls using combinations of scaffolds and culture techniques. Biomaterials, 2003, 24, 2277–2286.

Vakhrushev, I. V., Yarygin, N. V., & Yarygin, K. N. Tissue engineering of the bone by transplanting scaffolds seeded with mesenchymal stem cells. Khirurg (Surgeon), 2012, 4, 62–67 (in Russian).

Volova, T. G. Polyhydroxyalkanoates – Plastic Materials of the 21st Century: Production, Properties, and Application. Nova Science Pub., Inc. NY, USA, 2004, 282 p.

Volova, T. G., & Shishatskaya, E. I. Results of biomedical studies of PHAs produced in the Institute of Biophysics SB RAS and Siberian Federal University. In: *Polyhydroxyalkanoates (PHA): Biosynthesis, Industrial Production and Applications in Medicine*, Ch. 21. Nova Science Pub., Inc. NY, USA, 2014, 273–330.

Volova, T. G., Goncharov, D. B., Sukovatyi, A. G., Shabanov, A., Nikolaeva, E. D., & Shishatskaya, E. I. Electrospinning of polyhydroxyalkanoate fibrous scaffolds: effect on electrospinning parameters on structure and properties. J. Biomat. Sci. Polym. E., 2014b, 25, 370–393.

Volova, T. G., Goncharov, D. B., Sukovatyi, A. G., Shishatskaya, E. I. The influence of the electrospinning parameters on the properties of nano-scaffolds of the biodegradable polyester Bioplastotan. Plasticheskiye massy (Plastics), 2013, 7, 52–56 (in Russian).

Volova, T. G., Kiselev, E. G., Vinogradova, O. N. Nikolaeva, E. D., Chistyakov, A. A., Sukovatyi, A. G., & Shishatskaya, E. I. A glucose-utilizing strain, Cupriavidus eutrophus B-10646: growth kinetics, characterization and synthesis of multicomponent PHAs. PloS One, 2014a, 9, 87551–87566.

Volova, T. G., & Shishatskaya, E. I. Biorazrushayemyye polimery: sintez, svoistva, primeneniye (Biodegradable Polymers: Synthesis, Properties, Applications). Krasnoyarskii Pisatel. Krasnoyarsk, 2011, 49 p. (in Russian).

Volova, T. G., Shishatskaya, E. I., & Gordeev, S. A. Characterization of ultrafine fibers electrospun from a thermoplastic polyester, poly(3-hydroxybutyrate-co-3-hydroxyvalerate). Perspektivnyye materialy (Advanced Materials), 2006, 3, 25–29 (in Russian).

Volova, T. G., Shishatskaya, E. I., & Sinskey, A. J. Degradable Polymers: Production, Properties and Applications. Nova Science Pub., Inc. NY, USA, 2013, 380 p.

Wang, L., Wang, Z. H., Shen, C. Y., You, M. L., Xiao, J. F., & Chen, G. Q. Differentiation of human bone marrow mesenchymal stem cells grown in terpolyesters of 3-hydroxyalkanoates scaffold into nerve cells. Biomaterials, 2010, 31, 1691–1698.

Wang, M. Developing bioactive composite materials for tissue replacement. Biomaterials, 2003, 24, 2133–2151.

Wang, Y., Bian, Y. Z., Wu, Q., & Chen, G. Q. Evaluation of three-dimensional scaffolds prepared from poly(3-hydroxybutyrate-co-3-hydroxyhexanoate) for growth of allogeneic chondrocytes for cartilage repair in rabbits. Biomaterials, 2008, 29, 2858–2868.

Wang, Y., Cao, R., Wang, P. P., Jian, J., Jiang, X. L., Yan, C., Lin, X., Wu, L., Chen, G. Q., & Wu, Q. The differential effects of aligned electrospun PHBHHx fibers on adipogenic and asteogenic potential of MSCs through the regeneration of PPARr signaling. Biomaterials, 2012, 33, 485–493.

Williams, D. F., Martin, D. P., & Horowitz, D. M. PHA applications: addressing the price performance issue. I. Tissue engineering. Int. J. Biol. Macromol., 1999, 25, 111–121.

Yu, B. Y., Chen, P. Y., Sun, Y. M., Lee, Y. T., & Young, T. H. Response of Human Mesenchymal Stem Cells (hMSC) to the Topographic Variation of Poly(3-Hydroxybutyrate-co-3-Hydroxyhexanoate) (PHBHHx) Films. J. Biomater. Sci., 2012, 23, 1–26.

CHAPTER 4

DEGRADABLE POLYHYDROXYALKANOATES AS A BASIS FOR DRUG DELIVERY SYSTEMS

CONTENTS

4.1 INTRODUCTION

Designing and using new drug formulations is a priority task of pharmacology. Construction of sustained-release drug delivery systems is a promising and rapidly developing line of biotechnology and experimental pharmacology. New-generation drug delivery systems enable sustained release of the drugs, enhance their bioavailability, direct them to the site of a pathological process, and reduce possible side effects of the drugs (Pouton and Akhtar, 1996; Amass et al., 1998; Poncelet, 2005; Torchilin, 2006; Dutta et al., 2007; Panarin et al., 2014). Modern pharmacotherapy involves the use of micro- and nano-sized systems based on biodegradable polymers, which is a very promising approach, as by varying the structure of polymeric carriers, one can change pharmacological properties of the drug, including its biodistribution, permeability, and pharmacokinetics.

4.2   NEW-GENERATION SUSTAINED-RELEASE DRUG DELIVERY SYSTEMS

An ideal drug delivery system (DDS) should be inert, biocompatible, mechanically strong and stable, allowing the drug to be highly bioavailable and soluble, safe from accidental release, simple to administer and remove, capable of transporting the drug to the targeted site, without affecting intact tissues, therapeutically effective, comfortable for the patient, easy to fabricate and sterilize, and suitable for a wide range of drugs. By using different materials to prepare DDS, one can vary the rate of destruction of the polymer matrix and, hence, control drug release kinetics. Drug delivery systems have a number of advantages over free drugs: the drugs are retained in the patient's organism for longer periods of time; they have better pharmacokinetics, and many of the drugs become able to cross the membrane and blood-brain barriers; hydrophobic drugs become soluble; possible side effects of highly toxic drugs are reduced (Ambruosi et al., 2006). New drug delivery systems protect drugs from degradation and reduce initial drug concentrations and the number of administrations in the case of long-term medication (Kim et al., 2002; Yen et al., 2003).

As a rule, in the delivery system the drug resides in the pores of the construction in the free state (Freiberg et al., 2004; Medvesky et al., 2007; Panarin et al., 2014), but it can also be conjugated to polymer molecules through the chemical bond, which influences the drug release rate (Marcucci et al., 2004). Therefore, drugs are usually conjugated to short chemical sequences, so-called linkers (Furgeson et al., 2006). The most commonly used carriers in these systems are well-studied polymers such as dextran, poly-N-vinyl pyrrolidone, poly-N-(2-hydroxypropyl) methacrylamide containing links or functional groups used to conjugate drugs. Such substances as mono- and oligo-sugars, folic acid, lectins, poly-L-lysines, and antibodies can be used as linkers (Torchilin, 2006; Panarin et al., 2014). Controlled release of the drug from the system can be achieved in different ways: breaking of the chemical bond between the polymeric carrier and the drug; diffusion through the polymer layer; drug release due to degradation (erosion) of the polymer system; release of the active agent from the swelling hydrogel system (Kim et al., 2009; Olkhov et al., 2008). Based on the administration route, DDS can be divided into surgically implanted polymer systems (plates, films, sponges, tablets, microchips) and injected implants (gels, microspheres, microcapsules, erythrocytes, liposomes, nanospheres, and nanocapsules) (Jain et al., 2005). In recent years, there has been an increasing interest in injected micro- and nano-sized polymer systems shaped as microspheres, nanotubes, nanoparticles, and various carbon nanoparticles (Kingsley et al., 2006; Dutta et al., 2007; Bordes et al., 2009).

In contemporary feedback regulated drug delivery systems, the drug concentration at the drug target site is measured through a sensor and, depending on the ideal drug concentration, release is either increased or slowed down (Huo et al., 2005; Park et al., 2005). One of the major parameters of the body that changes during disease is temperature. That was the basis for proposing polymer-based hydrogels as drug carriers. Phase layering in solutions of such polymers occurs because with an increase in temperature, the swollen globule-like conformation changes into a coil-like structure. As this transformation occurs, the macromolecule shrinks or gets partially destroyed, and the drug readily diffuses into the solution (Cohn et al., 2009). Hydrogels sensitive to pH behave in a similar way (Ni et al., 2007; Cai et al., 2007).

The recent decade has seen growing interest in drug formulations with targeted delivery. The use of new-generation drug delivery systems, with drugs encapsulated in the polymer matrix that is degraded *in vivo*, makes the drugs bioavailable and ensures their sustained targeted release to organs and cells (Cheng et al., 2007; Wu et al., 2009). Sustained-release drug delivery systems are being developed in many fields of medicine: endocrinology, pulmonology, cardiology, oncology, etc. At the present time, biodegradable matrices are used to carry anesthetic drugs, antidepressants, hormones, contraceptive, anticancer, anti-inflammatory, and antiviral drugs, vaccines, and systems with encapsulated DNA for gene therapy (Kim et al., 2007; Vasir et al., 2007; Goforth et al., 2009). New drug formulations used as drug delivery systems based on polymer carriers have emerged on the global pharmaceutical market lately (Medvedeva et al., 2006).

Drug delivery systems are fabricated from well-known materials (phospholipids, hydrogels, polyesters, polyurethanes, polymers, polyamino acids, proteins, polysaccharides, chitosans) and novel materials, which meet the necessary requirements for degradation rates and physicochemical properties of drug delivery systems (composites, nanomaterials, fullerenes, dendrimers). Medical-grade materials must be perfectly harmless and functional. They must have necessary physical-mechanical properties, biodegrade without releasing toxic products, and be miscible with drugs in different phase states, without inactivating them (Liu et al., 2005). These materials should be easy to synthesize and their cost should not be high (Jain et al., 2005; Nair et al., 2007; Panarin et al., 2014).

The diversity of materials used as drug carriers enables construction of different dosage forms: films, compacted rods and pellets, microcapsules, microspheres (microparticles), nanoparticles, liposomes and micelles, which can be administered via different routes: subcutaneous, intramuscular, intravenous, intraperitoneal, transdermal, oral or as inhalers.

Both non-degradable and biodegradable polymer materials are used to construct modern drug formulations. The main drawback of non-degradable materials is the necessity to extract the polymer matrix from the implant site after the drug has been released. Biodegradable materials can be divided into hydrolytically degradable and enzymatically degradable ones. The majority of naturally occurring polymers are degraded enzymatically. Biodegradable matrices are gradually degraded in the body, slowly releasing the drug. Degradation rates of these materials are determined by their chemical structure, positions of monomers, the presence of ionic groups and random monomers or defects in the polymer chain, molecular weight, the presence of low-molecular-weight units, parameters of the process employed to produce them, sterilization parameters, storage conditions, a number of physicochemical factors of the medium (ion exchange, ionic strength, pH), the mechanism of hydrolysis, and implant shape and site. Chemical modification of the polymers can considerably influence their degradation rate (Vert et al., 1991). The most widely used materials for constructing drug delivery systems are polyesters such as polylactides, polylactide glycolides (Liu et al., 2005; Nair et al., 2007), polyanhydrides

(Goepferich et al., 2002; Jain et al., 2005), poly(ortho esters) (Heller et al., 2002); nonionic surfactants (Dutta et al., 2007); polysaccharides (starch, dextran, chitosan, gelatin) (Yun et al., 2004; Balthasar et al., 2005); and, in recent years, polyhydroxyalkanoates (Pouton and Akhtar, 1996; Sudesh et al., 2000; Piddubnyak et al., 2004, Amara, 2008; Duran et al., 2008).

Over the past decade, aliphatic esters of lactic and glycolic acids have become the most commonly used biodegradable polymers (Kuznetsova, 2013; Panyam, 2012; Regnier-Delplaceb, 2013; Sardushkin, 2013). The lifetimes of these materials are determined by their degree of crystallinity, molecular weight, and hydrophobicity of the monomer unit, ranging between several days and several years. These properties may influence the availability of water molecules to ester linkages, whose cleavage determines the degradation rate. The disadvantage of these polymers is that acid products of hydrolysis may affect the stability and biological activity of the encapsulated drug and cause patient discomfort (Cleland, 2001).

An important aspect of designing new-generation drug formulations is the development of technological processes for their production. The main requirement is the ease of fabrication. The diverse techniques used to fabricate polymer carriers and to load drugs into them can be classified into several major groups.

Large 3D constructs (such as pellets) are fabricated by cold or hot compression molding and by melt extrusion (El-Rrehim, 2005). The procedure may be reversed: compression may be followed by heating. In this case, polymer particles are not only compressed but also sintered, producing a strong 3D construct (Liu et al., 2005). Extrusion is a technique where molten polymer is forced through a die and is used to produce components of a fixed cross-sectional area. This technique is used to fabricate drug delivery systems in the form of films and fibers (Repka et al., 2003).

The most common techniques of embedding the drug in the polymer matrix are melt dispersion, spray drying, co-evaporation, co-precipitation, and co-milling. Melt dispersion techniques can be used to prepare systems with unstable molecules such as peptides and proteins, to stabilize them (Breitenbach, 2002). Micro- and nano-sized fibers are produced by electrospinning (Wang et al., 2010). Porous 3D constructs are fabricated by polymer foaming or by using high-pressure carbon dioxide (Sokolovsky-Parkov et al., 2007). This method is the method of choice when encapsulating protein compounds, as it does not involve the use of increased temperatures and organic solvents (Kanczler et al., 2007).

It has been generally accepted that the most promising drug delivery systems are sustained-release micro- and nano-particles, which can be administered subcutaneously or intramuscularly, adapted for oral administration or inhalations, and injected into the bloodstream (Freiberg, et al., 2004; Jain et al., 2008). Studies of silica, zinc oxide, and gold microparticles showed that particles of these materials smaller than 100 nm exhibit new, unusual properties and biological effects. However, several authors have reported adverse effects of nanocarriers, which, having penetrated into cells, can selectively accumulate in them and their compartments, become capable of transcytosis across epithelial and endothelial cells, can move along dendrites and axons, blood and lymphatic vessels, causing such adverse effects as oxidative stress and tissue inflammation (Semmler-Behnke et al., 2008; Napierska et al., 2009; Nair et al., 2009). Micro-and nano-particles can deliver drugs to different organs: lungs (Zhang et al., 2009), liver (Wu et al., 2009), intestine (Kong et al., 2009), and brain (Zensi et al., 2009). Polymer microparticles can be used to deliver toxic metabolites or factors suppressing the growth of blood vessels in the treatment of cancer (Charoenphol et al., 2010) and other diseases.

Encapsulation of drugs into the polymeric microparticles can be performed by three major processes: granulation, spray drying, and emulsification. Granulation is the simplest way to prepare particles/microcapsules with narrow size distribution. Granulation involves the use of electrostatic generators, nozzle resonance technology, jet cutter, and spinning disks (Poncelet, 2005; Jaworek et al., 2008). By application of an electrostatic potential on a pending droplet, charges accumulate on its surface creating a repulsion that opposes the surface tension. The resulting droplets will then have small sizes down to 20 μm for high voltage (Poncelet, 2005). Microfluidization can be regarded as granulation. In this case, capsule formation is based on the feeding of two immiscible fluids from the side channels to the central connecting channel (Choi et al., 2007). Another technique is microchannel emulsification, in which the dispersed phase (usually, a polymer solution) is passed through the perforated plate and into the stationary water phase (Nakagawa et al., 2004). The addition of surfactants decreases the surface tension and prevents the drops from coagulating (Morello et al., 2007). The spray-drying technique is a popular technique for preparing micro- and nano-particles. This is an economical and controllable process, which can be scaled-up (Freitas et al., 2005; Barbosa-Cánovas et al., 2005; Zhang et al., 2007). The drug is emulsified or dispersed in a solution of polymer in volatile solvent, fed to the heating chamber, and then spray-dried through a small nozzle. Spray-dried powder presents as microspheres of sizes below 100 μm. This method is suitable for encapsulation of thermostable compounds (Arpagaus et al., 2008). Encapsulation could also include agglomeration of fine particles or the coating of solid particles. Powders are fluidized and sprayed with the solution containing the drug and the binding suspension, which transform the powder into agglomerates. Once the agglomerates have reached the preset size, the spraying is stopped and the liquid evaporates. The spaces previously occupied by the liquid become empty, and the particles, which have changed their size and become porous, can be used for further processing, e.g., pelletizing (Poncelet, 2005).

A common method of microparticle preparation is emulsification. Emulsion can be prepared by using ultrasound, static mixing, or high-speed homogenization (Wang et al., 2007; Bazzo et al., 2008; Danhier et al., 2009). New methods have been proposed recently for encapsulation of drugs into nanoparticles. The nanoprecipitation technique is based on the interfacial deposition of polymers following displacement of a semi-polar solvent miscible with water from a lipophilic solution (Lassalle et al., 2007). An alternative to the nanoprecipitation procedure is the salting-out method. This method involves the use of a solution including the polymer and, eventually, the drug in a water-miscible solvent such as acetone or tetrahydrofuran (THF). The solution is emulsified under vigorous stirring in an aqueous gel containing the salting-out agent and, if required, a stabilizer. Nanoparticles with a narrow size distribution are produced, but the need of intensive purification of the nanoparticles is the principal limitation associated with this technique. The techniques listed above produce microcapsules that are then solidified by various stabilization techniques (hardening, evaporation, gelation, polymerization, coacervation, etc.).

Solvent evaporation from emulsion is another common method of microparticle preparation from biodegradable polymers. The solvent evaporation technique is employed to prepare microspheres from high-molecular-weight compounds, by using double or triple emulsions. The method of solvent evaporation from the water/oil/water (W/O/W) emulsion was proposed by Bisseri (1983) to prepare biodegradable microspheres. Microspheres can also be prepared by using oil/oil (O/O) and oil/water (O/W) emulsions; composite microparticles with ceramics are produced from the oil/solid/ water emulsions (Li et al., 2005; Wang et al., 2007). A W/O/W emulsion usually contains an aqueous solution of the drug, a polymer solution in organic solvent (dichloromethane, trichloromethane), and a surfactant solution (polyvinyl alcohol, polyoxyethylene sorbitan monooleate, sorbitan tri- and mono-oleate, polyethylene glycol, etc.).

The size and porosity of microparticles are the major parameters determining drug release kinetics. Porous microspheres can be prepared by increasing the proportion of the water phase in the emulsion (Yang et al., 2001), increasing the rate of solvent evaporation (Giovagnoli et al., 2007), using enzymatic reaction, e.g., between catalase and hydrogen peroxide in the water-oil emulsion (Bae et al., 2009), adding sucrose crystals or polyethylene glycol, and preparing composite microparticles (Parra et al., 2006; Naha et al., 2008). The diameter of microspheres can be controlled by the physical properties of the material, polymer concentration in the emulsion, surfactant type and concentration, and by changing process parameters, including emulsion agitation velocity, solvent evaporation rate, nozzle diameter, and emulsion flow rate (Champion et al., 2007; Karatas et al., 2009). Some authors reported an increase in the diameter of microspheres with the increase in polymer concentration in the emulsion, with a decrease in the emulsion agitation velocity, and at low surfactant concentrations (Morello et al., 2007; Berchane et al., 2006).

Researchers of the Institute of Biophysics SB RAS and the Siberian Federal University constructed microparticles of PHAs with different chemical structures and carried out integrated studies that showed how the chemical composition of PHAs and the production technique are related to the properties of resulting microparticles (Shishatskaya, 2013; Shishatskaya, Goreva, 2006; Shishatskaya et al., 2005, 2006; Goreva et al., 2012; Shershneva et al., 2014; Goreva et al., 2013; Eke et al., 2014).

Goreva et al. (2012) prepared microparticles by using two types of PHAs: the homopolymer of 3-hydroxybutyrate (P3HB) and the copolymer of 3-hydroxybutyrate and 3-hydroxyvalerate (P3HB/3HV) containing different amounts of hydroxyvalerate units (10.5, 20.0, and 37.0 mol.%). Microparticles were obtained via solvent evaporation from the double (water/oil) and triple (water/oil/water) emulsions. The double emulsion contained a 4% solution of the polymer and a 0.5% solution of PVA. The triple emulsions were prepared through addition of an aqueous solution of gelatin to the polymer solution. Polymer emulsions were mixed with a Heipolph RZR1 overhead-drive three-blade stirrer (Germany) at various stirring rates (300–1000 rpm), an IKA Ultra-Turrax T25 high-velocity homogenizer (Germany) (20,000 rpm), and a Sonicator S3000 ultrasonic generator (U.S.). The ultrasound power was varied from 12 to 20 V, and the treatment lasted between 60 and 300 s. Emulsions were stabilized with various types of surfactants: PVA, polyoxyethylene (20) sorbitan monooleate (Tween® 80), and sodium dodecyl sulfate (SDS). Emulsions were allowed to stay for 24 h under continuous mechanical stirring until full evaporation of the solvent was attained. The as-formed microparticles were collected through centrifugation (10,000 rpm, 5 min), washed six times with distilled water, and dried in a LS-500 lyophilizer (Russia).

In order to develop the technique of embedding the drug in polymer microparticles, the authors used Hoffmann's violet dye (triethylrosalinine hydrochloride, $M = 338$) and antibiotics with different molecular weights and degrees of solubility in aqueous solutions—rubomycin hydrochloride ($M = 564$, water solubility of 0.04 mg/mL), rifampicin ($M = 823$, water solubility of 1.4 mg/mL), vancomycin ($M = 1440$, water solubility of 100 mg/mL), and tienam ($M = 320$, water solubility of 10 mg/mL). The choice of the drugs was based on their demand in clinical practice for treatment of long-lasting diseases, their stability in solution, and the feasibility of mixing the drug with nonpolar solvents without any change in the properties of the drugs. In the case of the double emulsion, the drugs were preliminarily dissolved in dichloromethane and, then, the weighed portion of the polymer was added to the resulting solution. To obtain the triple emulsion, the polymer was dissolved in dichloromethane, and the aqueous solution of the antibiotic was added to the resulting solution. The system was stirred until the emulsion was obtained, and the solution of PVA was slowly added. Microparticles were loaded with rubomycin by using a triple emulsion composed of a P3HB solution in dichloromethane and polyethylene glycol with $M_w = 4 \times 10^4$ (PEG-40) taken at a ratio of P3HB: PEG

= 80: 20 (w/w). To the as-prepared solution, 1 mL of an aqueous solution of rubomycin (1% with respect to the weight of the polymer carrier) was added, and the resulting solution was homogenized for 1 min with the aid of ultrasound with a power of 12 W.

The most important parameters determining the properties of microparticles and the kinetics of drug release from them are their sizes and surface area. In contrast to particles formed from the double emulsion, which were loose and had small deformations of the surface in the form of cavities, microparticles prepared from the triple emulsion were regular spheres. The diameters of microparticles in both variants were between 0.5 and 70 μm (Table 4.1). The average diameter of microparticles prepared from the double emulsion was 14.31 ± 1.4 μm, which is close to the diameter of particles derived from the triple emulsion, 12.53 ± 1.1 μm.

Comparative analysis of the characteristics of the microparticles showed that the concentration of a polymer solution and the method of emulsion mixing are the major factors responsible for the diameters of microparticles. Thus, because of higher viscosity, an increase in the concentration of the polymer in the solution resulted in the enlargement of microparticles. At a solution density of 10 g/L, the average diameter of microparticles was 7.5 ± 0.6 μm; at 40 g/L, the average diameter of microparticles increased to 16.0 ± 0.9 μm. In addition, the solution density influenced the microparticle yield: with a decrease in the polymer concentration to 10–20 g/L, the microparticle yield increased to 73% (Table 4.1). The rate of mixing the polymer emulsion exerted the most pronounced effect on the sizes of microparticles. As the rate of emulsion mixing increased from 300 to 1000 rpm, the average diameter of microparticles decreased by a factor of nearly 3, to 5.57 ± 0.8 μm (Figure 4.1a, b).

Variations in the conditions of mixing of the polymer emulsion made it possible to prepare microparticles under 1 μm in diameter. At a high rate of mixing or a high power of UV treatment, the particles were more homogeneous in size and their average diameter was about 0.36–0.39 μm. Microparticles prepared by sonication had insignificant deformations of the surface, and their total yield was lower by a factor of 1.5 than the yield of particles from the emulsion under mechanical stirring (Table 4.1).

In this case, as opposed to the mechanical stirring, during which the shape of microparticles formed and stabilized gradually (Embelton and

TABLE 4.1 Preparation Conditions and Characteristics of Microparticles (A Solution Concentration of 4%, a PVA Concentration of 0.5%) (Goreva et al., 2012) (Reprinted with the permission of NAUKA.)

PHA	Dispersion technique		Average diameter, μm	Yield of micropar-ticles, %
	Ultrasound, W (dispersion duration, min)	Mechanical stirring, rpm		
P(3HB)*	–	300	7.5 ± 0.6	74.3 ± 2.4
P(3HB)**	–	300	12.2 ± 0.9	72.7 ± 1.5
P(3HB)	–	300	16 ± 0.86	68.3 ± 1.8
P(3HB)	–	500	9.61±0.9	73.5 ± 3.8
P(3HB)	–	1000	5.57 ± 0.8	78.6 ± 6.2
P(3HB)	–	1000	4.25 ± 0.27	67.7 ± 3.3
P(3HB)	–	1000	13.3 ± 0.7	58.0 ± 3.4
P(3HB)	–	1000	7.09 ± 0.4	75.4 ± 4.5
P(3HB)	–	20000	0.39 ± 0.08	84.6 ± 6.5
P(3HB)	12 (1 min)	–	2.5 ± 0.14	57.0 ± 4.5
P(3HB)	16 (1 min)	–	1.74 ± 0.13	62.0 ± 5.1
P(3HB)	20 (1 min)	–	1.2 ± 0.08	61.0 ± 5.4
P(3HB)*	20 (5 min)	–	0.36 ± 0.07	64.0 ± 4.8
P(3HB/3HV) (3HV10.5 mol.%))	–	1000	4.39 ± 0.2	78.0 ± 2.2
P(3HB/3HV) (3HV 20 mol.%))	–	1000	4.45 ± 0.3	76.4 ± 3.2
P(3HB/3HV) (3HV 37 mol.%)	–	1000	4.1 ± 0.2	74.4 ± 4.7
P(3HB)/PEG–40)	–	1000	4.73 ± 0.6	69.8 ± 4.6
P(3HB/3HV)/PEG–40 (3HV 10.5 mol.%))	–	1000	4.1 ± 0.6	73.7 ± 4.5
P(3HB/3HV)/PEG–40 (3HV 20 mol.%)	–	1000	3.54 ± 0.5	75.2 ± 4.7
P(3HB/3HV)/PEG–40 (3HV 37 mol.%)	–	1000	3.6 ± 0.6	64.0 ± 5.1

Note: In the first column, the content of P(3HB) units (mol %) is shown in parentheses.
* 1% P(3HB) solution.** 2% P(3HB) solution.*** Surfactant: 0.5% Tween®80.

Tighe, 1992), the worse quality of microparticles might be related to heating of the emulsion and quicker evaporation of the solvent. It was found that, during dissolution of PHAs and formation of emulsions, the molecular weight declined. Regardless of the technique of emulsion mixing, the molecular weight of the polymer for the microparticles was 30–35% lower than the corresponding value for initial polymer samples. Gel-permeation chromatography measurements showed that the molecular weight of the polymer matrix of particles prepared under mechanical stirring decreased to 54.6×10^4 (compared to 78.0×10^4 for the initial polymer). Sonication of the polymer emulsion had the same effect on the molecular weight of the polymer.

Taking into account the data on the advantageous effect of polymeric stabilizers on the properties of microparticles (Morella et al., 2007), various surfactants were investigated and it was shown that they substantially influenced the quality and yield of microparticles (Table 4.1). When PVA and SDS were added to the polymer emulsion, microparticles were regular spheres with an average diameter of 5.5–7.0 μm. In the case of Tween® 80, large conglomerates were formed, and the total yield of particles was lower (~60%). An increase in the concentration of PVA from 5 to 10 g/L improved the stability of the water/oil emulsion and entailed a decrease in the average diameter of particles to 4 μm.

In addition, the yield and diameter of microparticles were influenced by the chemical composition of the polymer. A comparative study of particles derived from a highly crystalline polymer of 3-hydroxybutyric acid and its copolymers containing different amounts of 3-hydroxyvaleric acid units was performed. Microparticles based on the P(3HB/3HV) copolymer were regular spheres with rough and porous surfaces; the surface became rougher as the content of 3-hydroxyvalerate in the copolymer was increased (Figure 4.1c, d). Although the content of 3-hydroxyvalerate units in the copolymer influenced the surface structure of microparticles, it had no effect on their diameters. The average diameter of microparticles derived from the P(3HB/3HV) copolymers, regardless of their comonomer ratio, was 4.10–4.45 μm.

These values were similar to the sizes of microparticles prepared from the homogeneous P3HB (5.57 ± 0.8 μm) (Table 4.1). As the second chemical compound was added to the polymer system (20% of PEG-40 was

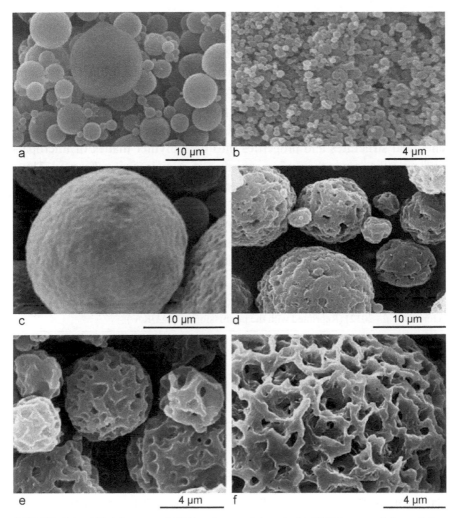

FIGURE 4.1 SEM images of PHA microparticles: (a, b) P3HB at emulsion stirring velocities of (a) 300 and (b) 20,000 rpm; (c) P3HB and (d) P(3HB/3HV) (3HB 37 mol.%) at emulsion stirring velocities of 1000 rpm; (e, f) microparticles prepared via mechanical stirring of the double emulsion containing 20% PEG-40 and either (e) P(3HB/3HV) (3HV 20 mol.%) or (f) P(3HB/3HV) (3HV 37 mol.%) (Goreva et al., 2012). (Reprinted with the permission of NAUKA.)

added based on the weight of the polymer carrier), the surface structure of microparticles changed but their sizes did not. Thus, when PEG-40 was added to the solution of a P(3HB/3HV) copolymer (the contents of 3-hydroxyvalerate units were 10.5, 20.0, and 37.0 mol.%), the surfaces of microparticles acquired an uneven porous structure with pore diameters between 1 and 3 μm (Figure 4.1e, f). The surfaces of microparticles prepared from P3HB and a small amount of PEG-40 were less rough and contained small cavities and single pores. An increase in the temperature of emulsion caused an increase in porosity. When the P3HB emulsion containing PEG-40 was heated to 30°C, porous particles with pore diameters of 3.5–4.0 μm were formed. During embedding of drugs with different chemical structures and properties into the polymer matrix of microparticles, no changes in the surface structure of particles were observed. Thus, microparticles loaded with different drugs had folded surfaces with small cavities and pores.

The drug encapsulation efficiency, EE, in the matrix of microparticles depended on their preparation technique and on the weight fraction of the drug in the polymer solution–emulsion (Table 4.2). In the case of Hoff-

TABLE 4.2   Characteristics of PHA-Based Microparticles With Encapsulated Drugs (Goreva et al., 2012) (Reprinted with the permission of NAUKA.)

Composition of microparticles	Composition of emulsion	Drug concentration, %	EE, %	Average diameter of microparticles, μm
P3HB/dye	Water/oil	1	78.3 ± 4.4	5.8 ± 0.8
		5	65.6 ± 5.08	7.21 ± 0.7
		10	56.3 ± 3.5	8.34 ± 1.1
P(3HB)/PEG-40/dye	Water/oil	1	64.0 ± 3.6	5.5 ± 0.8
P(3HB/3HV) (10 mol.%)	Water/oil	1	73.5 ± 5.2	4.3 ± 0.6
P(3HB/3HV) (37 mol.%)	Water/oil	1	69.3 ± 4.5	4.1 ± 0.9
P3HB/Rubomycin	Water/oil	1	82.3 ± 6.4	5.67 ± 0.7
P3HB/Rubomycin	Water/oil/water	1	65.2 ± 3.6	5.45 ± 0.4
		1	52.8 ± 3.2	8.9 ± 0.7
		5	57.5 ± 4.6	6.73 ± 0.9
P3HB/Rifampicin	Water/oil/water	5	65.5 ± 5.1	7.86 ± 0.9
P(3HB/3HV)/vancomycin	Water/oil/water	5	70.1 ± 3.5	8.91 ± 1.3
P3HB/Tienam	Water/oil/water	5	53.0 ± 4.5	6.43 ± 0.8
P3HB/Tienam				

mann's violet dye, with an increase in its amount in the polymer matrix of microparticles from 1 to 10% with respect to the weight of the polymer matrix, the average diameter of microparticles increased from 5.8 ± 0.8 to 8.3 ± 1.1 μm, while EE decreased from 78.0 ± 4.4 to $56.0 \pm 3.5\%$. Experiments with rubomycin hydrochloride showed that the value of EE was influenced by the technique of preparation of microparticles. In the case of the triple emulsion, the value of EE for this drug was almost 1.5 times lower than when microparticles were prepared from the double emulsion. Heating of the triple emulsion during embedding of rubomycin resulted in a decline in EE to $53.0 \pm 3.2\%$. The value of EE was influenced by the chemical composition of the drug. During encapsulation of high-molecular-weight antibiotics (for vancomycin, $M = 1440$ and, for rifampicin, $M = 823$), the values of EE were 70 and 65.5%, respectively. On average, these values were a factor of 1.3 higher than the corresponding parameters for low-molecular-weight antibiotics (for rubomycin, $M = 564$ and, for tienam, $M = 320$) (Table 4.2).

The yield of microparticles loaded with various drugs was generally lower (~65%) than the yield from the emulsion of the unloaded particles (85%) (Tables 4.2 and 4.3). In this case, a decline in the yield of microparticles may be associated with a decrease in the surfactant properties of the aqueous solution of PVA after addition of drugs to the polymer system and, possibly, with a reduction in the stability of the water/oil emulsion and an increase in the probability of merging of minute microparticles into coarser conglomerates (Morello et al., 2007).

The experimental curves of drug release from microparticles into the medium are shown in Figures 4.2–4.4.

All curves have a two-phase pattern. In the first phase (5–8 days), the concentration of the drug in the medium increased and, from 6–9 days onward, it changed insignificantly. As was shown for Hoffmann's violet dye, the release of the drug was influenced by the composition of the polymer used to synthesize microparticles.

The experiments showed that by varying the chemical composition of PHAs, one can prepare microparticles with different properties, which would be suitable for drug loading. The average diameter and ζ-potential of microparticles were found to be dependent on the level of loading (1, 5, and 10% of the polymer weight). None of the high-purity PHAs

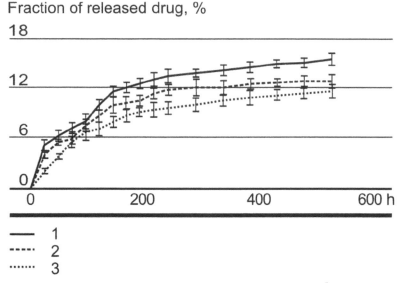

FIGURE 4.2 Dynamics of release of Hoffmann's violet dye from (3) PHA microparticles and (1, 2) P(3HB/3HV) copolymers containing (2) 10 and (3) 37 mol.% 3HV. The content of the dye is 1% with respect to the weight of the polymer matrix (Goreva et al., 2012) (Reprinted with the permission of NAUKA.)

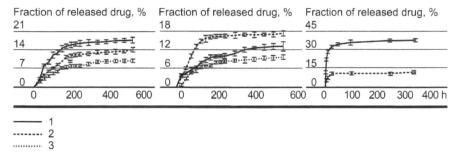

FIGURE 4.3 Dynamics of rubomycin release from (a) P3HB microparticles with different levels of antibiotic loading, (b) P3HB microparticles of different diameters, and (c) P3HB microparticles prepared with the addition of 20% PEG-40. (a) The contents of the antibiotic are (1) 1, (2) 5, and (3) 10% with respect to the weight of the polymer matrix; (b) the particle diameters are (1) 3.6, (2) 10.2, and (3) 15.6 μm; and (c) (1) P3HB/PEG-40 and (2) P3HB (Goreva et al., 2012). (Reprinted with the permission of NAUKA.)

directly contacting with NIH 3T3 fibroblast cells caused any toxic effect or impaired viability of these cells, i.e., all PHAs used in this study were biocompatible and suitable for biomedical applications.

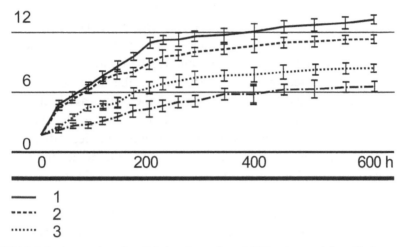

FIGURE 4.4   Dynamics of antibiotic release from P3HB microparticles: (1) tienam, (2) rubomycin, (3) rifampicin, and (4) vancomycin (Goreva et al., 2012). (Reprinted with the permission of NAUKA.)

Thus, during the study of dye release from microparticles prepared from P3HB, from P(3HB/3HV) (10 mol.%), and from P(3HB/3HV) (3HV 37 mol.%), 1.98 ± 0.29, 3.9 ± 0.18, and 5.17 ± 0.57% of the drug with respect to the incorporated amount of the drug were registered in the medium during the first day. Then (during the following six days), these differences were less pronounced: 7.85 ± 0.6, 10.0 ± 0.9, and 10.9 ± 0.6%, respectively. By the end of the experiment, the total release of the dye from the microparticles prepared from the copolymer was almost 1.5 higher than that from the microparticles prepared from P3HB (Figure 4.2).

The kinetics of drug release was influenced by the degree of loading of the matrix. As is seen from Figure 4.3, the higher the level of incorporation of the drug into the matrix, the more pronounced the release of rubomycin from P3HB microparticles. At the maximum loading and the minimum loading of the matrix, the levels of release of the antibiotic during Day 1 were 1.5 ±0.3 and 3.6 ± 0.4% (with respect to the weight of the incorporated antibiotic), respectively. However, from Day 5 onward, the curves reached a plateau. Then, the concentration of rubomycin in the model system remained practically unchanged. By the end of the experiment, the contents of the antibiotic in the medium were 9.5 ± 0.9, 13.6 ± 0.98, and 17.3 ±0.8%, respectively, with the initial levels of loading of microparticles of 1, 5, and 10% (with respect to the weight of the polymer carrier). The kinetics of rubomycin release from microparticles having different diameters but containing the same amounts of the antibiotic is presented in Figure 4.3b. During the initial period (Days 1–4 of the experiment), a gradual release of the drug was observed and, then, the release of the antibiotic remained practically unchanged (Days 6–22). The release of rubomycin into the medium depended on the microparticle diameter: the smaller the particles, the higher the release. For microparticles 3.6 μm in diameter, the release of rubomycin was 18.3 ± 1.1%; for particles 10.2 and 15.6 μm in diameter, the levels of release of rubomycin were 14.5 ± 0.9 and 10.3 ± 0.6% with respect to the incorporated drug, respectively. In addition, the release of rubomycin into the medium depended on the surface structure of microparticles: the higher the porosity of particles, the higher the release. As is seen from Figure 4.3c, during the first day, the release of rubomycin from porous particles (particles with pore diameters of 3.5–4 μm) was 28.4 ± 3.04%, whereas in the second variant (particles having the folded structure with small cavities), the release did not exceed 1.8 ± 0.06%. For 14 days, the release of rubomycin from porous particles was 49.03 ± 1.1%, while in the case of nonporous particles, the release of the antibiotic did not exceed 8.5 ± 0.26% with respect to the incorporated drug.

The dynamics of drug release from microparticles was influenced by the molecular weight of the drug: the higher the molecular weight, the slower and steadier the release of the drug (Figure 4.4). Higher-molecular-weight antibiotics (rifampicin, vancomycin) showed similar curves. After 24 h, the levels of release of these drugs were 1.02 ± 0.08 and 0.57 ± 0.06%, respectively, and by the end of the experiment (Day 22), 8.15 ± 0.7 and 6.8 ± 0.5%, respectively. For tienam and rubomycin, which have lower molecular weights, during Day 1, the variable values of release were 1.28 ± 0.1 and 1.20 ± 0.08% with respect to the weight of the drug incorporated into the matrix. At the end of the experiment, these values were

15.81 ± 1.1 and $13.06 \pm 0.80\%$, respectively. The rate of antibiotic release may depend also on the chemical nature of the drug. Thus, for tienam, the antibiotic having higher water solubility, the release was higher than that of rifampicin, whose release was almost seven times lower. A different picture was observed in the case of vancomycin and rubomycin. This result is apparently due to the fact that they contain functional groups capable of forming hydrogen bonds with the hydroxyl groups of the polymer. Similar data were obtained when the drugs diclofenac and gentamicin were embedded into the PHA polymer matrix (Poletto et al., 2007; Wang et al., 2007).

Analysis of the relationships shown in Figures 4.2–4.4 suggests that the release of encapsulated drugs from PHA microparticles occurs via diffusion. During the first phase, the drugs are released from the surface structures of the carrier; this period lasts between 5 and 8 days, depending on the properties of the polymer matrix. The second segment of the curve (from Days 6–9 to the end of the experiment) corresponds to the release of the drug from the internal structures of the matrix, including release due to a slow and insignificant change in the molecular weight of the polymer. Thus, our experiments showed that, during the experiment, M_w of the polymer used to prepare microparticles decreased from 54.6×10^4 to $(49.0\text{–}46.7) \times 10^4$. The average drop in the molecular weight of microparticles was no more than 10–15% with respect to the initial molecular weight of microparticles.

This study has defined factors influencing the properties of PHA-based microparticles: namely, the chemical composition of polymers, the type of the polymer system, and the technique of mixing employed. By varying these parameters, one can prepare microparticles having various diameters and degrees of porosity suitable for drug encapsulation. The absence of sharp bursts of the drug into the medium and the low rates of drug release from microparticles based on P3HB and P(3HB/3HV) copolymers suggest that microparticles are suitable for drug encapsulation. For microparticles with different diameters, the authors derived the mathematical description of the mechanism controlling the release of drugs from the polymer matrix in the form of PHA-based microparticles. Further variation in the conditions of designing of PHA carriers and improvement of available techniques will make it possible to control the rate of drug release from microparticles and to begin *in vivo* experiments.

Thus, construction of microcarriers of resorbable materials for embedding and delivery of medicinal drugs is a rapidly developing direction of experimental pharmacology. The important parameters of such medicinal systems are their size, surface properties, and electrokinetic potential (ζ-potential), which determine the process of distribution and functioning of microparticles in the body. The electrokinetic potential, as a major parameter of colloidal systems, is used as an indicator of stability and surface adsorption. This parameter characterizes the behavior of microparticles in aqueous media and essentially depends on the chemical structure of the material used, the method of preparation employed, and the size characteristics of microparticles. The paucity of literature data on the influence of conditions of preparation and characteristics of microparticles on the ζ-potential has determined the aim of the study below: to investigate the ζ-potential of microparticles prepared from a polymer of 3-hydroxybutyric acid (Shershneva et al., 2014).

Polymeric microparticles were prepared by two methods: emulsion and spray drying. The method of spray drying is based on rapid removal of water, which causes accelerated formation of microparticles. The Mini Spray Dryer B-290 (BUCHI Laboratory Equipment, Switzerland) is equipped with a nozzle (orifice of 0.7 mm in diameter), into which inert gas (argon) is fed; the centrifugal force acts on the dry microparticles, and they settle in high-production cyclones of the setup. The second method of preparing polymeric microparticles is based on evaporating the solvent from a double emulsion. For this, 1% solution of P3HB was gradually poured into 100 mL of 0.5% solution of polyvinyl alcohol under stirring with a high-speed homogenizer (IKA, Germany) at various speeds (4000 to 24,000 rpm with a step of 4000 rpm, for 3–5 min). The emulsion was then left to stay for one day under constant mechanical stirring until complete evaporation of solvent took place. Microparticles were collected by centrifugation (10,000 rpm for 5 min), washed with distilled water, and lyophilized (Alpha 1–2 LD plus, Christ®, Netherlands).

As a drug, we used an antibiotic of the cephalosporin type – Ceftriaxone ("LEKKO," Russia). Microparticles loaded with Ceftriaxone (1.5 and 10% of polymer weight) were prepared by the emulsion method. For this, 1, 5 and 10 mg of Ceftriaxone was homogenized with 1% P3HB with a Sonicator-S3000 (Misonix Incor., U.S.) at 10 W for 1–2 min. A 0.5% solution of polyvinyl alcohol (100 mL) was poured into the emulsion under stirring with a high-speed homogenizer (24,000 rpm, for 5 min). Other procedures were conducted by the method described above. Measurement of the ζ-potential of microparticles was conducted on a Zetasizer Nano ZS (Malvern, U.K.). We also measured the average size of microparticles, their size distribution and distribution range (polydispersity index, Đ). For investigating the stability of microparticles in aqueous media, a weighed sample of sterile dry particles was placed into balanced phosphate buffer (BPB) (pH 7.3–7.4) in small Falcon-type tubes. The tubes were incubated in a thermostat at 37°C for 30 days. In the course of observation, we measured the ζ-potential and the mean diameter of microparticles. Drug release from microparticles was determined photometrically (wavelength 240 nm) in a Cary 60 UV-Vis spectrophotometer (Agilent Technologies, U.S.).

SEM images of polymeric microparticles are presented in Figure 4.5. The microparticles prepared by spray drying had ζ-potential ranging between -87.1 ± 0.5 and 95.7 ± 0.6 mV (Figure 4.6a), which confirmed the physical stability of the samples. The literature data suggest that the absolute values of ζ-potential above 30 mV provide adequate physical stability and above 60 mV high physical stability (Müller, 1996).

Thus, spray drying produced microparticles of the average diameter between 3.0 and 6.5 µm (Figure 4.6a). Figure 4.6b shows the size distribution of the microparticles. The average diameter of microparticles produced at a temperature of 75°C varied between 4.9 ± 0.5 and 6.5 ± 0.4 µm, while at 85°C – between 5.1 ± 0.3 and 5.7 ± 0.7 µm. The microparticles prepared by the emulsion method at low stirring speeds (4000 rpm) had higher absolute values of ζ-potential (-26.4 ± 0.9 mV) and a larger average diameter (2.42 ± 0.17 µm) (Figure 4.7a, b). As the stirring speed was increased to 12,000 and 24,000 rpm, the average diameter of microparticles decreased to 1.36 ± 0.02 and 0.62 ± 0.02 µm and ζ-potential increased to -21.6 ± 0.6 and -11.8 ± 0.23 mV, respectively. The values of ζ-potential of P3HB microparticles prepared by spray drying were significantly lower than those of microparticles prepared by the emulsion method. A possible reason for this may be that surfactants that are used in the emulsion method

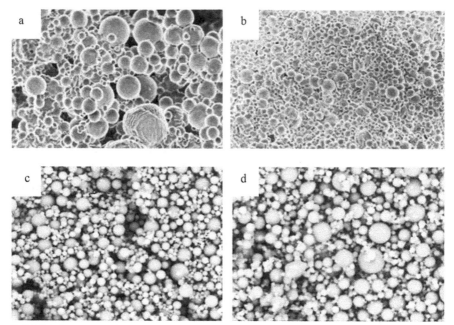

FIGURE 4.5   SEM images of P3HB microparticles prepared by the emulsion method (a, b) and by spray drying (c, d), where: (a) emulsion under stirring at 4000 rpm; (b) emulsion under stirring at 24000 rpm; (c) polymer solution drying with flow rate of 3.25 mL/min at 75°C; (d) polymer solution drying with flow rate of 3.25 mL/min at 85°C. Bar = 5 (a and b) and 30 μm (c and d) (Shershneva et al., 2014). (Reprinted with the permission of NAUKA.)

and are absent in preparing microparticles by spray drying adsorbed onto the surface of the particles. Thus, by using different methods of preparation, it is possible to influence the characteristics of microparticles (both ζ-potential and sizes).

For evaluating the influence of drug loading on the characteristics of microparticles, we constructed P3HB microparticles prepared by the emulsion method at the maximal stirring speed and loaded with antibacterial drug Ceftriaxone to different degrees (Table 4.3, Figure 4.8). As the degree of loading was increased, ζ-potential decreased and the average diameter of microparticles increased. The fraction of microparticles of the average diameter between 400 and 700 nm declined with an increase in the degree of drug loading (from 1 to 10% of polymer weight), and the fraction of microparticles sized between 700 and 1200 nm increased. The fraction of microparticles of diameter between 200 and 400 nm did not

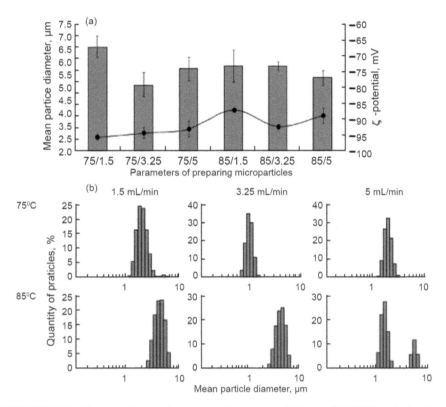

FIGURE 4.6 Characteristics of microparticles prepared from a 1% P3HB solution by spray drying: (a) mean diameter (μm) and ζ-potential of microparticles (mV); (b) size distribution. Parameters of preparing microparticles: temperature of inert gas at the system input (75 and 85°C) and solution feed rate (1.5, 3.25, and 5.0 mL/min) (Shershneva et al., 2014). (Reprinted with the permission of NAUKA.)

change and constituted about 15% in all samples. Ð, as a parameter of size distribution range, had rather low values in all samples, which confirmed the monodispersity of the microparticles (Table 4.3).

It is well-known that the value of ζ-potential may influence the resistance of particles residing in solutions to aggregation (Kovacevic et al., 2014). This phenomenon is related to formation of a double electric layer on the surface of microparticles and the adjacent layer of water, carrying charges of the opposite signs. Absolute values of ζ-potential of about 20 mV provide only short-term physical stability, while values between –5 mV and 5 mV lead to rapid aggregation of particles (Müller, 1996).

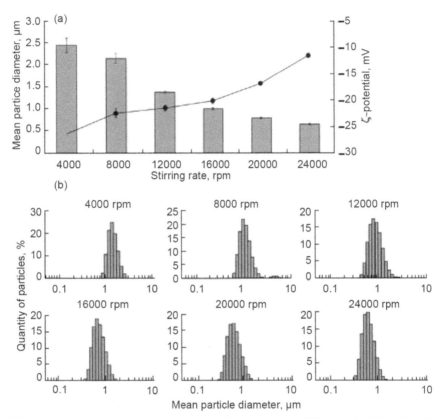

FIGURE 4.7 Characteristics of microparticles prepared at different velocities of stirring the emulsion of 1% P3HB: (a) mean diameter (μm) and ζ-potential of microparticles (mV); (b) size distribution (Shershneva et al., 2014). (Reprinted with the permission of NAUKA.)

TABLE 4.3 Characteristics of P3HB Microparticles Loaded with Ceftriaxone to Different Concentrations (Shershneva et al., 2014) (Reprinted with the permission of NAUKA.)

Ceftriaxone concentration in emulsion, % of polymer weight	Encapsulation effectiveness, %	Mean diameter, μm	Polydispersity index Đ	ζ-potential, mV
No loading				
1	–	0.627 ± 0.005	0.226 ± 0.022	−11.8 ± 0.23
5	81.5 ± 3.1	0.684 ± 0.017	0.160 ± 0.026	−18.3 ± 0.4
10	74.6 ± 4.9	0.752 ± 0.032	0.234 ± 0.055	−22.3 ± 0.3
	62.3 ± 6.5	0.812 ± 0.024	0.258 ± 0.032	−23.5 ± 0.1

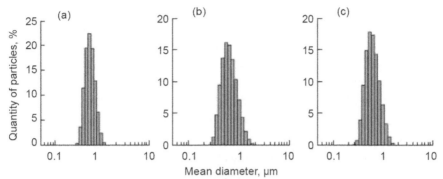

FIGURE 4.8   Size distribution of microparticles loaded with Ceftriaxone: (a) 1%, (b) 5% and (c) 10% of polymer weight (Shershneva et al., 2014). (Reprinted with the permission of NAUKA.)

To estimate the stability of polymeric microparticles of P3HB in liquid media we investigated changes in their characteristics upon incubation in the BPB during 30 d (Figure 4.9). During the first 24 h, the size of microparticles increased somewhat, and the most probable reason for that was rather high ζ-potential of microparticles, which caused their aggregation. The average size of the agglomerates was 1.1 μm, which was practically two times more than the mean diameter of initial microparticles. Further observation (to 744 h) did not show significant changes in ζ-potential.

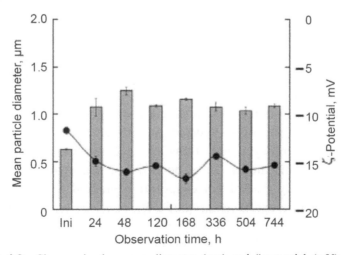

FIGURE 4.9   Changes in the mean diameter (μm) and ζ-potential (mV) of P3HB microparticles upon incubation in the BPB (Shershneva et al., 2014). (Reprinted with the permission of NAUKA.)

The dynamics of change in the ζ-potential of drug-loaded microparticles was different upon incubation in the BPB (Figure 4.10a). From Day
6 (144 h) onward, we noted some reduction of ζ-potential as the drug
was released into the medium, and by the end of experiment, the values
of the ζ-potential were -21.9 ± 0.4, -23.0 ± 0.8, and -24.1 ± 0.5 mV for
microparticles loaded to 1, 5 and 10% of polymer weight, respectively.
With the degree of loading of microparticles increased from 1 to 10%
of polymer weight, we noticed an increase in Ceftriaxone release (Figure
4.10b). Drug release after three days (72 h) constituted $1.33 \pm 0.04\%$, 3.1
$\pm 2.4\%$, and $9.3 \pm 0.9\%$ of the encapsulated drug for microparticles loaded
to 1, 5 and 10% of polymer weight, respectively. By the end of experiment
(744 h) the release of Ceftriaxone was $9.4 \pm 1.2\%$, $23.8 \pm 1.5\%$ and 45.9
$\pm 1.4\%$ of its initial content in microparticles. The changes in size and
ζ-potential of the microparticles loaded with Ceftriaxone (Figure 4.10a,
c) must have been caused by primary adsorption of antibiotic on the surface of microparticles during microparticle preparation. When microparticles interacted with the model medium, the drug was washed away from
the surface on the first day of observation, which subsequently led to the
stable state of the system in the BPB. Thus, the method of preparing microparticles influenced their size and the value of ζ-potential. Regardless of
the preparation technique employed, the microparticles based on P3HB
had negative values of ζ-potential, between -20 mV (emulsion method)
and -95 mV (spray drying). The value of ζ-potential was also influenced
by the loading of the microparticles with the drug, increasing the physical
stability of microparticles in the BPB as compared with microparticles that
did not carry a drug.

The study investigated ζ-potentials of microparticles prepared
from biodegradable poly-3-hydroxybutyrate as related to the method
of preparation employed and taking into account the size of particles
incubated in liquid media. The ζ-potential of microparticles prepared
from emulsion by the solvent evaporation method was -20 mV; the
ζ-potential of microparticles prepared by spray drying was lower: -95
mV. The value of ζ-potential was influenced by drug loading into microparticles; the drug-loaded microparticles incubated in the balanced
phosphate buffer for 30 days had higher physical stability than those
without drug loading.

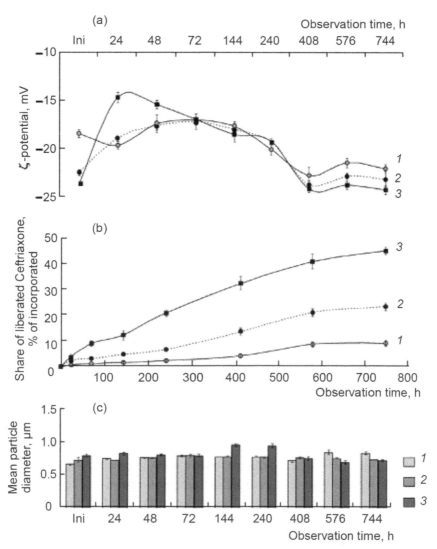

FIGURE 4.10 Changes in ζ-potential (a) and mean diameter (c) of P3HB microparticles loaded with Ceftriaxone: 1 – 1%; 2 – 5% and 3 – 10% of polymer mass, upon exposure in BPB during drug release into the medium (b) (Shershneva et al., 2014). (Reprinted with the permission of NAUKA.)

Until recently, the published studies mainly described prepara-tion of microcarries of drugs from the P3HB homopolymer and 3-hydroxybutyrate/3-hydroxyvalerate copolymers. Murueva et al. (2013)

were the first to report a study of microparticles prepared from PHAs with different chemical compositions. The microparticles were prepared at the Institute of Biophysics SB RAS (Volova et al., 2008, 2013; 2014). They were prepared from the homopolymer of 3-hydroxybutyric acid, P3HB, and 3-hydroxybutyric/4-hydroxybutyric, 3-hydroxybutyric/3-hydroxyvaleric, 3-hydroxybutyric/3-hydroxyhexanoic acid copolymers. The copolymers contained different molar fractions of the second monomer unit and differed in their physicochemical properties, mainly, the degree of crystallinity.

High-purity PHA specimens—a homopolymer of 3-hydroxybutyric acid P3HB and 3-hydroxybutyric/4-hydroxybutyric acid P(3HB/4HB), 3-hydroxybutyric/3-hydroxyvaleric acid (P3HB/3HV), and 3-hydroxybutyric/ 3-hydroxyhexanoic acid P(3HB/3HHx) copolymers were produced in the Institute of Biophysics SB RAS by cultivation of hydrogen-oxidizing microorganisms (Table 4.4).

Microparticles were prepared by the solvent evaporation technique, using double (water/oil) emulsions. The double emulsion contained 0.4 g PHA in 10 mL of dichloromethane and 100 mL 0.5% (w/v) PVA. The resulting double emulsion was mechanically agitated at 24,000 rpm [IKA Ultra-Turrax T25 digital high-performance homogenizer (Germany)] until the solvent had completely evaporated. All emulsions were continuously mixed mechanically for 24 h, until the solvent had completely evaporated. Microparticles were collected by centrifuging (at 10,000 rpm, for 5 min), rinsed 6 times in distilled water, and freeze-dried in an Alpha 1–2 LD plus (Christ®, Germany). Microparticles loaded with doxorubicin (DOX) were prepared using the solvent evaporation technique from the double emulsion. DOX (4, 20 or 40 mg) was dissolved in 10 mL of dichloromethane containing 0.4 g P3HB or P(3HB/3HV) (6.5 mol.%). Aqueous phase as a dispersion medium for the microparticles production was prepared by using 100 mL of a 0.5% (w/v) PVA aqueous solution. The emulsion was agitated at 24,000 rpm [IKA Ultra-Turrax T25 digital high-performance homogenizer (Germany)] until the solvent had completely evaporated. Microparticles were collected by centrifuging (at 10,000 rpm, for 5 min), rinsed six times in distilled water, and freeze-dried in an LS-500 lyophilic dryer (Russia).

TABLE 4.4 Biodegradable PHAs of Different Chemical Compositions Used to Prepare Microparticles

Polymer composition, mol. %	Structural formula	Polymer properties		
		M_n, kDa	M_w, kDa	Đ
P3HB 100	CH₃ O +O-CH-CH₂-C+ₓ	710	1 200	1.71
P(3HB/3HV) 93.5/6.5	CH₃ O CH₂ O +O-CH-CH₂-C+ₓ+O-CH-CH₂-C+ᵧ (CH₃)	890	1700	1.91
P(3HB/3HV) 89.5/10.5		840	1500	1.79
P(3HB/3HV) 80/20		663	1500	2.27
P(3HB/3HV) 63/37		1026	2000	2.00
P(3HB/4HHx) 93/7	CH₃ O O +O-CH-CH₂-C+ₓ+O-CH₂-CH₂-CH₂-C+ᵧ	270	950	3.52
P(3HB/4HB) 93.9/6.1	CH₃ O CH₂ O +O-CH-CH₂-C+ₓ+O-CH-CH₂-C+ᵧ (CH₃ / CH₂)	140	410	2.93
P(3HB/4HB) 84/16		370	970	2.62

Differences in the basic physical properties of the polymers under study (Table 4.4) influenced the characteristics of the microparticles. SEM images of the surface microstructure of microparticles prepared from PHAs that differed in their chemical composition and physicochemical properties showed certain dissimilarities (Figure 4.11).

Whatever the PHA composition, microparticles were heterogeneous in their shape, and their surface structures were different. Microparticles prepared from P3HB and the P(3HB/3HV) containing the lowest molar

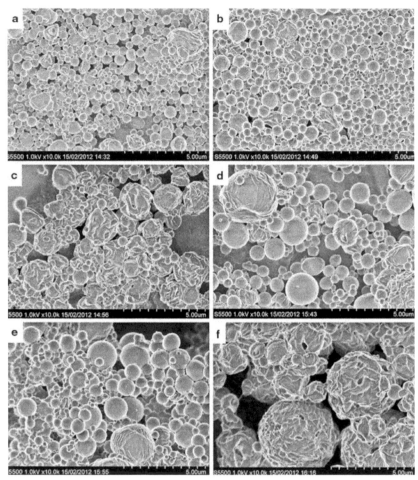

FIGURE 4.11 SEM images of the microparticles prepared from PHAs of different chemical compositions: a – P3HB, b – P(3HB/3HV) (3HV 6.5 mol.%), c – P(3HB/3HV) (3HV 37 mol.%), d – P(3HB/3HHx) (3HHx 7 mol.%), e – P(3HB/4HB) (4HB 6.1 mol.%), f – P(3HB/4HB (4HB 16 mol.%). Bar = 5 μm (Murueva et al., 2013). (Reprinted with kind permission from Springer Science+Business Media: Murueva, A. V., Shishatskaya, E. I., Kuzmina, A. M., Volova, T. G., Sinskey, A. J. Microparticles prepared from biodegradable polyhydroxyalkanoatesas matrix for encapsulation of cytostatic drug. J. Mater. Sci. Mater. Med., 2013, 24, 1905–1915.)

fraction of 3HV (6.5%) were practically smooth and of a regular spheri-cal shape, without surface deformation. Microparticles prepared from the P(3HB/3HV) with a high molar fraction of 3HV (37%) and P(3HB/4HB) (16% 4HB) had a rough surface; some of the particles were irregularly shaped. Visual estimation showed that P(3HB/4HB) microparticles were of larger size. Microparticles prepared from P(3HB/3HHx) (7 mol.% HHx) and P(3HB/4HB) (6.1 mol.% 4HB) had a spherical shape and smooth surface.

Important parameters determining the tissue specificity of the particles and their ability to cross biological barriers are their size and size distribution. Nanoparticles generally vary in size between 10 and 1,000 nm (Soppimatha et al., 2001). Microparticulated drug delivery systems of bigger size are very promising with various methods of administration: oral (osmotic minipumps), parenteral (nanoparticles and nanocapsules), subcutaneous (implants), intracavitary (intrauterine inserts and various suppositoria), buccal, etc. (Vasilyev et al., 2001).

The average diameter of microparticles prepared from PHA copolymers was larger than that of the particles prepared from the homopolymer (Figure 4.12a) although the matrices prepared from P(3HB/3HV) (10 mol.% 3HV) and those prepared from 3HB were similarly sized, and their average diameter was about 750–700 nm. The average diameter of microparticles prepared from the P(3HB/3HV) with the molar fraction of 3HV amounting to 37% was almost twice greater, reaching 1.25 μm. The average diameter of P(3HB/3HHx) particles did not differ significantly from that of P(3HB/3HV) (37 mol.% 3HV) ones: 1.14 μm. Microparticles prepared from P(3HB/4HB) containing 6.1 and 16 mol.% 4HB were significantly larger than other copolymer microparticles, and their diameters were 2.3 and 2.6 μm, respectively (Figure 4.12a). Another important parameter is ζ-potential of microparticles, which characterizes stability or coagulation of the particles in the dispersion medium (Maia, Santana, 2004). Determination of the ζ-potential of microparticles prepared from PHAs with different chemical composition gave the following results (Figure 4.12b): the lowest ζ-potential was recorded for P(3HB/3HHx) microparticles (−32.2 mV); the second-lowest values of this parameter were recorded for P(3HB/4HB) (about −29–27 mV). P3HB microparticles had the highest ζ-potential (about -11 mV). Microparticles prepared from P(3HB/3HV) with different molar fractions of 3HV had a lower ζ-potential, which varied between −23 and −26 mV and was not influenced by the molar fraction of 3HV.

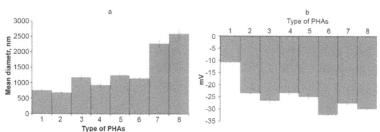

FIGURE 4.12 Mean diameter (a) and ζ-potential (b) of the microparticles prepared from PHAs of different chemical compositions: 1 – P3HB, 2 – P(3HB/3HV) (3HV 6.5 mol.%), 3 – P(3HB/3HV) (3HV 10.5 mol.%), 4 – P(3HB/3HV) (3HV 20 mol.%), 5 – P(3HB/3HV) (3HV 37 mol.%), 6 – P(3HB/3HHx) (3HHx 7 mol.%), 7 – P(3HB/4-HB) (4HB 6.1 mol.%), 8 – P(3HB/4HB) (16 mol.%) (Murueva et al., 2013). (Reprinted with kind permission from Springer Science+Business Media: Murueva, A. V., Shishatskaya, E. I., Kuzmina, A. M., Volova, T. G., Sinskey, A. J. Microparticles prepared from biodegradable polyhydroxyalkanoatesas matrix for encapsulation of cytostatic drug. J. Mater. Sci. Mater. Med., 2013, 24, 1905–1915.)

Figure 4.13 shows results of MTT assay: determination of viability of cells cultured in the presence of PHA microparticles treated with H_2O_2 plasma or by autoclaving on direct contact with fibroblast NIH 3T3 cells. At 3 d after seeding, counts of attached cells showed that the number of cells on microparticles treated with H_2O_2 plasma was higher. The largest numbers of cells (up to 28–33 in the field of view) were attached to microparticles prepared from P3HB and P(3HB/3HV) with 20 mol.% 3HV.

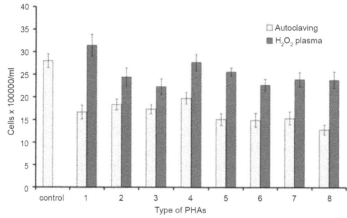

FIGURE 4.13 The number of cells attached to the microparticle surface in 3 h after seeding (numbers as in Figure 4.12). Reference – polystyrene (Murueva et al., 2013). (Reprinted with kind permission from Springer Science+Business Media: Murueva, A. V., Shishatskaya, E. I., Kuzmina, A. M., Volova, T. G., Sinskey, A. J. Microparticles prepared from biodegradable polyhydroxyalkanoatesas matrix for encapsulation of cytostatic drug. J. Mater. Sci. Mater. Med., 2013, 24, 1905–1915.)

That number was 1.4–1.8 times higher than the number of cells attached to the microparticles sterilized by autoclaving. The number of cells attached to autoclaved microparticles prepared from P(3HB/3HV) (6.5, 10 and 37 mol.% 3HV), P(3HB/3HHx), and P(3HB/4HB) (6.1 and 16 mol.% 4HB) was half that recorded on the corresponding microparticles treated with H_2O_2 plasma. A possible explanation for this might be that treatment of polymer devices by physical methods (laser cutting or plasma) strengthens interphase adhesion joints, increasing surface hydrophilicity and, hence, improving its adhesion properties.

MTT assay did not reveal any cytotoxic effect of autoclaved or plasma-treated PHA microparticles. The number of viable cells adhering to the surface of the matrices treated with H_2O_2 plasma was higher than on the surface of the autoclaved ones in all treatments (Figure 3). Results of the cell counts obtained using the fluorescent DAPI DNA stain were as follows: at 3 d after fibroblast NIH 3T3 cells were seeded onto microparticles, the number of cells on PHA microparticles treated with H_2O_2 plasma was significantly higher than on autoclaved ones (Figure 4.14). On plasma-treated microparticles prepared from PHAs with different chemical composition, cells spread well and formed a monolayer. On the corresponding PHA microparticles sterilized by autoclaving, the number of cells was 1.5–2 times lower, and they showed an irregular shape. As differences in the number of cells proliferating on microparticles prepared from PHAs of different types are insignificant, all of the polymers investigated in this study are of good quality, showing high biocompatibility.

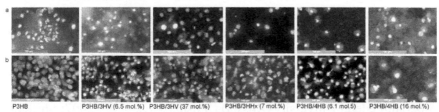

FIGURE 4.14 DAPI staining of fibroblast NIH 3T3 cells on microparticles of different types sterilized with autoclaving (a) and H_2O_2 plasma (b), 7 days after seeding: P3HB, P(3HB/3HV) (3HV 6.5 mol.%), P(3HB/3HV) (3HV 37 mol.%), P(3HB/3HHx) (3HHx 7 mol%), P(3HB/4HB) (4HB 6.1 mol.%), P(3HB/4HB) (4HB 16 mol.%) (Murueva et al., 2013). (Reprinted with kind permission from Springer Science+Business Media: Murueva, A. V., Shishatskaya, E. I., Kuzmina, A. M., Volova, T. G., Sinskey, A. J. Microparticles prepared from biodegradable polyhydroxyalkanoatesas matrix for encapsulation of cytostatic drug. J. Mater. Sci. Mater. Med., 2013, 24, 1905–1915.)

The experiments showed that by varying the chemical composition of PHAs, one can prepare microparticles with different properties, which would be suitable for drug loading. The average diameter and ζ-potential of microparticles were found to be dependent on the level of loading (1, 5, and 10% of the polymer mass). None of the high-purity PHAs directly contacting with NIH 3T3 fibroblast cells caused any toxic effect or impaired viability of these cells, i.e., all PHAs used in this study were biocompatible and suitable for biomedical applications.

4.3   BIOLOGICAL COMPATIBILITY OF PHA-BASED MICROPARTICLES

Microspheres are considered to be the best form for injection-delivered prolonged-action drugs. However, microspheres, though containing a small amount of biomaterial, have a large surface area and, being injected subcutaneously or intramuscularly, contact with an extensive area inside the host organism, which may cause stronger tissue reaction (Anderson, 1997). Rather few results have been reported on *in vivo* biocompatibility of P3HB, including data on local tissue reaction to poly-3-hydroxybutyrate and poly-3-hydroxybutyrate/3-hydroxyvalerate implants (Gogolewski et al., 1993; Shishatskaya et al., 2004). Investigations of PHA biocompatibility were performed using large implants, in the form of sutures, film grafts or pins (Williams and Martin, 2002; Qu et al., 2006). It is well-known, though, that biocompatibility of biomaterials depends not only on the chemical structure and purity of the specimens but, to a large extent, on the size and shape of the implant and the implantation site (Anderson and Shive, 1997).

Studies of biocompatibility of PHA microparticles were performed *in vitro* and *in vivo* by researchers of the Institute of Biophysics SB RAS and Siberian Federal University (Russia) (Shishatskaya et al., 2007; Shishatskaya et al., 2008; Shishatskaya et al., 2011; Murueva et al., 2013: Eke et al., 2014).

The purpose of the study was to test poly-3-hydroxybutyrate microparticles for cytotoxicity *in vitro*, in fibroblast culture, and to investigate tissue response to intramuscular implantation of P3HB *in vivo* (Shishatskaya et al., 2008). PHA-based microspheres were prepared by the solvent evaporation technique, using a triple emulsion. 600 mg of the polymer (P3HB) and 200 mg of polyethylene glycol (PEG40, molecular mass 40 kDa) were dissolved in 10 mL of dichloromethane at 40°C. Then, 1 mL of a 6% gelatin solution (40°C) was added and the mixture was shaken vigorously. The resulting double water/oil (W/O) emulsion was allowed to cool to room temperature and, then, it was gradually poured into 150 mL of a 0.5% PVA solution, which was stirred with a three-blade propeller (at 700 rpm, for 20 min), to obtain a triple (water/oil/water, W/O/W) emulsion. The emulsion was continuously mixed mechanically for 24 h, until the solvent was completely evaporated. Microspheres were collected by centrifuging (at 10,000 rpm, for 5 min), rinsed 7–8 times in distilled water, and freeze-dried in an LS-500 lyophilic dryer (Russia).

In the first phase of the study, extracts of polymer microparticles were investigated for toxicity in the culture of mouse fibroblasts NIH 3T3 cells according to the ISO 10993-5 standard. As a positive control, we used red technical rubber GOST (the Russian Federation state standard) – 5496-78. This standard corresponds to the European analog of the cytotoxic positive control – Para rubber (Freshney, 2000). As a negative control, polystyrene "Greiner bio-one" extract was used. Cells were cultured under standard conditions in the presence of extracts for 72 h, and then they were examined to analyze their morphology, viability (by live staining with trypan blue), and metabolic activity (using MTT assay) (Shishatskaya et al., 2008b).

Mouse fibroblast cells (NIH 3T3 line) cultured in the presence of polymeric P3HB microsphere extracts retained the morphology of normal cells, like those grown in the control, on polystyrene. Cell viability test, performed by the method of live staining with trypan blue, showed that $99.8\pm0.2\%$ of the cultured cells did not incorporate the dye, i.e., remained highly viable, in contrast to the positive control (rubber extract), in which most of the cells died. The doubling time of fibroblasts corresponded to the generation time of the cells cultured on standard medium and in the treatment culture – 25.1 ± 1.8 and 25 ± 2 h; and 24.9 ± 2.1 and 168 ± 21.3 on the negative and positive controls. The MTT assay did not indicate any toxic effect of the polymer extract on the metabolic activity of fibroblasts, either. Cell concentration of the treatment culture was comparable to those of the negative control and the standard medium culture ($190\pm23 \times 10^6$), and that was almost an order of magnitude higher than in the positive control ($200\pm24 \times 10^5$). It is well-known that only active dehydrogenases of living cells will convert MTT into insoluble formazan crystals. This study showed that extracts of polymeric P3HB microparticles did not produce any toxic effect on proliferation and metabolic activity of mouse fibroblast NIH 3T3 cells.

The *in vitro* study of biocompatibility of PHA microparticles was followed by experiments on laboratory animals (Shishatskaya et al. 2007, 2008, 2011). All animals in the treatment group, which had been injected with a microsphere suspension, were healthy and ate well. Their body mass and masses of their internal organs were similar to those of the rats in the control group. At 24 h after the injection of polymeric microparticles, the microscopic picture at the injection site was characterized by a slight tissue edema and leukocyte infiltration. The cluster of microspheres injected intramuscularly may be considered as an open porous implant.

The nature and extent of the tissue reaction to the implant is characterized by the presence of specific cell types (Anderson and Shive, 1997). After injury due to surgical intervention and implantation of the material or device, aseptic tissue inflammation develops at the site of implantation. It is usually divided into several phases: alteration (injury), exudation, and proliferation; the last phase is at the same time the first phase of tissue repair. The phases of exudation and proliferation are sometimes subdivided into neutrophil, macrophage, and fibroblast phases. The neutrophil phase occurs during the first few hours of implantation: polymorphonuclear leukocytes (PNLs) migrate from the vessels to the site of damage, surround it, and, after 6–12 h, form a leukocyte barrier. The lifetime of PNLs is short; within 24 h, neutrophilic leukocytes stop migrating and their decomposition starts.

In our experiment, the initial inflammatory response was observed in 24 h after the injection of P3HB microparticles: up to 20–25 polymorphonuclear leukocytes could be seen in the field of view. This tissue reaction was of short duration; after 1 week, the number of polymorphonuclear leukocytes decreased significantly. The low number of leukocytes during that time suggests that tissue reaction to the implantation of polymeric microparticles can be described as a mild inflammatory response. After 1 week of implantation, we observed infiltration of fibroblast cell elements and formation of a thin fibrous capsule at the interface between the microspheres and the intact muscular tissue, as a foreign body reaction of the tissue.

The foreign body reaction is expressed as the presence of macrophages at the microsphere/tissue interface (Anderson and Shive, 1997). In this phase (the macrophage phase), macrophages penetrate into the leukocyte barrier and phagocytize cell detritus and degradation products from the tissue and implanted material. Macrophages surround the foreign body and form neutrophil-macrophage→ macrophage → macrophage-fibroblast barriers, which prevent the formation of granulation tissue (Kao et al., 2007). When biodegradable materials are implanted, the macrophage response, which follows the neutrophil response, increases rather than decreases. The reason for this is that macrophages and foreign body giant cells phagocytize and resorb biodegradable materials. The main contribution to the formation of foreign body giant cells (FBGCs) is made by macrophages, which phagocytize the implanted material. It is well-known that FBGCs, formed on the surface of or around the particles of the implanted material by the fusion of macrophages and by nuclear division of macrophages without cytokinesis, resorb the material (Dijkhuizen-Radersma et al., 2002).

Figure 4.15 shows the tissue at the microsphere implantation site at 2 weeks. One can see mononuclear secretory-phagocytic macrophages and a few FBGCs with 2–3 nuclei. In addition to their phagocytic activity, macrophages initiate the formation of granulation tissue. The granulation tissue response is characterized by fibroblast infiltration and the development of blood capillaries. If an implant is made of non-resorbable material, a thick fibrous capsule is usually formed around it. If biocompatible and biodegradable implants are used, the tissue response may not involve the formation of a thick fibrous capsule or, if the capsule is formed, it is later involuted (Shishatskaya et al., 2004). At 2 weeks, the cluster of the injected microspheres was encapsulated by thin connective tissue. At that time, the percentage of large particles (over 10–15 μm) in the cluster of microspheres was significantly reduced and did not exceed 14% in the field of view. Active fibroblast elements were identified in the structure of the thin capsule; its mean thickness was no more than 50–60 μm; it consisted of 4–6 fibroblast layers. The reaction was almost 3 times

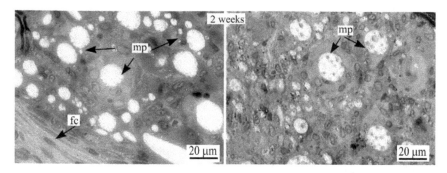

FIGURE 4.15 Microscopic picture of tissue at the site of P3HB microsphere implantation (2 weeks after implantation): mp – microparticles, fc – fibrous capsule. The bar is 20 μm (Shishatskaya et al., 2008b). (Reprinted with permission from the Journal of Siberian Federal University.)

less intensive than the reaction of muscular tissue to the implantation of sutures that we reported previously. This result is in good agreement with the data reported by Fournier et al. (2006), who found that the intramuscularly injected microsphere cluster was isolated from the muscular tissue by a very thin fibrous capsule, which was much less developed than the capsule formed in the event of subcutaneous injection. At the site of injection of P3HB microparticles, the newly formed tissue was significantly vascularized, indicating high biocompatibility of the polymeric microspheres and favorable tissue reaction. The area of the edema at the implantation site was considerably decreased, and there were no necrotic regions. There were a few leukocytes among the microspheres. We recorded an increase in the number of mature secretory-phagocytic macrophages (up to 6.36±0.42 in the field of view) on the inner side of the capsule adhering to the implant and at the cluster of the microspheres. Later in the experiment, we observed more intensive macrophage infiltration at the site of injection of the microspheres; the numbers of both mononuclear macrophages and FBGCs increased.

At 5 weeks, the number of mono- and poly-nuclear macrophages at the site of implantation of the microspheres increased (Figure 4.16a). The fraction of large particles became smaller and did not exceed 4–5% of the total amount of the particles in the field of view, suggesting that the polymeric matrix was being degraded. FBGCs with 6–8 nuclei were observed to aggregate around large (more than 10 μm in diameter) microspheres.

At the interface between the intact muscular tissue and the microspheres there still was a fibrous capsule, although very thin, consisting of 2–3 layers of mature fibroblasts (Figure 4.16b). The macrophage response grew more intense at 7–9 weeks; the number of FBGCs increased and so did the number of their nuclei. At the site of implantation of the microspheres, we clearly saw macrophages grouped around large microparticles. In some cases, we could see penetration of cell elements into the surface matrix of the microspheres and degraded (resorbed) parts around the circumference of the particles at the implant/tissue interface (Figure 4.17a), which indicated that the polymeric matrix of the microparticles was being degraded. There were no fibrous capsules either at the microspheres/tissue interface or around any of the microspheres. This is a very important fact because the fibrous capsule around a polymeric particle could significantly influence the release behavior of the encapsulated drug and resorption of the polymeric matrix.

At the end of the experiment (11–12 weeks), the tissue reaction remained essentially the same: we observed pronounced macrophage infiltration with a large number of poly-nuclear FBGCs, which either surrounded polymeric particles or were grouped around a cluster of smaller microparticles; there were rather large fusions of FBGCs with 10–12 and even more nuclei (Figure 4.17a). There were no fibrous capsules at the interface between the intact muscular tissue and the cluster of the microspheres. In some parts of the microsphere cluster, there was polymeric

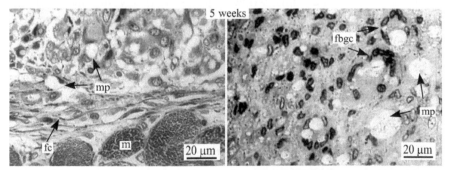

FIGURE 4.16 Microscopic picture of tissue at the site of P3HB microsphere implantation (5 weeks after implantation): mp – microparticles, mph – macrophages, FBGCs – foreign body giant cells, tf – tissue fibers, fc – fibrous capsule. The bar is 20 µm (Shishatskaya et al., 2008). (Reprinted with permission from the Journal of Siberian Federal University.)

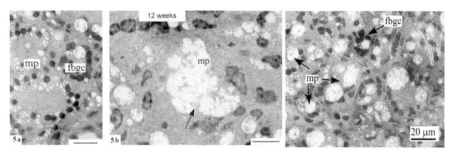

FIGURE 4.17 Histology of the in vivo P3HB degrading microspheres implanted intramuscularly, at 12 weeks. The bar is 10 μm (Shishatskaya et al., 2008). (Reprinted with permission from the Journal of Siberian Federal University.)

detritus, as a degradation product of larger particles. The fraction of microspheres larger than 10–15 μm in the field of view significantly decreased (to about 2–3%) and fragmented microparticles were seen (Figure 4.17 b). For quite a long time, however, most of the microspheres persisted in the tissue undegraded, suggesting that *in vivo* bioresorption of microparticles must be rather a long process, which makes poly-3-hydroxybutyrate a good candidate for fabricating a prolonged-action drug formulation intended for intramuscular injection.

These experiments showed that poly-3-hydroxybutyrate microspheres of diameter between 0.9 and 35.0 μm were biocompatible; they neither inhibited growth and metabolic activity of fibroblasts nor caused negative response in animals that were intramuscularly injected with them. The tissue response to the implantation of poly-3-hydroxybutyrate microparticles was characterized by mild inflammatory reaction of short duration, pronounced and ever increasing macrophage infiltration involving foreign body giant cells that resorbed the polymeric matrix, and the granulation response, involving the formation of a highly vascularized thin fibrous capsule at the microspheres/tissue interface, which was resorbed after several weeks of implantation. No fibrous capsules were formed around polymeric microparticles; neither necrosis nor any other adverse morphological changes and tissue transformation in response to the implantation of the polymeric microparticles were recorded. The results of the study suggest that poly-3-hydroxybutyrate is a good candidate for fabricating prolonged-action drugs in the form of microparticles intended for intramuscular injection.

Finding materials for intravenous sustained-release drug delivery systems presents a special challenge. Although polymer-based microparticles have recently received much attention as drug carriers, there are very few data in the available literature regarding biocompatibility of microparticles injected into the bloodstream. Blood-compatible materials must not induce thromboses, thromboembolisms, antigenic response, destruction of blood constituents and plasma proteins; they must also retain their mechanical-physical properties (Kim et al., 2002; Yen and Huang, 2003). There are very few literature reports on injecting polymer microparticles into the bloodstream, and the available data do not provide a solution to this issue. The distribution of 3-μm to 12-μm diameter [141]Ce-labeled polystyrene/divinylbenzene microparticles intravenously injected to Beagle dogs was found to depend on the size of the particles: larger particles (over 7 μm) were retained for long periods of time (up to 4 weeks) in the lungs, while smaller (3–5 μm) ones – in the liver and spleen (Kanke et al., 1982; Kon et al., 1993). The authors observed vascular occlusions around clusters of microspheres and pinpoint necrotic sites without tissue destruction. Studies published in the mid-1980s – early 1990s and more recent studies (Semmler-Behnke et al., 2008; Saatchi et al., 2009) that investigated the distribution of microparticles prepared from different polymeric materials in tissues of internal organs of animals (mainly rats or mice) did not address the effects of administration of the microparticles.

Krasnoyarsk researchers (Shishatskaya et al., 2009) performed *in vivo* studies addressing consequences of intravenous administration of polymer microparticles. Particles for intravenous administration were prepared from ^{14}C-labeled poly-3-hydroxybutyrate. The ^{14}C-labeled microparticles were prepared by the solvent evaporation technique, using a triple emulsion. 600 mg of the P3HB, and 200 mg of polyethylene glycol (PEG40, molecular mass 40 kDa) were dissolved in 10 mL of dichloromethane. Then, 1 mL of a 6% gelatin solution (40°C) was added, and the mixture was shaken vigorously. The resulting double water/oil (W/O) emulsion was allowed to cool to room temperature and then it was gradually poured into 150 mL of a 0.5% PVA solution, which was stirred with a three-blade propeller (at 2000 rpm, for 20 min) (Heidolph, Germany), to obtain a triple (water/oil/water, W/O/W) emulsion. The emulsion was continuously mixed mechanically for 24 h, until the solvent was completely evaporated. Microspheres were collected by centrifuging (at 10000 rpm, for 5 min), rinsed 7–8 times in distilled water, and freeze-dried in the LS-500 lyophilizer (Prointex, Russia). The size of microspheres with a diameter larger than 3 μm was determined using the Automatic Particle Counter + Analyzer system (Casy TTC, Scharle System GmbH, Germany). The obtained size distribution was used to describe the particle size. For intravenous injections, a fraction of microspheres with a diameter smaller than 3.8 μm was selected, using a track filter. The size of the particles was estimated using the Scan master program of the CMM-2000 scanning multi-viewer microscope, taking into account the parameters that determined the image size.

Microspheres of this size are considered to be the best for injections. The fraction of the particles between 5 and 10 μm in diameter constituted 30.4±1.1%, and the fraction of the 10 to 20 μm particles – 34.4±2.3%. The largest of the harvested particles were 35 μm in diameter, and their fraction did not exceed 3.5%. The average diameter of microspheres was 10±0.23 μm. Particles taken for intravenous injection were 0.5–3.8 μm. The microparticles used in the experiment had an average diameter of 2.40 ± 0.21 μm.

Experiments were conducted on adult female Wistar rats (200–240 g each) in accordance with the international and Russian guidelines for ethical treatment of laboratory animals. The rats were kept in an animal facility and fed a standard diet in accordance with the directive on maintaining animals and experimenting on them (Genin et al., 2001). A sterile suspension of microspheres (5 mg in 0.5 mL of physiological saline, 5×104 cpm/g) with a diameter no more than 3.8 μm was injected to rats through the tail vein, without any anesthesia. Intact animals were used as control. Three hours after the injection, 24 hours after the injection, and then every week, three animals were sacrificed by using an overdose of a volatile anesthetic. Their internal organs were removed, examined macroscopically, weighed, dried and ground; then, radioactivity counts were performed. A 100 mg sample was placed into a plastic vial (PerkinElmer/Packard, U.S.) containing 15 mL of dioxane scintillation solution; 1 L dioxane contained 2,5-diphenyloxazole, 10 g; 1,3-di-2,5-phenyloxazolyl benzene, 0.25 g; naphthalene, 100 g. Radioactivity counts were performed in a TRI-CARB 2100TR scintillation counter (Packard BioScience Company, U.S.). Radiocarbon was measured in the heart, lungs, liver, kidneys, bone marrow, and blood, without taking into account the radioactivity of soft and hard tissues and excreted metabolic products. To study the resorption of the polymer scaffold and accumulation of ^{14}C-containing PHA degradation products, samples of the dried tissues were subjected to methanolysis, and fatty acid methyl esters were determined using the GCD plus chromatograph mass spectrometer (Hewlett Packard, U.S.). To determine high-molecular- weight (undecomposed) polymer in the organs, it was extracted from the tissues with chloroform and precipitated with hexane. Then, the polymer was methylated and chromatographed as described above; the sensitivity was 10–^{11} g. The general tissue reaction to implanted microspheres was investigated using conventional histological techniques. The samples were fixed in 10% formalin and embedded in paraffin; 5–10 μm thick microtome sections were stained with hematoxylin and eosin.

All animals that had been injected with a microsphere suspension were healthy and ate well throughout the experiment. Their body mass and masses of their internal organs were similar to those of the rats in the control group. Macroscopic examination of the rats' internal organs and histological studies of the tissue sections did not show any adverse changes in them throughout the observation period. The dynamics of ^{14}C activity concentrations in tissues of internal organs in the course of the experiment is shown in Figure 4.18. Three hours after the injection of the microparticles into the bloodstream of the animals, the largest ^{14}C activity concentration (14680±417 cpm×g) was registered in heart tissues. The second largest concentration of the label was determined in kidney tissues (7520±81 cpm×g) and the third – in lung tissues (5280±65 cpm×g). Liver and spleen tissues contained similar ^{14}C activities – (4400±58 cpm×g). The lowest radioactivity levels were registered in blood and bone marrow. One month later, ^{14}C activity concentration in the heart tissues was 4.5 times lower than its initial level, amounting to 3240±57 cpm×g. Radioactivity of liver tissues dropped to 15120±90 cpm/g. Radioactivity levels decreased in the tissues of kidneys and lungs to 4720±69 cpm/g and 940±97 cpm/g, respectively. At the same time, ^{14}C concentration in the spleen tissues somewhat rose, which may be accounted for by the accumulation of polymer resorption products in this organ, where, in addition to hydrolytic enzymes, there are active macrophage-type cells, which

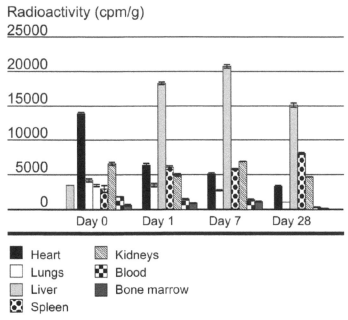

FIGURE 4.18 Dynamics of radiocarbon concentration in tissues of internal organs of the animals intravenously injected with 14C-labeled P3HB microparticles (Shishatskaya et al., 2009). (Used with permission of the Journal of Siberian Federal University and the Journal "Bulletin of Experimental Biology and Medicine.")

resorb cell elements. Macrophages are known to actively degrade poly-3-hydroxybutyrate (Shishatskaya et al., 2009).

The level of radioactivity in blood and bone marrow did not exceed 300 cpm/g. Thus, the tissues of internal organs sorbed different amounts of intravenously injected microparticles. This is even more obvious from the analysis of the data in Figure 4.19, which shows *in vivo* accumulation of the label, taking into account the masses of the organs. Data in Figure 4.19 correspond to the data in Figure 4.18, suggesting the following conclusion: the main target for microparticles was the liver tissue; the average level of ^{14}C accumulated in the liver during the experiment was approximately 60% of the injected dose. During the period of microparticle circulation, a reliable decrease in the organ radioactivity was detected, which, we believe, is due to the process of polymer matrix biodegradation and release of the label part with carbon-containing P3HB biodegradation products. Thus, three hours after the injection of microparticles to the animals, the

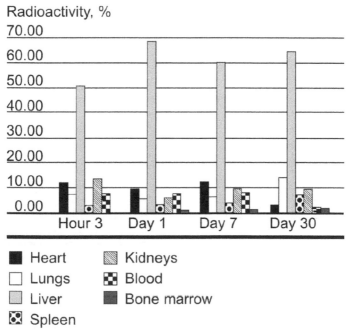

FIGURE 4.19 Dynamics of radiocarbon accumulation in internal organs of the animals intravenously injected with 14C-labeled P3HB microparticles (Shishatskaya et al., 2009). (Used with permission of the Journal of Siberian Federal University and the Journal "Bulletin of Experimental Biology and Medicine.")

total radioactivity of the organs was 38720 ± 2575 cpm/g, i.e., 80% of the injected dose (Figure 4.20). 120 days after the intravenous injection, the total radioactivity dropped to 28210 ± 2054 cpm/g, which could be attributed to polymer biodegradation.

The amounts of radiocarbon detected in the rats' organs did not show the true accumulation of microparticles in the organs: PHAs are biodegradable polyesters and are bioresorbed *in vivo* by the enzymes and cell elements of blood and tissues, and the polymer degradation products are removed with metabolites (Jendrossek et al., 2001). The ^{14}C activities recorded in the tissues comprise the radiocarbon of the high-molecular-weight polymer matrix, i.e., undecomposed microparticles, and the radiocarbon of polymer degradation products.

To compare the biodegradation rates of the polymer in different organs and to determine the lifetime of polymer microparticles, tissues were subjected to chromatographic analysis. During the experiment, the amounts

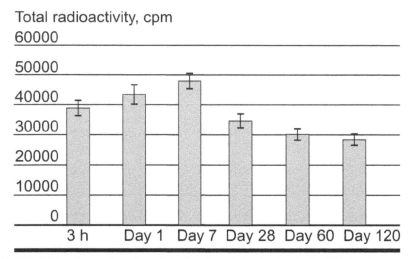

FIGURE 4.20 Dynamics of total 14C radioactivity in the body (Shishatskaya et al., 2009). (Used with permission of the Journal of Siberian Federal University and the Journal "Bulletin of Experimental Biology and Medicine.")

of the polymer substance varied in every organ. At 8 weeks, polymer content in the heart and lungs dropped almost by a factor of ten compared to week 1. However, in our opinion, this cannot be accounted for by the degradation of polymer matrix in these organs, but rather by the bloodstream washing out the initially introduced microparticles from the pulmonary circulation. Eight weeks later, polymer content in the liver and kidneys decreased insignificantly: by a factor of 1.5 in the kidneys and by a factor of 1.2 in the liver. A more significant, 2.5-fold, decrease was recorded in the spleen. 12 weeks later, at the end of the experiment, polymer content decreased as compared to the values measured after an 8 week interval: almost twofold in the heart and lungs, threefold in the kidneys, 18-fold in the spleen, and 25-fold in the liver (Table 4.5).

The results are indicative of different PHA biodegradation rates in animal tissues. They also show that at the end of the experiment, the organs still contained high-molecular-weight polymer and, thus, some of the microparticles could have remained undecomposed. However, spectrometric methods used to detect high-molecular-weight PHAs are based on registering fatty acid methyl esters (i.e., polymer-forming monomers) – products of preliminary hydrolysis of the polymer to monomers and their

TABLE 4.5 Results of Chromatographic Analysis of Methyl Esters of 3-Hydroxybutyrate Monomers Contained in Tissues of Rats' Internal Organs (Shishatskaya et al., 2009) (Used with permission of the Journal of Siberian Federal University and the Journal "Bulletin of Experimental Biology and Medicine.")

Organ	P3HB content (1×10^{-4} mg/organ)		
	1 week	8 weeks	12 weeks
Heart	1527	132	75
Lungs	1694	111	56
Liver	5861	4889	197
Spleen	606	251	14
Kidneys	189	124	43

subsequent methylation. Thus, the values obtained may comprise both monomers of hydroxybutyric acid resulting from methanolysis of the high- molecular-weight polymer and monomer products of its natural biological degradation caused by enzymes and cells of the organs.

The high-molecular-weight (undecomposed) polymer was extracted from the samples and analyzed. That was done to determine the concentration of the undecomposed high-molecular-weight polymer in internal organs at different time points after the injection of microparticles (Table 4.6). One week after the injection, the polymer contents in the lungs, heart, and liver were larger than those registered in the spleen and kidneys. Eight weeks after the injection, the polymer contents in the lungs and heart dropped dramatically (similarly to the data in Table 4.5), but that was rather due to particle washout than degradation processes. This conclusion was confirmed by the measurements of tissue radioactivity. The data at the end of a 12-week interval given in Tables 4.1 and 4.2 differ significantly: chromatography of the tissues determined much higher amounts of the polymer substance (Table 4.5) than the amounts of the high-molecular-weight polymer recovered from the organs (Table 4.6). This leads to the conclusion that at the end of the experiment, the major portion of the ^{14}C-labeled polymer was present *in vivo* as monomers of hydroxybutyric acid and its low-molecular-weight degradation products.

The second important conclusion based on the data of Table 4.2 is that some polymer microparticles remained undecomposed in the animals' organs at different time points of the experiment. Finally, the fact that the

TABLE 4.6 Residual High-Molecular-Weight Polymer in Tissues of Rats' Internal Organs (Shishatskaya et al., 2009) (Reprinted with permission from the Journal of Siberian Federal University.)

Organ	P3HB content (1×10^{-4} mg/organ)		
	1 week	8 weeks	12 weeks
Heart	420	30	5
Lungs	600	5	6
Liver	360	110	80
Spleen	6	10	5
Kidneys	16	20	7

examined organs, especially the liver and the spleen, contained relatively low amounts of the polymer, while radioactivity of the tissues was high, is indicative of the rapid degradation of the polymer matrix in them.

In conclusion, this study shows that poly-3-hydroxybutyrate microparticles are suitable for intravenous administration; tissues of internal organs accumulate different concentrations of polymer microparticles, and polymer matrix is degraded in them at different rates. The presence of high-molecular-weight polymer matrix in internal organs, which is indicative of microparticles remaining undecomposed, suggests that poly-3-hydroxybutyrate can be used to construct systems for slow (up to 12 weeks) drug delivery to tissues of internal organs via intravenous administration.

4.4   THERAPEUTIC EFFICACY OF PHA-BASED MICROPARTICLES

Research involving construction of microparticles for sustained release of rubomycin, an anthracycline antibiotic, by encapsulating the drug into a PHA matrix, and investigation of its antitumor effectiveness was conducted in a number of studies (Shishatskaya et al., 2008, 2012; Murueva et al., 2013). As anticancer drugs of cytotoxic action have low selectivity and often produce an adverse effect on the body, it is particularly important to design controlled-release delivery systems for them. Among these drugs are anthracycline antibiotics, including rubomycin and its oxidized form – doxorubicin, which are widely used clinically in treatment of various cancers. Rubomycin is an effective anticancer drug (Giewirtz, 1999). As on contact with tissue rubomycin induces necrosis, it is only administered intravenously; moreover, being very toxic, it may cause grave side effects in the patients (Lin et al., 2005). In the early 1990s, in order to alleviate the toxic effect of rubomycin and to increase the time of its *in vivo* circulation, it was proposed to entrap it in erythrocytes (Ataullakhanov et al., 1993) and to conjugate it with high-molecular-weight compounds such as polyethylene glycol, an *N*-(2-hydroxypropyl/methacrylamide) copolymer (Husseini et al., 2002).

The investigations of the past few years have confirmed the high efficiency of using polymer nano- and micro-carriers as drug delivery systems. Several studies have been conducted to compare the toxicity of free and encapsulated doxorubicin (DOX) (Lee et al., 2005; Zhao et al., 2007). Doxorubicin-loaded chitosan microspheres were tested in cell culture (Kim et al., 2007). *N*-(2-hydroxypropyl/methacrylamide) copolymer-DOX conjugates with pH-controlled drug release were found to be effective for suppressing *in vitro* cell proliferation (Mrkvan et al., 2005). Copolymer of styrene-maleic acid (SMA) was used to construct micelles containing doxorubicin by means of a hydrophobic interaction between the styrene moiety of SMA and DOX. The SMADox micelle preparation was less cytotoxic to the SW480 human colon cancer cell line compared with free doxorubicin (Greisha et al., 2004). Park and coauthors described surface-modified polylactide nanoparticles and reported a study of drug cytotoxicity in A20 murine lymphoma cells in culture. Doxorubicin encapsulated in surface-modified nanoparticles killed up to 60% of the total cell population in culture (Park et al., 2009).

The cytostatic effect of microparticles loaded with DOX was estimated in the culture of tumor cells—HeLa (Murueva et al., 2013). Figure 4.21 shows results of evaluation of the inhibiting effect produced by DOX-loaded microparticles on HeLa cell culture versus the effect of the free drug (Figure 4.22). The effect of small particles (0.2 μm), which contained the highest DOX concentration (0.6 μg/mL), was similar to the effect of the free drug: they became effective within the same time interval and had the same inhibitory action on the cells. Particles containing the medium and the lowest concentrations of DOX (3.2 μg/mL and 0.6 μg/mL) began inhibiting the growth of tumor cells at Day 3 of the experiment, similarly to the free drug, but the strongest inhibitory effect was recorded at Day 4. The reason is that drug release from the polymer matrix during the first

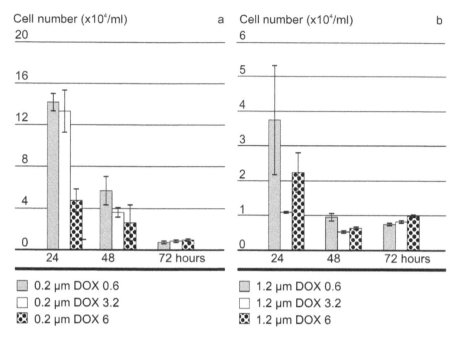

FIGURE 4.21 MTT assay: the effect of DOX encapsulated in 0.2 μm (a) and 1.2 μm (b) polymer microparticles on the number of viable cells in HeLa cell culture (Shishatskaya's data).

two days was slow, resulting in low DOX concentrations in the cell culture (about 0.09 μg/mL and 0.07 μg/mL in the experiments with the highest and lowest concentrations of encapsulated DOX, respectively), and this concentration was not high enough to suppress HeLa cell growth.

The effect of DOX encapsulated in larger polymer particles was more pronounced (Figure 4.21b). The delay of the inhibitory effect was only recorded at Day 1 and only for the particles with the lowest and the medium DOX concentrations (DOX concentrations in the culture were 0.08 μg/mL and 0.28 μg/mL, respectively). At Day 2, however, the cytostatic effect of the encapsulated DOX was comparable with the effect of the free drug.

These results showed the effectiveness of the cytostatic drug encapsulated in microparticles prepared from resorbable polymers in HeLa cell culture.

Results of *in vitro* studies were used as a basis for experiments on animals. Experiments aimed at evaluation of medicinal effectiveness of

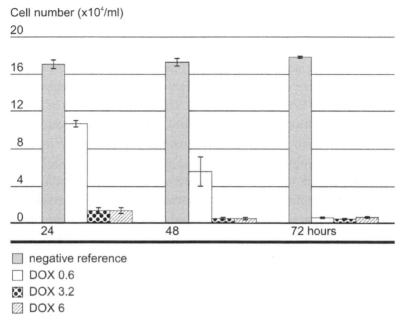

Cell number (x10⁴/ml)

■ negative reference
□ DOX 0.6
▨ DOX 3.2
▨ DOX 6

FIGURE 4.22 The effect of free DOX concentration on the number of viable cells in HeLa cell culture: negative control, drug-free culture (DOX concentration)—0.6, 3.2, 6 µg/mL (Shishatskaya's data).

PHA microparticles were preceded by the studies that proved high biological compatibility of PHA microparticles and showed that they could be administered to animals intramuscularly, intravenously, and intraperitoneally, causing no adverse effects (Volova et al., 2013; Volova and Shishatskaya, 2011).

Experiment was performed on laboratory animals (BALB/c mice, 20–23 g, from Krasfarma Breeding Center; 30 animals in the control and 30 in the treatment group) with transplanted Ehrlich ascites carcinoma (EAC), supplied by the Institute of Pharmacology, Tomsk Research Center, Siberian Branch of the Russian Academy of Medical Sciences (Shishatskaya et al., 2008). Microspheres with diameters of <2 µm, 2–10 µm, and >10 µm constituted 33.3±2.2, 38.0±1.9, and 29.0±2.3%, respectively. The largest shells were 35 µm in diameter, their percentage not exceeding 3–5%. Microspheres about 10 µm in diameter were taken for the experiment. Tumor cells were transplanted intraperitoneally to all animals in a dose of

3×10^6 cells/mouse in 0.2 mL saline. Animal morbidity and mortality after this EAC dose is 100%, the maximum life span after cell injection is <2–3 weeks (Emmanuel, 1977). The animals in the treatment group received 50 mg DOX (0.16 mg/animal, or 8 mg/kg) intraperitoneally simultaneously with EAC. One more group of animals received DOX in the same concentration 5 days after EAC transplantation. The antitumor effect of DOX was evaluated by reduction in animal mortality, tumor volume reduction (ml), and reduction in cell concentration in ascitic fluid (counted in Goryaev chamber), as the volume of ascitic fluid and proliferative pool of tumor suspension cells are the parameters determining tumor growth rate. The percentage of necrotic cells was evaluated by trypan blue staining. The microspheres prepared from the triple polymer emulsion were character-ized by a regular spherical shape and well-developed "wrinkled" porous surface; they had different diameters.

Seven days after EAC transplantation, the tumor volume in the control animals reached 0.62 ± 0.08 mL; in the treatment animals, it was lower by one order of magnitude (0.020 ± 0.004 mL; Table 4.7). Total cell counts

TABLE 4.7 Effect of DOX on EAC Development in Mice With Tumors (M±m) (Shishatskaya et al., 2008) (Reprinted with the permisison of the Journal "Bulletin of Experimental Biology and Medicine.")

Param-eter	Day of observation			
	7		14	
	Control	Treatment	Control	Treatment
Tumor volume, mL	0.62 ± 0.08	0.020 ± 0.004	5.26 ± 0.49	0
Count of tumor cells/mL	$(3.49 \pm 0.41) \times 10^8$	$(1.25 \pm 0.31) \times 10^8$	$(6.41 \pm 0.89) \times 10^8$	$0 \ (1.56 \pm 0.23) \times 10^8$
Count of necrotic cells per mL	$(0.68 \pm 0.01) \times 10^7$	$(0.10 \pm 0.01) \times 10^8$	$(0.32 \pm 0.09) \times 10^7$	$(0.07 \pm 0.01) \times 10^8$
% of total cell count	1.88 ± 0.42	3.04 ± 0.27	0.490 ± 0.004	$3.59 \pm 0.52*$

Note: *$p < 0.05$ compared to the control.

were comparable in the control and treatment groups: $(3.49\pm0.41)\times10^8$ and $(1.25\pm0.31)\times10^8$ cells/mL ascitic fluid, respectively, while the number of necrotic tumor cells was significantly higher in animals injected with the drug, 3.04% of total cell count *vs.* 1.88% in the control. Seven more days later, the antitumor effect of DOX was even more pronounced. Tumor volume in the control animals increased by one order of magnitude and reached 5.26 ± 0.49 mL, while treatment animals exhibited no signs of ascites. Therefore, in order to collect tumor cells, 1 mL saline was injected into the abdominal cavity of treatment animals. The total number of EAC cells increased in the control animals and was higher by a factor of 4 compared with the treatment animals. The number of necrotic cells in the control animals decreased in comparison with day 7 postinjection, while in treatment animals, this parameter increased. These results indicate inhibition of proliferative activity of EAC by DOX.

In the control group, mass dying of mice started at Day 14, and by Day 21, mortality had reached 100%. All dead animals had large ascitic tumors and died with symptoms characteristic of the late stage of the disease (dyspnea, poor mobility, food refusal). The average life span of mice with tumors in the control group was 8 days. The mortality curve of mice injected with DOX simultaneously with EAC differed significantly from the control. At Day 13 and later on, just a few deaths occurred in this group, and by Day 21, the survival of the mice in this group reached 80%, the mean life span for that period being 16 days (twice longer than in the control). Mortality rates in the treatment group decreased over 30 days, and the mortality curve looked smooth. No signs of EAC were observed in the 40% survivors on Day 55 (Figure 4.23).

Injection of DOX 5 days after EAC transplantation (not simultaneously with it) did not influence the development of tumor process and the shape of the mortality curve.

Mathematical processing of the data on mortality of mice with tumors by the least-squares method provided kinetic parameters characterizing antitumor activity of DOX. The kinetic coefficients of mortality rate (c) in control mice with tumors were 1.96 *vs.* 0.23 in the treatment group. The kinetic coefficient χ (the coefficient of superposition of kinetic curves, named the coefficient of inhibition of tumor development) directly shows how many times slower (in comparison with the control)

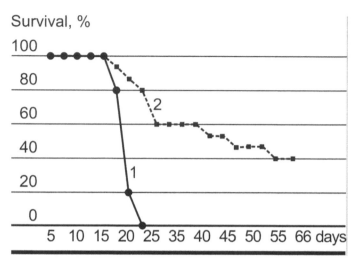

FIGURE 4.23 Survival of mice with transplanted EAC. (i) control group; (ii) treatment animals injected with a single dose of DOX (Shishatskaya et al., 2008). (Reprinted with the permisison of the Journal "Bulletin of Experimental Biology and Medicine.")

the tumor process develops in animals treated with antitumor drugs, and it is suggested for evaluating the antitumor drug activity. The kinetic coefficient indicates that DOX is characterized by pronounced antitumor activity and reduces mortality in animals with EAC transplanted in a lethal dose virtually by one order of magnitude in comparison with the control. Moreover, the encapsulated DOX dosage form can be injected locally, without causing adverse reactions, which are an obstacle to the use of free rubomycin.

Then, the efficacy of rubomycin encapsulated in polymer micropar-ticles was evaluated in experiments with mice inoculated with Ehrlich solid carcinoma (ESC) (Shishatskaya et al., 2012). The polymer of β-hydroxybutyric acid (poly-3-hydroxybutyrate), synthesized by the tech-nology developed at Institute of Biophysics, with molecular weight of 800 kDa, 75% crystallinity, and melting temperature of 168°C, served as the polymeric carrier. Anthracycline cytostatic doxorubicin (Lance-Pharm) was used in the study. Experimental microparticles loaded with doxorubi-cin (antibiotic content 5% of the polymeric matrix weight) were prepared by evaporation of the emulsion solvent. The average diameter of doxoru-bicin-loaded microparticles was 0.34±0.03 μm.

Cancerous process in the form of ESC (Institute of Pharmacology, Tomsk Research Center) was created in laboratory animals (adult BALB/c mice, 20 g, from Krasfarma Breeding Center). Before the experiment, the cells were passaged in the ascitic EC form. The resultant tumor cell culture was injected subcutaneously into the right hip in a dose of 2.5×10^6 cells per mouse in 0.2 mL saline. Day 4–7 after ESC cell injection (subcutaneous transplantation) is the optimal period for the beginning of drug therapy (Sofyina, 1979). Tumor size during this period (measured by the diameter of the hip at the site of injection) was about 8 mm. Seven days after ESC transplantation, the animals were divided into 5 groups, 8 per group. The initial diameter of the hip in animals of all groups before therapy was about the same (7.5 mm). Group 1 were intact animals, group 2 were animals with ESC without drug therapy (negative control). Group 3 animals received intravenous doxorubicin on Day 7 after ESC transplantation and then weekly (positive control). Animals of Groups 4 and 5 (treatment) were injected with one and two cycle doses of encapsulated doxorubicin (equivalent to 75 and 150 mg/m^2 of doxorubicin) in 0.2 mL saline, respectively, into the site of tumor palpation. Body surface area of adult mice was calculated from standard values. The experiment lasted 28 days, including 21 days of the treatment. The antitumor effect of experimental doxorubicin form was compared with the effect of intravenous injection of free doxorubicin and was evaluated by the state of animals, peripheral blood composition, and daily measurements of the hip diameter at the site of tumor transplantation (Wong et al., 2007). Microscopic examination of the tumor was carried out on histological sections over the course of experiment. The tumor area and structure were evaluated by tissue necrosis, capillary necrosis and plethora, and lymphocytic infiltration.

No adverse effect of injection of microparticles loaded with doxorubicin was recorded. Blood tests in Groups 2 and 3 showed no changes beyond the normal range of values or appreciable differences in comparison with intact animals. Group 3 animals developed a 2-fold reduction of the leukocyte count by the end of experiment. This was presumably caused by the antibiotic toxicity for hemopoietic organs. Measurements of the average diameter of the hip showed different dynamics of the parameter in different groups. The parameter increased significantly by day 28 after ESC transplantation in the negative control group (1.7 ± 0.01 cm), in comparison with 1.05 ± 0.03 cm in the positive control group. The values were higher in animals of Groups 4 and 5, receiving doxorubicin in microparticles: 1.48 ± 0.05 and 1.77 ± 0.06 cm, respectively, which was comparable to the values in animals with tumor process without therapy. However, higher values of the average diameter of the hip in animals injected with encapsulated doxorubicin in comparison with the hip size in animals injected with the free antibiotic could be explained by edema of muscle tissue because of its contact with doxorubicin released from the particles. Hence, in addition to measurements of the hip diameter at the site of tumor transplantation, the tumor was resected and weighed, its diameter was measured, and the tumor morphology was examined.

Formation of slit-like necrosis in tumor tissue was found in all groups after 1 week (Figure 4.24). In the positive control group, the necroses occupied large areas close to the central part of the tumor, while the tumor tissue was retained at the periphery, in zones of infiltrative growth. In treatment groups, the necroses were grouped in large fields in the central compartments, while smaller ones were chaotically scattered at the periphery. The area of necrotic zones in treatment and positive control groups was about 30% of tumor tissue area (Figure 4.25). Two weeks after the beginning of therapy the antitumor effect was significantly higher in the positive control group than in the treatment groups. However, after 3 weeks, the maximum tumor growth inhibition (area of necrotic zones reaching 78% of tumor tissue area) was recorded in Group 5 (treatment) (in response to double cycle dose of the drug). In Group 3 animals (intravenous drug weekly) and in Group 4 (a single cycle dose of the drug encapsulated in microparticles), the necrotic zones occupied 63–64% of tumor tissue area during this period; the values in these groups virtually did not differ. The necrotic zone area in the negative control group varied between 18 and 30% during the experiment.

As doxorubicin could cause irreversible cardiomyopathy and congestive heart failure, special histological studies of myocardial tissue were carried out. They included standard staining of heart preparations with hematoxylin and eosin and additional staining by Regaud's method, detecting the pathological changes in the myocardium. Histological studies of myocardial tissues of intact animals, animals from positive control group, and two treatment groups showed no changes in the myocardial structure.

The data were in good agreement with the results of Russian and foreign authors. The experimental form of nanosomal doxorubicin, loaded into

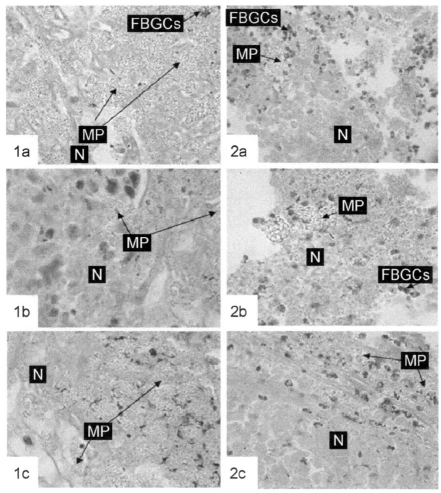

FIGURE 4.24 Morphological structure of tumor tissue: weeks 1 (a), 2 (b), and 3 (c) of treatment. N: necrosis; M: muscle; T: tumor tissue; FBGCs: foreign body giant cells; MP: microparticles. Hematoxylin and eosin staining (Shishatskaya et al., 2012). (Reprinted with the permisison of the Journal "Bulletin of Experimental Biology and Medicine.")

biodegraded polybutylcyanoacrylate particles, modified by polysorbate 80, led to tumor growth inhibition, reduced cell proliferation and density of vessels (Shishatskaya et al., 2005). Thermosensitive liposomes exhibited a pronounced cytotoxic effect *in vitro* and *in vivo* and selectively transported doxorubicin to the tumor (Ataullakhanov et al., 1993). Moreover, the developed doxorubicin formulations were characterized by a lower cardiotoxicity in comparison with the standard drug (Husseini et al., 2002).

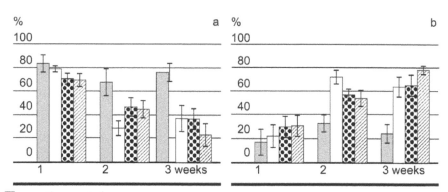

FIGURE 4.25 Areas of tumor tissue (a) and necrosis (b) in sections. Light bars: Group 2 (negative control); vertically hatched bars: Group 3 (positive control); dark bars: Group 4 (treatment); horizontally hatched bars: Group 5 (treatment) (Shishatskaya et al., 2012). (Reprinted with the permisison of the Journal "Bulletin of Experimental Biology and Medicine.")

   Thus, these experiments with the models of ascites and solid EC showed that the doxorubicin encapsulated in microparticles prepared from resorbable poly-3-hydroxybutyrate is a good candidate for local administration to the tumor development site. The inhibitory effect of a single or double administration of the encapsulated drug is similar to that of free doxorubicin injected intravenously every week, but doxorubicin-loaded microparticles do not have any adverse effects on the blood system.

KEYWORDS

- **embedding of drugs**
- **polymer matrix**
- **microparticles**
- **drug release kinetics**
- **drug administration in vivo**
- **tumor growth arrest**

REFERENCES

Amara, A. Polyhydroxyalkanoates: from basic research and molecular biology to application. IUM Engineering Journal, 2008, 9, 37–72.

Amass, W., Amass, A., Tighe, B. A Review of Biodegradale Polymers: Uses, Current Developments in the Synthesis and Characterization of Biodegradable Polyesters, Blends of Biodegradable Polymers and Recent Advances in Biodegradation Studies. Polymer Int., 1998, 47, 89–144.

Ambruosi, A., Gelperina, S., Khalansky, A., Tanski, S., Theisen, A., Kreuter, J. Influence of surfactants, polymer and doxorubicin loading on the anti-tumor effect of poly(butyl cyanoacrylate) nanoparticles in a rat glioma model. J. Microencapsul., 2006, 23, 582–592.

Amirah, M. G., Amirul, A. A., Wahab, H. A. Formulation and characterization of rifampicin-loaded P(3HB-co-4HB) nanoparticles. Int. J. Pharm. Pharm. Sci., 2014, 6, 141–146.

Anderson, J., Shive, M. Biodegradation and biocompatibility of PLA and PLGA microspheres. Adv. Drug. Deliv. Rev., 1997, 28, 5–24.

Arpagaus, C., Schafroth, N. Spray drying of biodegradable polymers in laboratory scale. Book of abstractsof XVI International conference of bioencapsulation. Ireland, Dublin, 2008, 1–4.

Ataullakhanov, F. I., Vitvitsky, V. M., Batasheva, T. V., Komarova, S. V. The effect of glutaric aldehyde treatment of murine erythrocytes loaded with rubomycin on rubomycin and hemoglobin release. Biotekhnologiya (Biotechnology), 1993, 2, 40–43 (in Russian).

Bae, S., Son, J., Park, K., Han, D. K. Fabrication of covered porous PLGA microspheres using hydrogen peroxide for controlled drug delivery and regenerative medicine. J. Control. Release, 2009, 133, 34–43.

Balthasar, S., Michaelis, K., Dinauer, N., Van Briesen, H., Kreuter, J., Langer, K. Preparation and characterization of antibody modified gelatin nanoparticles as drug carrier system for uptake in lymphocytes. Biomaterials, 2005, 26, 2723–2732.

Barbosa-Cánovas, G., Ortega-Rivas, E., Juliano, P., Yan, H. Encapsulation Processes. Food Powders: Physical Properties, Processing, and Functionality. Kluwer Academic Publishers. NY, 2005, 199–218.

Bazzo, G., Lemous-Senna, E., GoncalvesM., Pires, A. T. N. Effect of preparation conditions onmorphology, drug content and release profiles of poly(hydroxybutyrate) microparticles containing piroxicam. J. Braz. Chem. Soc., 2008, 19, 914–921.

Berchane, N., Jebrail, F., Carson, K., Rice-Ficht, A. C., Andrews, M. J. About mean diameter and size distributions of poly(lactide-co-glycolide) (PLG) microspheres. J. Microencapsul., 2006, 23, 539–5552.

Bissery, M. C., Puisieux, F., Thies, C. A study of process parameters in the making of microspheres by the solvent evaporation procedure. Third Exp. Cong. Int. Technol. Pharm., 1983, 3, 233–239.

Bordes, P., Pollet, E., Averous, L. Nano-biocomposites: Biodegradable polyester/nanoclay systems. Prog. Polym. Sci., 2009, 34, 125–155.

Breitenbach, J. Melt extrusion: from progress to drug delivery technology. Eur. J. Pharm. Biopharm., 2002, 54, 107–117.

Cai, J., Guo, J., Ji, M. Preparation and characterization of multiresponsive polymer composite microspheres with core-shell structure. Colloid. Polym. Sci., 2007, 285, 1607–1615.

Cai, S., Thati, S., Bagby, T. R., Diab, H. M., Davies, N. M., Cohen, M. S., Forrest, M. L. Localized doxorubicin chemotherapy with a biopolymeric nanocarrier improves survival and reduces toxicity in xenografts of human breast cancer. J. Control. Release, 2010, 146, 212–218.

Champion, J., Katare, Y., Mitragotri, S. Particle shape: A new design parameter for micro- and nanoscale drug delivery carriers. J. Control. Release, 2007, 121, 3–9.

Charoenphol, P., Huang, R., Eniola-Adefeso, O. Potential role of size and hemodynamics in the efficacy of vascular-targeted spherical drug carriers. Biomaterials, 2010, 31, 1392–1402.

Cheng, J., Teply, B., Shrifi, I., Sung, J., Luther, G., Gu, F. X., Levy-Nissenbaum, E., Radovic-Moreno, A. F., Langer, R., Farokhzad, O. C. Formulation of functionalized PLGA-PEG nanoparticles for in vivo targeted drug delivery. Biomaterials, 2007, 28, 869–876.

Choi, C., Jung, J. H., Rhee, Y. W., Kim, D. P., Shim, S. E., Lee, C. S. Generation monodisperse alginate microbeads and in situ encapsulation of cell in microfluidic device. Biomed. Microdevices, 2007, 9, 855–862.

Chokshi, R., Zia, H. Hot-melt extrusion technique: a review. Iran. J. Pharm. Res., 2004, 3, 3–16.

Cleland, J. L., Duenas, E. T., Park, A., Daugherty, A., Kahn, J., Kowalski, J., Cuthbertson, A. Development of poly-(D, L-lactide-coglycolide) microsphere formulations containing recombinant human vascular endothelial growth factor to promote local angiogenesis. J. Control. Release, 2001, 72, 13–24.

Cohn, D., Sagiv, H., Benyamin, A., Lando, G. Engineering thermoresponsive polymeric nanoshells. Biomaterials, 2009, 30, 3289–3296.

Conway, B. R., Eyles, J. E., Alpar, H. O. Immune response to antigen inmicrospheres of different polymers. Proc. Int. Symp. Control. Release Bioact. Mater., 1996, 23, 335–336.

Danhier, F., Lecouturier, N., Vroman, B. Paclitacxel-loaded PEgylated PLGA-based nanoparticles: In vitro and in vivo evaluation. J. Control. Release, 2009, 133, 11–17.

Dijkhuizen-Radersma, R., Hesseling, S. C., Kaim, P. E., de Groot, K., Bezemer, J. M. Biocompatibility and degradation of poly(ether-ester) microspheres: in vitro and in vivo evaluation. Biomaterials, 2002, 23, 4719–4729.

Duran, N., Alvarenga, M., Silva, E., Melo, P. S., Maecato, P. D. Microencapsulation of antibiotic rifampicin in poly(3-hydroxybutyrate-co-3-hydroxyvalerate). Arch. Pharm. Res., 2008, 31, 1509–1516.

Dutta, R. C. Drug carriers in pharmaceutical design: promises and progress. Curr. Pharm. Des., 2007, 13, 761–769.

Eke, G., Kuzmina, A. M., Goreva, A. V., Shishatskaya, E. I., Hasirci, N., Hasirci, V. In vitro and Transdermal Penetration of PHBV Micro/Nanoparticles. J. Mater. Sci. Mater. M., 2014, 25, 1471–1481.

Eldridge, J. H., Hammond, C. J., Meulhrock, J. A. Controlled vaccine release in the gut-associated lymphoid tissues. I. Orally administered biodegradable microspheres target the patches. J. Control. Release, 1990, 11, 209–214.

El-Rehim, H. A. A. Properties and biotic hydrolysis of radiation cross-linked poly(e-caprolactone). Nucl. Instrum. Meth. B, 2005, 229, 293–301.

Embelton, J. K., Tighe, B. J. Polymers for biodegradable medical devices. X: Microincapsulation studies: control of polyhydroxybutyrate-hydroxyvalerate microcapsule porosity via polycaprolactone blending. Biomaterials, 1993, 10, 341–352.

Experimentalnaya otsenka protivoopukholevykh preparatov v SSSR i SShA (Experimental evaluation of anticancer drugs in the USSR and the USA). In: Z. P. Sofyina, ed. Moscow, 1979, 296 p. (in Russian).

Fournier, E., Passirani, C., Colin, N., Sagodira, S., Menei, P., Benoit, J. P., Montero-Menei, C. N. The brain tissue response to biodegradable poly(methylidene malonate 2.1.2)-based microspheres in the rat. Biomaterials, 2006, 27, 4963–4974.

Freiberg, S., Zhu, X. Polymer microspheres for controlled drug release. Int. J. Pharm., 2004, 282, 1–18.

Freitas, S., Merkle, H., Gander, B. Microencapsulation by solvent extraction/evaporation: reviewing the state of the art of microsphere preparation process technology. J. Control. Release, 2005, 102, 313–332.

Furgenson, D., Dreher, M., Chilkoti, A. Structural optimization of a "smart" doxorubicin-polypeptide conjugate for thermally targeted delivery to solid tumors. J. Control. Release, 2006, 110, 362–369.

Giovagnoli, S., Blasi, P., Schoubben, A., Rossi, C., Ricci, M. Preparation of large porous biodegradable microspheres by using a simple double-emulsion method for capreomycin sulfate pulmonary delivery. Int. J. Pharm., 2007, 333, 103–111.

Goepferich, A., Tessmar, J. Polyanhydride degradation and erosion. Adv. Drug Deliv. Rev., 2002, 54, 911–931.

Goforth, R., Salem, A., Zhu, X. Miles, S., Zhang, X. Q., Lee, J. H., Sandler, A. D. Immune stimulatory antigen loaded particles combined with depletion of regulatory T-cells induce potent tumor specific immunity in a mouse model of melanoma. Cancer Immunol. Immunother., 2009, 58, 517–530.

Gogolewski, S., Javanovic, M., Perren, S., Dillon, J. G., Hughes, M. K. Tissue response and in vivo degradation of selected polyhydroxyacids: Polylactides PLA), poly(3-hydroxybutyrate) (PHB), and poly(3-hydroxybutyrate-co-3-hydroxyvalerates (PHB/PHV). J. Biomed. Mater. Res., 1993, 27, 1135–1148.

Goreva, A. V., Shishatskaya, E. I., Volova, T. G., Sinskey, A. J. Characterization of polymer microparticles based on resorbable polymers of hydroxyalkanoic acids as a platform for drug delivery. Vysokomolekulyarnyye soedineniya (High-molecular-weight compounds), Series A, 2012, 54, 224–236 (in Russian).

Goreva, A. V., Shishatskaya, E. I., Kuzmina, A. M., Volova, T. G., Sinskey, A. J. Microparticles prepared from biodegradable polyhydroxyalkanoates as matrix for encapsulation of cytostatic drug. J. Mater. Sci.: Mater. Med., 2013, 24, 1905–1915.

Greisha, K., Sawa, T. SMA-doxorubicin, a new polymeric micellar drug for effective targeting to solid tumors. J. Control. Release, 2004, 97, 219–230.

Heller, J., Barr, J., Ng, S. Y. Poly(ortho esters) synthesis, characterization, properties and uses. Adv. Drug Deliv. Rev., 2002, 54, 1015–1039.

Huo, D., Deng, S., Li, L., Ji, J. Studies of poly(lactic-*co*-glicolic) acid microspheres of cisplatin for lung-targeting. Int. J. Pharm., 2005, 289, 63–67.

Husseini, G. A., Rapoport, N. Y., Christensen, D. A., Pruitt, J. D., Pitt, W. G. Kinetics of ultrasonic release of doxorubicin from pluronic P105 micelles. Colloid. Surface. B., 2002, 24, 253–264.

Jain, J., Modi, S., Domb, A., Kumar, N. Role of polyanhydrides as localized drug carriers. J. Control. Release, 2005, 103, 541–563.

Jain, K. Drug Delivery Systems – An Overview. In: K. Jain. Drug Delivery Systems. Humana Press, USA, 2008, 1–50.

Jaworek, A. Electrostatic micro- and nanoencapsulation and electroemulsification: A brief review. J. Microencapsul., 2008, 25, 443–468.

Kanazler, J., Barry, J., Ginty, P. Supercritical carbone dioxide generated vascular endothelial growth factor encapsulated poly(DL-lactic acid) scaffolds induce angiogenesis *in vitro*. Biochem. Biophys. Res. Commun., 2007, 352, 135–141.

Kao, W. J., Zhao, Q. H., Hiltner, A., Anderson, J. M. Theoretical analysis of *in vivo* macrophage adhesion and foreign body giant cells formation on polydimethylsiloxane, low density polyethylene and polyetherurethane. J. Biomed. Mater. Res, 1994, 28, 73–79.

Karatas, A., Sonakin, O., Kilicarslan, M. Poly(e-caprolactone) microspheres containing Levobunolol HCl prepared by a multiple emulsion (W/O/W) solvent evaporation technique: Effect of some formulation parameters on microparticle characteristics. J. Microencapsul., 2009, 26, 63–74.

Khalansky, A. S., Hekmatara, T., Bernreuther, C., Rubtsov, B. V., Kondakova, L. I., Matschke, J., Kreuter, J., Glatzel, M., Gelperina, S., Shvets, V. I. Morphological evaluation of antitumor effect of nanosomal doxorubicin formulation toward the experimental glioblastoma in rats. Biofarmatsevticheskiy zhurnal (Biopharmaceutical journal), 2011, 3(2), 41–50 (in Russian).

Kim, J., Kwak, B., Shim, H. Preparation of doxorubicin-containing chitosan microspheres for transcatheter arterial chemoembolization of hepatocellular carcinoma. J. Microencapsul., 2007, 24, 408–419.

Kim, K. Pack, D. Microspheres for Drug Delivery. In: M. Ferrari. BioMEMS and Biomedical Nanotechnology. Springer Science and Business Media. NY, 2006, 19–50.

Kim, T. H., Lee, H., Park, T. Regylated recombinant human epidermal growth factor (rhEGF) for sustained release from biodegradable PLGA microspheres. Biomaterials, 2002, 23, 2311–2317.

Kingsley, J., Dou, H., Morehesd, J., Rabinow, B., Gendelman, H. E., Destache, C. J. Nanotechnology: a focus on nanoparticles as a drug delivery system. J. Neuroimmune. Pharmacol., 2006, 1, 340–350.

Kong, H., Lee, Y., Hong, S., Han, J., Choi, B., Jung, Y., Kim, Y. M. Sulfate-conjugated methylprednisolone as a colontargeted methylprednisolone prodrug with improved therapeutic properties against rat colitis. J. Drug Target., 2009, 17, 450–458.

Kostanski, J. W., Thanoo, B. C., DeLuca, P. P. Preparation, characterization, and *in vitro* evaluation of 1- and 4-month controlled release orntide PLA and PLGA microspheres. Pharm. Dev. Technol., 2000, 5, 585–596.

Kovačević, A. B., Müller, R. H., Savić, S. D., Vuleta, G. M., Keck, C. M. Solid lipid nanoparticles (SLN) stabilized with polyhydroxysurfactants: Preparation, characterization and physical stability investigation. Colloid. Surface. A, 2014, 444, 15–25.

Kuznetsova, I. G., Severin, S. E. Using a copolymer of lactic and glycolic acids to prepare nano-sized drug formulations. Farmatsevticheskaya tekhnologiya (Pharmaceutical technology), 2013, 5(4), 30–36 (in Russian).

Lassalle, V., Ferreira, M. PLA Nano- and Microparticles for Drug Delivery: An Overview of the Methods of Preparation. Macromol. Biosci., 2007, 7, 767–783.

Lee, E. S., Na, K., Bae, Y. H. Doxorubicin loaded pH-sensitive polymeric micelles for reversal of resistant MCF-7 tumor. J. Control. Release, 2005, 103, 405–488.

Lin, R., Ng, L. S., Wang, C. *In vitro* study of anticancer drug doxorubicin in PLGA-based microparticles. Biomaterials, 2005, 26, 4476–4485.

Liu, S. J., Tsai, Y. E., Ueng, S. A novel solvent-free method for the manufacture of biodegradable antibiotic-capsules for a long-term drug release using compression sintering and ultrasonic welding techniques. Biomaterials, 2005, 26, 4662–4669.

Marcucci, F., Lefoulon, F. Active targeting with particulate drug carriers in tumor therapy: fundamentals and recent progress. Drug Discov. Today, 2004, 9, 219–228.

Medvecky, L., Stulajterova, R., Briancin, B. Study of Controlled Tetracycline Release from Porous Calcium Phosphate/Polyhydroxybutyrate Composites. Chem. Pap., 2007, 61, 477–484.

Medvedeva, N. V., Ipatova, O. M., Ivanov, Y. D., Drozhzhin, A. I., Archakov, A. I. Nanobiotechnologies and nanobiomedicine. Biomeditsinskaya khimiya (Biomedical chemistry), 2006, 52(6), 529–546 (in Russian).

Morello, A., Forbes, N., Mathiowitz, E. Investigating the effects of surfactants on the size and hydrolytic stability of poly(adipic anhydride) particles. J. Microencapsul., 2007, 24, 40–56.

Mrkvan, T., Sirova, M., Etrych, T. Chemotherapy based on HPMA copolymer conjugates with pH-controlled release doxorubicin triggers anti-tumor immunity. J. Control. Release, 2005, 64, 241–246.

Murueva, A. V., Shishatskaya, E. I., Kuzmina, A. M., Volova, T. G., Sinskey, A. J. Microparticles prepared from biodegradable polyhydroxyalkanoatesas matrix for encapsulation of cytostatic drug. J. Mater. Sci. Mater. Med., 2013, 24, 1905–1915.

Murueva, A. V., Shershneva, A. M., Shishatskaya, E. I., Volova, T. G. The Use of Polymeric Microcarries Loaded with Anti-Inflammatory Substances in the Therapy of Experimantal Skin Wounds. Bulletin of Experimental Biology and Medicine, 2014, 157(5), 614–619 (in Russian).

Müller, R. H. Zeta potential und Partikelladung – Kurze Theorie, praktische Me durchfuhrung, Dateninterpretation, WissenschaftlicheVerlagsge-sellschaft. Stuttgart, 1996, 26 p.

Naha, P., Kanchan, V., MannaP., Panda, A. K. Improved bioavailability of orally delivered insulin using Eudragitt-L30D coated PLGA microparticles. J. Microencapsul., 2008, 25, 248–256.

Nair, A., Sasidharan, Rani, V. D., Menon, D., Nair, S., Manzoor, K., Raina, S. Role of size scale of ZnO nanoparticles and microparticles on toxicity toward bacteria and osteoblast cancer cells. J. Mater. Sci: Mater Med. Suppl. 1, 2009, 20, 235–241.

Nair, L., Laurencin, C. Biodegradable polymers as biomaterials. Prog. Polym. Sci., 2007, 32, 762–798.

Nakagava, K., Iwamotoa, S., Nakajima, M., Shono, A., Satoh, K. Microchanel emulsification using gelatin and surface-free coacervate microencapsulation. J.Colloid Interface Sci., 2004, 278, 198–205.

Napierska, D., Thomassen, L., Rabolli, V., Lison, D., Gonzalez, L., Kisch-Volders, M., Martens, J. A., Hoet, P. H. Size-Dependent Cytotoxicity of Monodisperse Silica Nanoparticles in Human Endothelial Cells. Small, 2009, 5, 846–853.

Ni, H., Kawaguchi, H., Endo, T. Preparation of pH-sensitive hydrogel microspheres of poly(acrylamide-co-methacrylic acid) with sharp pH-volume transition. Colloid. Polym. Sci., 2007, 285, 819–826.

Olkhov, A., Iordansky, A., Kosenko, R. Diffusion properties of novel biomedical materials based on poly(3-hydroxybutyrate) and cellulose. Plasticheskiye massy (Plastics), 2008, 11, 44–48 (in Russian).

Panarin, E. F., Lavrov, N. A., Solovsky, M. V., Shalnova, L. I. Polimery – nositeli biologicheski aktivnykh veshchestv (Polymers as Carriers of Biologically Active Substances). In: E. F. Panarin, N. A. Lavrov. Professiya Publishers. St. Petersburg, 2014, 304 p. (in Russian).

Panyam, J., Labhasetwar, V. Biodegradable nanoparticles for drug and gene delivery to cells and tissue. Adv. Drug Deliver. Rev., 2012, 64, 61–71.

Park, E. K., Lee, S., Lee, Y. Preparation and characterization of methoxy poly(ethylene glycol)/poly(e-caprolactone) amphiphilic block copolimeric nanospheres for tumor-specific folate-mediated targeting of anticancer drugs. Biomaterials, 2005, 26, 1053–1061.

Park, J., Fong, P., Lu, J., Russell, K. S., Booth, C. J., Saltzman, W. M., Fahmy, T. M. PEGylated PLGA nanoparticles for the improved delivery of doxorubicin. Nanomedicine, 2009, 5, 410–418.

Parra, D. F., Fusaro, J., Gaboardi, F., Rosa, D. S. Influence of poly (ethylene glycol) on the thermal, mechanical, morphological, physical-chemical and biodegradation properties of poly (3-hydroxybutyrate). Polym. Degrad. Stabil., 2006, 91, 1954–1959.

Piddubnyak, V., Kurcok, P., Matuszowicz, A. Oligo-3-hydroxybutyrates as potential carriers for drug delivery. Biomaterials, 2004, 25, 5271–5279.

Poletto, F., Jager, E., Re, M., Guterres, S. S., Pohlmann, A. R. Rate-modulating PHBV/PCL microparticles containing weak acid model drugs. Int. J. Pharm., 2007, 345, 70–80.

Poncelet, D. Microencapsulation: Fundamentals, methods and applications. In: J. Blitz. Surface Chemistry in Biomedical and Environmental Science. Springer: The Netherlands, 2005, 23–34.

Pouton, C. W., Akhtar, S. Biosynthetic polyhydroxyalkanoates and their potential in drug delivery. Adv. Drug Deliver. Rev., 1996, 18, 133–162.

Qu, X., Wu, Q., ZhangK. In vivo studies of poly(3-hydroxybutyrate-co-3hydroxyhexanoate) based polymers: Biodegradation and tissue reactions. Biomaterials, 2006, 27, 3540–3548.

Regnier-Delplaceb, C., Thillaye du Boullay, O., Siepmann, F., Martin-Vaca, B., Demonchaux, P., Jentzer, O., Danède, F., Descamps, M., Siepmann, J., Bourissou, D. PLGAs bearing carboxylated side chains: Novel matrix formers with improved properties for controlled drug delivery. J. Control. Release, 2013, 166, 256–267.

Repka, M. A., Prodduturi, S., Stodghill, S. P. Production and characterization of hot-melt extruded films containing clotrimazole. Drug Dev. Ind. Pharm., 2003, 29, 757–765.

Rhee, H. J., Birgh-De Winter, S. D., Dafms, W. H. The differentiation of monocyte into macrophages, epithelioid cells in subcutaneous granulomass. Cell Tis. Res., 1979, 198, 355–378.

Saatchi, K., Häfeli, U. O. Radiolabeling of biodegradable polymeric microspheres with (99mTc(CO)3)+ and *in vivo* biodistribution evaluation using MicroSPECT/CT imaging. Bioconjug. Chem., 2009, 20, 1209–1217.

Sardushkin, M. V., Kienskaya, K. I., Il'yushenko, E. V., Avramenko, G. V. Fabrication of rifampicin microcapsules with a polylactide shell. Russ. J. Appl. Chem., 2013, 86(5), 782–786.

Semmler-Behnke, M., Kreyling, W., Lipka, J., Fertsch, S., Wenk, A., Takenaka, S., Schmid, G., Brandau, W. Biodistribution of 1.4 – and 18-nm gold particles in rats. Small, 2008, 4, 2108–2111.

Shershneva, A. M., Murueva, A. V., Shishatskaya, E. I., Volova, T. G. A study of the electrokinetic potential of drug microcarriers of resorbable polymers Bioplastotan. Biofizika (Biophysics), 2014, 59(4), 684–691 (in Russian).

Shishatskaya, E. I. Biomedical investigation, application of PHA. Macromol. Symposia, 2008, 269, 65–81.

Shishatskaya, E. I. Contemporary reconstruction technologies in medicine: the contribution of natural polymers. Izvestiya VUZov "Fizika" (Bulletin of universities "Physics"), 2013, 56(13/3), 58–63 (in Russian).

Shishatskaya, E. I., Goreva, A. V. Microparticles from biodegradable poly(3-hydroxybutyrate) as a matrix for rubomycin encapsulation. Perspektivnyye materialy (Advanced materials), 2006, 4, 65–70 (in Russian).

Shishatskaya, E. I., Volova, T. G., Efremov, S. N., Puzyr, A. P., Mogilnaya, O. A., Efremov, S. N. Tissue response to the implantation of biodegradable polyhydroxyalkanoate sutures. J. Mater. Sci. Mater. Med., 2004, 15, 719–728.

Shishatskaya, E. I., Zhemchugova, A. V., Volova, T. G. A study of biodegradable polyhydroxyalkanoates as anticancer drug carriers. Antibiotiki i khimioterapiya (Antibiotics and Chemotherapy), 2005, 2–3, 4–13 (in Russian).

Shishatskaya, E. I., Goreva, A. V., Voinova, O. N., Volova, T. G. Tissue response to intramuscular implantation of microparticles prepared from resorbable polymers. Byulleten experimentalnoi biologii i meditsiny (Newsletter of Experimental Biology and Medicine), 2007a, 12, 635–639 (in Russian).

Shishatskaya, E. I., Goreva, A. V., Voinova, O. N., Volova, T. G.. Resorbable polyhydroxyalkanoates as a carrier of anti-tumor drugs. J. Biotechnol., 2007, 131, 50.

Shishatskaya, E. I., Goreva, A. V., Inzhevatkin, E. V., Voinova, O. N., Khebopros, R. G., Volova, T. G. Anticancer effectiveness of rubomycin encapsulated in resorbable polymer matrix. Byulleten experimentalnoi biologii i meditsiny (Newsletter of Experimental Biology and Medicine), 2008, 3, 333–336 (in Russian).

Shishatskaya, E. I., Voinova, O. N., Goreva, A. V., Mogilnaya, O. A., Volova, T. G. Biocompatability of polyhydroxybutyrate Microspheres: in vitro and in vivo evaluation. J. of Siberial Federal University. Biology., 2008 b,1, 66–77 (in Russian).

Shishatskaya, E. I., Goreva, A. V., Kalacheva, G. S., Volova, T. G. Distribution and resorption of intravenously administered polymer microparticles in tissues of internal organs of laboratory animals. J. Siberial Federal University. Biology, 2009, 2, 453–465 (in Russian).

Shishatskaya, E. I., Goreva, A. V., Kalacheva, G. S., Volova, T. G. Distribution and resorption of intravenously administered polymer microparticles in tissues of internal organs of laboratory animals. Byulleten experimentalnoi biologii i meditsiny (Newsletter of Experimental Biology and Medicine), 2009, 11, 542–546 (in Russian).

Shishatskaya, E. I., Voinova, O. N., Goreva, A. V., Mogilnaya, O. A., Volova, T. G. Biocompatability of polyhydroxybutyrate Microspheres: in vitro and in vivo evaluation. J. Mat. Sci. Mat. Med., 2008, 19, 2493–2502.

Shishatskaya, E. I., Goreva, A. V., Kalacheva, G. S., Volova, T. G. Distribution and resorption of intravenously administered polymer microparticles in tissues of internal organs of laboratory animals. Byulleten experimentalnoi biologii i meditsiny (Newsletter of Experimental Biology and Medicine), 2009, 11, 542–546 (in Russian).

Shishatskaya, E., Goreva, A., Kalacheva, G., Volova, T. G. Biocompatibility and resorption of intravenously administered polymer microparticles in tissues of internal organs of laboratory animals. J. Biomater. Sci. Polym. Edn., 2011, 22, 2185–2203.

Shishatskaya, E. I., Goreva, A. V., Kuzmina, A. M. A study of medicinal effectiveness of doxorubicin loaded into microparticles prepared from resorbable bioplastotan in experiments with laboratory animals with solid Ehrlich carcinoma. Byulleten experimentalnoi biologii i meditsiny (Newsletter of Experimental Biology and Medicine), 2012, 154, 741–745 (in Russian).

Sokolovsky-Parkov, M., Agashi, K., Olaye, A. Polymer cariers for drug delivery in tissue engineering. Adv. Drug Deliver. Rev., 2007, 59, 187–206.

Sudesh, K., Abe, H., Doi, Y. Synthesis, structure and properties of polyhydroxyalkanoates: biological polyesters. Prog. Polym. Sci., 2000, 25, 1503–1555.

Tazina, E. V., Oborotova, N. A. Selective drug delivery to the tumor using thermosensitive liposomes and local hypothermia. Rossiiskiy bioterapevticheskiy zhurnal (Russian biotreatment journal), 2008, 3, 4–12 (in Russian).

Torchilin, V. Multifunctional nanocarriers. Adv. Drug Deliver. Rev., 2006, 58, 1532–1555.

Treshchalin, I. D., Pereverzeva, E. R., Bodyagin, D. A. A comparative experimental toxicological study of doxorubicin and its nanosomal formulations. Rossiiskiy Bioterapevticheskiy Zhurnal (Russian Biotreatment Journal), 2008, 3, 24–32 (in Russian).

Vasir, J., Labhasetwar, V. Biodegradable nanoparticles for cytosolic delivery of therapeutics. Adv. Drug Deliver. Rev., 2007, 59, 718–728.

Vert, M., Li, S., Garreau, H. More about the degradation of LA/GA-derived matrices in aqueous media. J. Control. Release, 1991, 16, 15–26.

Volova, T. G., Shishatskaya, E. I. Razrushayemyye biopolimery: polucheniye, svoistva, primeneniye (Degradable biopolymers: production, properties, applications). Krasnoyarskii Pisatel. Krasnoyarsk, 2011, 392 p. (in Russian).

Volova, T. G., Shishatskaya, E. I., Sinskey, A. J. Degradable Polymers: Production, Properties and Applications. Nova Science Pub., Inc. NY, USA, 2013, 380 p.

Volova, T. G., Kalacheva, G. S., Steinbuchel, A. Biosinthesis multi-component polyhydroxyalkanoates by the bacterium *Wautersia eutropha.* Macromol. Symposia, 2008, 269, 1–7.

Volova, T., Zhila, N., Kalacheva, G., Brigham, C., Sinskey, A. J. The effects of the intracellular poly(3-hydroxybutyrate) reserves on physiological-biochemical properties and growth of *Ralstonia eutropha.* Res. Microbiol., 2013, 164, 164–171.

Volova, T. G., Kiselev, E. G., Vinogradova, O. N. Nikolaeva, E. D., Chistyakov, A. A., Sukovatyi, A. G., Shishatskaya, E. I. A glucose-utilizing strain, Cupriavidus eutrophus B-10646: growth kinetics, characterization and synthesis of multicomponentPHAs. Plos One, 2014, 9, 1–15.

Wang, Y., Wang, X., Wie, K., Zhao, N., Zhang, S., Chen, J. Fabrication, characterization and long-term *in vitro* release of hydrophilic drug using PHBV/HA composite microspheres. Mater. Lett., 2007, 61, 1071–1076.

Williams, S. F., Martin, D. P. Applications of PHAs in Medicine and Pharmacy. In: A. Steinbüchel. Series of Biopolymers in 10 vol. Willey-VCY Verlag GmbH, 2002, 4, 91–121.

Wu, Q., Wang, Y., Chen, G. Medical application of microbal biopolyesters polyhydroxyalkanoates. Artif. Cell Blood Sub., 2009, 37, 1–12.

Yakovlev, S. G. Biologicheski aktivnyye mikro- i nanochastitsy iz poli(3-oxibutirata), ego sopolimerov i kompozitov (Biologically active micro- and nanoparticles of poly(3-hydroxybutyrate), its copolymers and composites). PhD thesis in biology. Specialty 03.01.06: Biotechnology. Moscow, A. N. Bakh Institute of Biochemistry RAS, 2013, 182 p. (in Russian).

Yang, Y., Chung, T., Ng, N. Morphology, drug distribution, and *in vitro* release profiles of biodegradable polymeric microspheres containing protein fabricated by double-emulsion solvent extraction/evaporation method. Biomaterials, 2001, 22, 231–241.

Yen, H., HuangY. Injectable biodegradable polymeric implants for the prevention of post-operative infection. Am. J. Drug Deliv., 2003, 1, 1–8.

Yun, Y. H., Goetz, D. J., Yellen, P. Hyaluronan microspheres for sustained gene delivery and site-specific targeting. Biomaterials, 2004, 25, 147–157.

Zensi, A., Begley, D., Pontikis, C., Legros, C., Mihoreanu, L., Wagner, S., Büchel, C., von Briesen, H., Kreuter, J. Albumin nanoparticles targeted with Apo E enter the CNS by transcytosis and are delivered to neurons. J. Control. Release, 2009, 137, 78–86.

Zhang, H., Bei, J., Wang, S. Multi-morphological biodegradable PLGE nanoparticles and their drug release behavior. Biomaterials, 2009, 30, 100–107.

Zhang, J., Li, X., Zhang, D., Xiu, Z. Theoretical and experimental investigations on the size of alginate microspheres prepared by dropping and spraying. J. Microencapsul., 2007, 24, 303–322.

Zhao, J., Liu, C., Tao, X., Shan, X. Q., Sheng, Y., Wu, F. Preparation of hemoglobin-loaded nano-sized particles with porous structure as oxygen carriers. Biomaterials, 2007, 28, 1414–1422.

PART III

IMPLANTS AND CELL GRAFTS OF PHAs FOR TISSUE REGENERATION

CHAPTER 5

POTENTIALS OF POLYHYDROXYALKANOATES FOR REPAIR OF SKIN DEFECTS

CONTENTS

5.1 INTRODUCTION

Reconstructive medicine needs novel materials for effective regeneration of skin injured due to burns, traumas, and surgical interventions. There are hundreds of surgical and therapeutic devices for covering and healing skin defects and a great number of materials and drugs used to prepare these devices (Minchenko, 2003; Abaev, 2006). Principles and methods of replacing the integrity of the skin are determined by such factors as the depth and severity of the injury, the phase of the wound healing process, the wound location, the degree of microbial invasion, the patient's other diseases, and the drugs taken by the patient (Struchkov, 1975; Tumanov and German, 2000). The basic principle of wound therapy is wound debridement and creation of the optimal conditions for wound healing.

For many centuries, wound dressings have been used to stop bleeding and protect the wound against microbial invasion. In practical surgery, treatment of infected wounds under dressings remains the major clinical approach, as it is very convenient and economical. Disruption of the skin's integrity causes massive fluid loss: loss by evaporation may reach $60-100$ mg/cm^2/h, which is comparable with the evaporation rate of water from the open surface. One of the most important requirements for wound dressings, especially ones used to cover vast wounds, is their ability to limit exudate evaporation. The rate of fluid evaporation through the bandage must be higher than through the intact skin but lower than through the eschar, about 1400 g/m^2/d (Tumanov and German, 2000). The wound dressing must effectively remove excessive wound exudate and its toxic components, enable adequate gas exchange between the wound and the air, prevent the wound from secondary infection and avoid contamination of the environment, wet the wound surface adequately, have anti-adhesive properties, and be sufficiently strong (Abaev, 2006).

Materials for wound dressings must have specific properties for a particular type of injury and be suitable for different phases of the healing process. For instance, during debridement, the cover must absorb excessive exudate and microorganisms populating the wound. During the granulation phase, the wound dressing material must wet the wound surface, and facilitate angiogenesis and filling of the defect site by collagen fibers. During epithelialization, it must favor and induce cell proliferation and tissue regeneration (Kaem, 1990).

The following conditions are required for wound healing:
- moisturized wound surface;
- sufficient oxygen tension in wound tissues;
- the absence of excessive wound exudate;
- protection from external injuries;
- prevention of secondary infections; and
- prevention of excessive heat loss.

Wound dressings are commonly used to treat skin defects. Many new types of wound dressings, which differ in their chemical compositions and drugs contained in them, have been designed recently. At the same time, there is no universal dressing that could be used to cover burns of various depths in different phases of healing.

An ideal wound dressing must:
- create optimal microenvironment for wound healing;
- effectively absorb wound exudate;
- protect the wound against microbial invasions;
- be sufficiently permeable for gases (oxygen, carbon dioxide) to enable repair processes;
- be permeable for water vapor but keep the wound bed moist;
- be supple and conform to the wound surface in all areas;
- not exhibit pyrogenicity, antigenicity, and toxicity; and
- not cause local inflammation and allergy.

In addition to that, wound dressings should:
- be transparent, to enable wound observation;
- be able to carry antibacterial drugs and drugs for wound repair;
- be sterilization tolerant;
- be convenient for use by medical staff and the patient; and
- be easily removable from skin surface.

Wound dressings can be functionally divided into the following groups:
- protective (preventing external contamination and protecting the wound from mechanical injuries);
- absorptive (capable of absorbing the wound exudate and preventing its accumulation under the dressing);
- curative (anesthetic, hemostatic, preventing infection, facilitating wound healing, compatible with drugs);
- transporting (air permeable, preventing exudate evaporation); and
- technological (costly, complex, challenging to manufacture, sterilizable).

Biodegradable wound dressings are mainly prepared from various naturally occurring polymers, and bioinert ones – from synthetic materials. Therapeutic wound dressings have different properties and applications. They can be subdivided into natural and synthetic skin replacements.

Natural wound dressings are mainly represented by various preserved epidermis and dermis grafts. The "gold standard" of wound dressing is natural skin. For short periods, skin flaps, which include the epidermal layer and the underlying dermis, are stored at a temperature between 0 and 8°C in the aqueous medium (in growth medium), which contains amino acids and glucose as well as antioxidants and cryoprotectants. For longer storage, skin flaps are freeze-dried, preserved in glycerol, or deep-frozen. The quality of skin grafts is certainly influenced by the method and conditions of storage. The commercially produced cell-free dermis templates are usually preserved by freeze-drying. Sviderm and Alloask D are porcine dermis grafts. Commercially produced allografts include AlloDerm, Integra, and Dermagraft.

Wound dressings can be classified in various ways. According to their properties, they may be: (i) absorptive, (ii) protective, (iii) drug-eluting, and (iv) atraumatic. According to the mechanism of their effect, they can be divided into the following types:

- absorptive wound dressings;
- covers preventing exudate evaporation;
- non-adherent dressings;
- resorbable dressings; and
- insulating dressings.

The main functional property of absorptive biocompatible dressings is their ability to absorb wound exudate, which may be significant. The loss of fluid from the wound surface in the case of the second- or third-degree burn may reach 3.5 mL/kg/% of the burned surface area for 24 h. The absorptive ability of the polymer dressing is mainly determined by the free volume of the pores, while the origin of the polymer determines the rate of fluid absorption. The optimal absorption ability has not been determined yet. For instance, absorptive ability of hygroscopic cotton wool is 2000–2500% and that of cellulose bandages reaches 3,400%. The purpose of using wound dressings preventing exudate evaporation is not to allow tissue fluids to evaporate, as the rate of evaporation through the injured skin may reach 0.5–2.2 mL/cm²/h. The generally accepted optimal rate of evaporation through the wound dressing is between 6 and 12 mg/cm².

Wound dressings may be in the form of sponges, gels, films, and spray coatings. The porous structure of sponges makes them highly absorptive and permeable for gases and oxygen vapor. Sponges are fabricated from natural polymers (collagen, chitosan, alginates, cellulose, etc.) and synthetic polymers (polyurethane etc.). Sponges are treated with antibiotics, proteolytic enzymes, hemostatic agents, etc. Natural polymer-based sponges are mainly used in the second phase of the wound healing process. These are formulations based on collagen – Kolaspon, Combutec-2, Oblekol (with sea buckthorn oil), on alginates – Algipor, and other polysaccharides – Aubazidan, Aubazipor. Combutec-2 contains not only collagen but also antibacterial components – glutaraldehyde, chinosolum, and boric acid. It is used for permanent wound closure of debrided second- and third-degree burns; for temporary wound closure of the third-degree burns after eschar removal, flat granulated posttraumatic wounds, trophic ulcers and bedsores; for preparing the wounds for autograft implantation and closing the donor sites. Combutec-2 should be used after the necrotic layer of the tissue has been removed and granulation has started (i.e., when the first phase of healing is over and the second begins) and during epithelialization of small wounds. In the treatment of trophic ulcers and bedsores, Combutec-2 is used after necrotic tissue has been removed from the wound surface and bacterial contamination has been reduced. Some wound dressings consist of two types of polymers – collagen and chitosan. Collagen sponge with antibacterial shikonin, Geshispon, is a relatively new type of skin graft.

Wound dressings in the form of gels are generated when the exudate wets the powdered substance on the wound surface. Functionally, these dressings are usually drainage sorbents. They are intended to absorb not only wound exudate but also microorganisms. Wound dressings of this type are fabricated from various synthetic and natural polymers (methyl methacrylate derivatives, dextran, acrylamide, agar-agar, etc.). They may be produced as powders that undergo gelation when wetted: Geliperm, Haidron, Debrizan, Dejisan, Gelevin, Celosorb, Collasorb, Colladisorb, etc. Another form is hydrogel: Inerpan (produced by polymerization of amino acids leucine and glycine), Galacton and Galagran (antibacterial formulations based on pectin, functioning as sorbents and facilitating repair).

Skin replacements in the form of films have been increasingly used recently. They are fabricated from various natural and synthetic materials: collagen, polyvinylchloride, polyethylene, polypropylene, polyethylene terephthalate, poly-epsilon-caprolactone, etc. Many of the films are sufficiently strong and supple; they are convenient to use. However, these wound dressings do not have sufficient vapor permeability. Polycaprolactone films show the best vapor permeability: $2 \text{ g/(h} \times \text{cm}^2)$ with the film thickness of about 0.15 mm. Hydrophilic polyurethane films show good promise; the best known representatives of these films are Op-Site, Tegaderm, and Cutinova hydro. These wound dressings are convenient to use, elastic, and transparent; they adhere well to the wound surface. Fixomull Stretch is a highly adhesive, water and vapor permeable wound dressing, consisting of a white nonwoven polyester substrate coated with a skin-friendly polyacrylate adhesive. Some films (Tegaderm, Foliderm) are impermeable to bacteria but permeable to air and water vapor, which makes them good dressings for burns. At the same time, vapor permeability of the majority of such dressings is too low. Omiderm, which is made of elastic hydrophilic polyurethane, has much better vapor permeability and, what is very important, can be used in combination with antibacterial drugs. Films of natural polymers – collagen (Biocol), chitosan, and bacterial cellulose (Baccelasept) – are also commonly used wound dressings. Baccelasept is impregnated with polymeric antiseptics (catapolum, cygerolum), which render it antibacterial.

A special group of wound dressings used in management of burn injuries comprises non-adhesive dressings, which are subdivided into the following varieties: metallized dressings, paraffin gauze, gauze impregnated with liniments or emulsions. Paraffin gauze dressings include Para-nett, Bactigras, Tulle-gras, and Sofra-tulle, with or without antibacterial drugs (gentamycin, soframycin, etc.) impregnated into them. Non-adhesive wound dressings Adaptic (Johnson & Johnson), Damor (Damor America Inc.), Fucidin, and Betadine are impregnated with non-smearing ointments, emulsions or creams. Skintact (Robinson) is a wound dressing that shows low adherence and high absorptive ability.

Very commonly used atraumatic dressings, which do not adhere to the wound and protect granulations and epithelium, are ointment dressings. They keep the wound moist and reduce scarring. Adhesive but hypotraumatic absorbing dressings based on natural and synthetic polymers are also used in clinical practice. Such dressings are represented by Algipor (Russia) – hermetically sealed 10-mm thick porous plates. Collagen is used to prepare resorbable wound dressings because it can facilitate fibrillogenesis and because it can be lysed and replaced by connective tissue. Combutec (Russia) is sponge based on water-soluble collagen, which has high absorptive ability. The stimulating effect of chitin and the chitin-based chitosan is associated with the activation of biological treatment of the wound (Collahit FA, Russia). Studies of the effect of chitosan preparations on healing of purulent wounds showed that the use of chitosan ascorbate osmogel facilitated wound decontamination and healing.

Wound dressings are used in the treatment of various wounds, but traditional bandages are going out of use. Scientific achievements enabled creation of modern wound dressings, which have different properties depending on the phase of wound healing in which they are used. Resorbable wound dressings meet all biomedical requirements and can be useful in both early and late phases of wound healing. Prolonged-action formulations are usually based on high-molecular-weight polymer materials, which are degraded in the body without releasing toxic substances. Designing resorbable polymer wound dressings with different biodegradation rates is one of the important objectives of biomedical research. There is great activity in searching for new functional materials and matrices.

The review of the literature and experimental results suggested good potential of the materials based on the biodegradable PHAs, which

- do not swell and are not hydrolyzed in the aqueous medium; do not dramatically acidify tissues; are thermostable; can be sterilized by conventional methods; are permeable to water vapor and oxygen; are biocompatible; do not cause local irritant or allergic effect;
- can be processed into hydrophobic non-adhesive wound dressings in the form of film and/or membrane composed of ultrafine polymer fibers; are transparent, enabling wound observation; can be used as drug carriers.

5.2   PHA MICROPARTICLES AS TRANSDERMAL SYSTEMS FOR TREATING SKIN DEFECTS

Transdermal drug applications are frequently used because of the effectiveness of the localized treatment, low cost, relatively low side effects, and maximum drug availability at the target site, avoiding systemic circulation. Many bioactive agents, however, do not have the necessary physicochemical properties for satisfactory efficacy when applied topically (Sloan et al., 2006). Transdermal skin treatment requires the absorption of drug through the skin into the body. One of the biggest challenges in developing an effective system is to transfer the drug through the tightly structured stratum corneum when the skin is not compromised (Yow et al., 2009). There are basically three layers of skin that need to be passed: epidermis, dermis and subcutaneous tissue. The epidermis is the outer layer and serves as a barrier between the body and the environment (Gawkrodger, 2002). The dermis gives the skin its mechanical strength. The subcutaneous tissue is the lowermost, insulating and cushioning, layer. The hair follicles and sweat glands serve as a route of entry for nanoparticles. Encapsulation of the drug in a carrier allows the drug to diffuse into hair follicles and sweat glands, where drug release can occur in the deeper layers of the skin (Prausnitz et al., 2008; Arora et al., 2008). The carriers and dissolved substances are transported via the capillary system of the subcutaneous tissue. This is the basis for the development of drug-carrying nanoparticles for targeted follicular delivery (Wosicka et al., 2010; Prow et al., 2011).

Recently, considerable research effort has been directed towards development of drug formulations that would safely cross the skin. One of the approaches is the use of micro- and nano-sized carriers. New drug delivery systems such as nanocapsules, nanospheres, liposomes, dendrimers, and emulsions have been designed. Micro- and nano-sized carriers fabricated from biodegradable polymers can be used to construct sustained-release drug delivery systems that can be administered via different routes, including transdermal administration. Murueva et al. (2014) were the first to investigate therapeutic efficacy of PHA microparticles loaded with anti-inflammatory diclofenac and hormone dexamethasone in the treatment of model skin defects in laboratory animals. That study was preceded by another one, which investigated interactions of PHA microparticles with skin cells and tissues. The study was performed by researchers of Department of Micro and Nano Technology at Middle East Technical University (Ankara, Turkey) in cooperation with researchers of Biotechnology Department at Siberian Federal University (Eke et al., 2014). The purpose of that study was to develop microparticles from poly-3-hydroxybutyrate/3-hydroxyvalerate and study the cell and skin penetration of these particles. Particles of three different mean diameters – 1.9 µm, 426 nm, and 166 nm – were produced for the study. Interaction of the microparticles with cells was investigated in the culture of mouse fibroblast L929 cells. The cells were counted with a hemocytometer (Blau Brand, Germany). Nile Red loaded microparticles were suspended in the culture medium and added into 24-well plates. Cells were seeded in the wells and incubated for 24 h. Then, the medium was removed, wells were washed twice with sterile PBS, MTT solution was added to determine the number of viable cells. After staining, the particles and the cells were observed with confocal laser scanning microscopy (CLSM) (Leica DM2500, Germany).

MTT assay showed that the size of polymer microparticles did not significantly influence the viability and proliferation of fibroblasts. At the same time, the increase in the concentration of microparticles in the cell culture considerably influenced fibroblast proliferation and concentration (Eke et al., 2014). When the concentration was increased from 0.5 to 1 mg/mL, L929 cell proliferation rate decreased. For 1.9-µm and 426-nm particles, the proliferation rates increased in a statistically significant manner throughout the whole duration of the test. For the low 166-nm particles, however, no statistically significant difference could be observed. These cells did not show a change in their shape implying that the decrease is not due to a negative effect like toxicity. It is stated that particles with a size up to about 100–200 nm can be internalized by receptor-mediate endocytosis, while larger particles have to be taken up by phagocytosis (Yin et al., 2005; Xiong et al., 2010). Thus, it is expected that the small size particles could improve efficacy of the particle based topical drug delivery systems because of more availability within the cell. The literature data about the effects of penetration of the particles into cells are contradictory: both increased and decreased proliferation rates have been reported. In the study by Eke et al. (2014), larger particles (1.9 µm) caused considerably lower proliferation rates than smaller ones. The cells took up 166-nm and 426-nm particles. The larger particles (1.9 µm), however, were unable to penetrate into the cells. Microparticles were generally located in the cytoplasm near the nuclei. The particles appeared to be intact, otherwise, lysis by lysosomal enzymes would release the

Nile Red into the medium causing it to stain all the organelles with lipoid components.

Intracellular uptake of particulates is proposed to be either by phagocytosis or by endocytosis (Sahoo et al., 2002). It was reported that nanoparticles around 500 nm are taken up by macrophages by phagocytosis (Foster et al., 2001). However, for the smaller nanoparticles, the main route of cellular entry is through fluid phase endocytosis and it leads to entrapment inside intracellular vehicles, such as endosomes or lysosomes (Benfer et al., 2012). Desai et al. (1997) reported intracellular uptake of nanoparticles and showed that this depends on the size and hydrophobicity. This supports the observation about the inability of the 1.9-μm-diameter particles to penetrate inside the cells (Eke et al., 2014).

Eke et al. (2014) studied the penetration of P(3HB/3HV) microparticles through skin in the *in vivo* experiments. Experiments were conducted on male BALB/c mice, 20–25 g each. Dorsal sections (1 cm^2) of the mice were shaved and swabbed with the particle suspension. During the experiment, the animals were sacrificed and skin samples were removed and used in histology. For the determination of the P(3HB/3HV), gas chromatograph-mass spectrometer (GCD Plus, Hewlett Packard, U.S.) equipped with a fused silica capillary column was used.

Four hours after application of the polymer microparticles, polymer was found in the skin; the amount of the polymer depended on the size of the particles applied (Eke et al., 2014). Soon after application, about 1% was found in the skin and this gradually decreased by tenfold in 10 days. In none of the skin sections the presence of the particles could be detected, which probably can be explained by the size of the particles being lower than the resolution of the light microscopic examination. However, chromatography showed the polymer was in the skin (up to 100 μg) indicating that the particles penetrated the skin. Results of that study suggested that PHAs were suitable for designing transdermal drug delivery systems. Particles of the average diameter of 426 nm appear to be promising therapeutic tools for the treatment of compromised skin.

The use of microparticles of degradable PHA to repair model skin defects was reported by researchers of the Siberian Federal University in a study conducted within the framework of mega project "Biotechnology of Novel Biomaterials" (Decree of the RF Government No. 220 of April

9, 2010) (Murueva et al., 2012). The authors investigated microparticles prepared from the polymer of P3HB by using the previously developed process (Goreva et al., 2012). The particles were loaded with diclofenac and dexamethasone (17.5 and 1.2% of the mass of the polymer matrix, respectively). The average diameter of the particle was 630±10.3 nm; zeta potential was 19.2±0.21 mV.

Experiments were conducted on Balb/c mice, 20–23 g each. Dorsal sections of mice were shaved, and a 2.5% solution of potassium dichromate was applied. As reported in the literature (Raben et al., 1970), the best time to start the treatment of the skin defect is on Day 5–8 after potassium dichromate application. The severity of skin injury was scored as follows: 0 – no response; 0.5 – isolated red spots; 1 – moderate diffuse hyperemia; 2 – distinct hyperemia and edema; 3 – pronounced redness and edema; 4 – small erosions; 5 – hemorrhagic crust and vast ulcers (Eremenko et al., 1999). Scoring of skin wounds (second/third-degree chemical burns) is given in Table 5.1.

At Day 2 after potassium dichromate application, moderate hyperemia was observed on the skin of all animals (1–1.3 scores); edemas were observed on some of the animals. At Day 6 after potassium dichromate application, hemorrhagic crust was observed on the skin (4–5 scores). In 3 days after application, the thickness of the skinfold increased to 0.83 mm, while the skinfold of intact mice was 0.58 mm thick. By Day 7, the thickness of the skinfold of the treatment mice had reached 1.7 mm. The histology of tissue samples taken from the wounds showed indications of several phases of the wound development: traumatic necrosis (alteration), inflammation, and regenerative histogenesis. Histologic examination defined the primary necrosis areas and the perinecrotic areas, which

TABLE 5.1   Scoring of the Severity of Animals' Skin Damage

Parameters	Measurement unit	n	Days				
			2	3	5	6	7
Degree of skin manifestations	score	10	1	2	3.7	4.2	4.6
Skinfold thickness	mm	10	0.78	0.83	0.87	1.64	1.7

*According to Eremenko et al. (1999).

were the sites of regenerative histogenesis and adaptive restructuring of tissue elements during the subsequent treatment. In the traumatic necrosis phase, at Day 8, the animals showed epidermal necrosis, and necrosis of the stratum papillare and the reticular layer of the dermis. The average area of the burn wound was 5.01 cm² (Figure 5.1).

The duration of the experiment was 20 days, including 12 days of the treatment. The treatment was started 8 days after the skin was compromised. The mice were divided into three groups: Group 1 (negative control), no treatment; Group 2 (reference), conventional treatment by applying diclofenac and dexamethasone (0.08 mg diclofenac and 0.02 mg dexamethasone per mouse) daily, for 11 days; Group 3 (treatment), daily applications of microparticles carrying the drugs in the form of water-organic gel, which contained water-soluble drugs in the same amounts as those administered to Group 2 mice, for 11 days. The process of regenerative histogenesis and adaptive restructuring of tissue elements during the treatment was monitored by histological examination and planimetry techniques. Changes in the area of the wound surface were monitored by Popova's method (Popova, 1942), to determine the wound healing rate. At Days 4, 8, and 12 of the treatment, the animals were sacrificed and skin samples were removed and used in histology. Examination included evaluation of the inflammatory reaction and acanthosis (an indicator of increased proliferation of basal and spinous cells); counting of horn cysts (a marker of

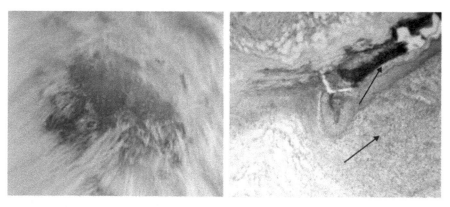

FIGURE 5.1   Model wound of the mice's skin at Day 8 after application of potassium dichromate: to the left is a photograph; b is a micrograph (hematoxylin and eosin staining, 100×): n – necrotic area, i – inflammatory infiltration area (Shishatskaya's data).

the lagging cell differentiation in the epidermis under the conditions of accelerated regeneration); quantification of sebaceous glands and hair follicles (a source of undifferentiated epithelial cells that can be involved in regeneration of the epidermis in the case of vast shallow burns).

The following results were obtained. The Group 1 animals (no treatment) showed symptoms of intoxication, including the reduced appetite and motion. In the reference and treatment groups, the animals showed moderate intoxication symptoms for the first 2–3 days of observation, which disappeared by Days 6–8. Figures 5.2 and 5.3 show results of wound treatment: changes in the burn wound area and wound healing rate.

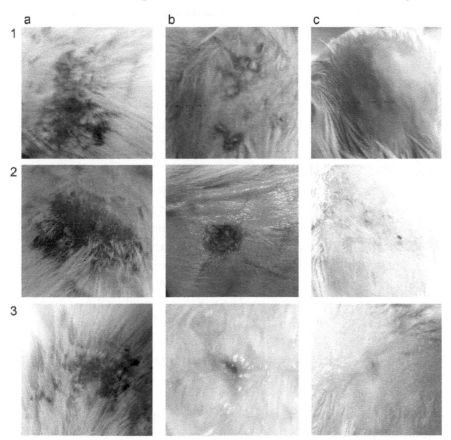

FIGURE 5.2 The skin wounds of the animals at different time points of the treatment: 1 – no treatment; 2 – the reference group; 3 – the treatment group at Days a – 4, b – 8, c – 12 (Murueva et al., 2014). (Reprinted with the permisison of the Journal "Bulletin of Experimental Biology and Medicine.")

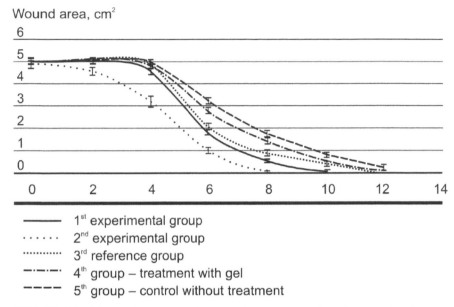

FIGURE 5.3 Changes in the wound area in laboratory animals of the control, reference, and treatment groups (Murueva et al., 2014). (Reprinted with the permisison of the Journal "Bulletin of Experimental Biology and Medicine.")

The process of tissue regeneration in the wounds treated with drugs embedded in microparticles occurred at a faster rate than in the reference group. By Day 8 of the treatment, the burn area on the animals of the treatment group had reduced to 0.5 cm², or almost by a factor of 10; in the reference group, it had reduced to 0.875 cm², or by a factor of 5.7; in the no-treatment group, the area of the burn had decreased to 1.75 cm², or by a factor of 2.75 (Figure 5.3). The average wound healing rate for the treatment animals was 0.113 cm²/d in the first four days and 0.50 cm²/d between Days 4 and 8, which was higher than the healing rates in the reference group animals: 0.103 and 0.33 cm²/d, respectively. By Day 12 of the treatment, the crusts had separated, revealing healthy skin, in the animals of all groups.

At Days 2–4 of the treatment, microscopic examination of the defect site in all animals showed similar histological patters: edema of all layers of the dermis, inflammatory cell infiltration, multifocal necrosis of all skin layers, small vessel spasms (Figure 5.4). At Day 8, the wounded skin of Group 1 animals had areas where the eschar had separated and

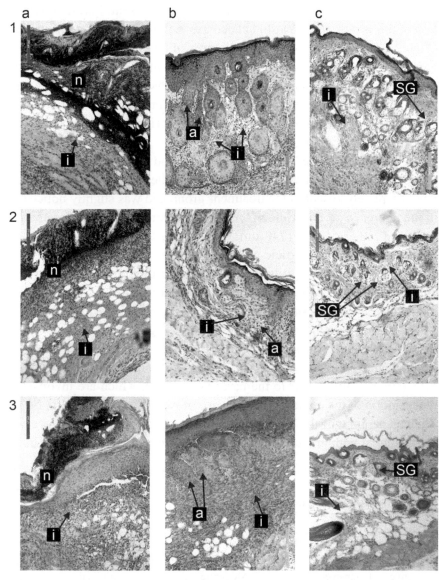

FIGURE 5.4 Results of histology of the wound healing process in the animals: 1 – negative control, 2 – reference, 3 – treatment; at Days a – 4, b – 8, c – 12 of the treatment. Notations: a – acanthosis, n – necrosis, i – inflammatory infiltration, SG – sebaceous glands. Hematoxylin and eosin staining, 100× (Murueva et al., 2014). (Reprinted with the permisison of the Journal "Bulletin of Experimental Biology and Medicine.")

suppuration occurred, with no upper skin layers present. Further observation of this group showed a slight decrease in the inflammatory infiltration (Figure 5.3c). The morphological pictures of wound healing in the treatment and reference groups were different: wound epithelialization was observed in many of the animals; granulation tissue was moderately vascularized; diffuse and weakly pronounced infiltration of granulation tissue was observed. By the end of the observation period, the skin in the defect site was healthy-looking (Figure 5.2).

Inflammation tended to decrease by the end of the experiment in all groups, but to different extents (Table 5.2). At the end of the experiment, it was completely absent in the treatment group and was slightly noticeable in the reference group, while in the no-treatment group, the inflammation remained rather deep – 1740 µm. Only in the treatment group animals, which received microparticle-embedded drugs, there was no necrosis at the end of the experiment. In the negative control group and in the reference group, although the skin was healthy-looking at the end of the experiment, there were very small areas of loose epithelial tissue.

Results of counting hair follicles, sebaceous glands, and horn cysts, as objective indicators of the healing process are given in Table 5.3. Sebaceous glands and hair follicles are sources of undifferentiated epithelial cells (structurally similar to the basal layer cells), which are involved in regeneration of the epidermis in the case of extensive shallow skin burns. The regeneration of the skin was first observed in the group that received the treatment with microparticle-encapsulated drugs. At Day 8, hair bulbs

TABLE 5.2 The Area of Necrosis and the Average Depth of Inflammation in Animals' Skin Wounds (Murueva et al., 2014) (Reprinted with the permisison of the Journal "Bulletin of Experimental Biology and Medicine.")

	Day 4		Day 8		Day 12	
Group	Necrosis	Inflamma-tion, µm	Necrosis	Inflamma-tion, µm	Necrosis	Inflamma-tion, µm
Negative control	Extensive (++++)	1712.6	Moderate (++)	2303.1	Minimal (+)	1740.6
Reference	Extensive (++++)	1520.7	Extensive (++++)	3081.8	Moderate (++)	700.8
Treatment	Extensive (++++)	2078.7	Minimal (+)	1427.6	–	467.0

and sebaceous glands were observed in the defect sites of the treatment group animals, while the reference group animals only had horn cysts as an indicator of the slow differentiation of the epidermis cells.

Another very important indicator of the repair of skin defects is acanthosis – the thickening and increase in the number of rows in the layers of spinous cells and granulosa cells, with the epidermal appendages penetrating deep into the skin. Results of examination of the acanthosis are listed in Table 5.4, suggesting that at Day 4 of the treatment, proliferative acanthosis was observed on the periphery of the wound in the treatment group (12 strands penetrating to a depth of 300 μm); in the reference group, no signs of acanthosis were detected. At Day 8 of the treatment, the number of the strands in the treatment group increased to 18, and they penetrated to a depth of 660 μm; in the reference group, only 4 strands penetrating to a depth of 312.5 μm were observed. By Day 12, in both the treatment group and the reference group, the wounds had healed completely, the acanthosis had disappeared, and the normal morphological structure of the epidermis had been recovered.

Although the effects of the free drugs and the microparticle-encapsulated ones were comparable, some parameters (the burn wound area, the healing rate, the number of acanthotic strands, and the numbers of hair

TABLE 5.3 The Numbers of Hair Follicles and Sebaceous Glands During the Treatment of Skin Defects with the Free Drugs and the Drugs Encapsulated in Polymer Microparticles (Murueva et al., 2014) (Reprinted with the permisison of the Journal "Bulletin of Experimental Biology and Medicine.")

	The number of hair follicles, sebaceous glands, and horn cysts		
Group	Day 4	Day 8	Day 12
Negative control	–	–	8 hair bulbs
			4 sebaceous glands
			8 horn cysts
Reference	–	8 horn cysts	22 hair bulbs
			17 sebaceous glands
			no horn cysts
Treatment	–	5 hair bulbs	20 hair bulbs
		9 sebaceous glands	15 sebaceous glands

TABLE 5.4 The Number of Acanthotic Strands and the Average Depth to Which They Penetrated into the Dermis During the Treatment of Skin Defects in the Animals (Murueva et al., 2014) (Reprinted with the permisison of the Journal "Bulletin of Experimental Biology and Medicine.")

Group	The number of acanthotic strands and the average depth to which they penetrated into the dermis		
	Day 4	Day 8	Day 12
Negative control	–	7 strands 295.6 µm	16 strands 324.7 µm
Reference	–	4 strands 312.5 µm	no acanthosis
Treatment	12 strands 300.6 µm	18 strands 659.7 µm	no acanthosis

follicles, sebaceous glands, and horn cysts) were indicative of the more pronounced reparative processes in the skin of the animals that received drugs encapsulated in microparticles. Moreover, while free drugs (ointments) were applied to the wound surface every day, during 11 days, microparticles were applied every 3 days.

These results suggest that PHA microparticles loaded with anti-inflammatory drugs can be used for regeneration of skin defects as sustained-release systems.

5.3   A STUDY OF EFFECTIVENESS OF PHA FILMS AS WOUND DRESSINGS FOR RECONSTRUCTING MODEL SKIN INJURIES

A disadvantage of many wound dressings is that they stick to the wound surface, injuring the regenerating tissues and causing pain. The injurious daily bandaging reduces the advantages of the dressing materials. The adhesion of the dressing to the wound surface may be caused by various reasons, the most frequent of which is "gluing" of the dressing to the wound surface. The "glue" is the exudate, which dries and forms an eschar. As the hydrophilicity of the polymer material constituting the dressing is increased, it adheres to the wound surface more firmly. During the granulation phase, the adhesion of the wound dressing is associated with the growth of the granulation tissue into the pores of the dressing material. This problem is usually resolved by fabricating wound dressings from hydrophobic polymer materials or by using hydrophobic synthetic polymer to make the wound-facing surface of the dressing (Tumanov and German, 2000). However, such dressings do not adhere properly to the wound surface and their absorption rate is low, causing exudate accumulation on the wound.

Atraumatic absorptive wound dressings can be designed by using resorbable materials. This line of research has not been sufficiently developed yet, as very few biodegradable materials meet all the requirements for wound dressing materials. Researchers of the Siberian Federal University and the Institute of Biophysics SB RAS have constructed wound dressings from degradable hydrophobic PHAs and tested them in experiments on closing and treating skin defects. The team of researchers constructed cell scaffolds and tested their effectiveness for regeneration of skin defects. Nonwoven membranes composed of randomly oriented ultrafine fibers prepared by electrospinning from PHA terpolymers of two types – PHAs containing major molar fractions of 3-hydroxybutyrate (3HB), 4-hydroxybutyrate (4HB), and 3-hydroxyvalerate (3HV) monomers and PHAs containing major molar fractions of 3-hydroxybutyrate (3HB), 3-hydroxyvalerate (3HV), and 3-hydroxyhexanoate (3HHx) monomers – and flexible films prepared by solution casting and nonwoven membranes made from PHA copolymers of 3HB and 4HB, P(3HB/4HB), were used as cell scaffolds. Various types of cells were seeded onto the scaffolds to evaluate the effects of the type of the polymer and scaffold on cell attachment and proliferation, cell morphology and ability of the stem cells to differentiate to specified cell types in specialized media. Cells were cultured on the DMEM medium with 10% fetal bovine serum and a solution of antibiotics, on matrix of P(3HB/4HB), and P(3HB/3HV/4HB) and P(3HB/3HV/3HHx) terpolymers.

Polymer films and electrospun nonwoven membranes shaped as 10-mm-diameter disks and 3D porous constructs (5×5×5 mm) were placed into 24-well culture plates (TPP Techno Plastic Products AG, Switzerland) and sterilized in a Sterrad NX medical sterilizer (Johnson & Johnson, U.S.). MSCs harvested from the bone marrow and adipose tissue of Wistar rats were used to test the ability of the PHA scaffolds to favor cell attachment and facilitate cell proliferation and directed cell differentiation. The MSC suspension was seeded onto Petri dishes; the nutrient medium was refreshed after 24 h, to remove the unattached cells. Cells were cultivated in the incubator at a temperature of 37°C, in the CO_2 environment, on the DMEM medium with 10% fetal bovine serum and a solution of antibiotics. The adipose tissue cells were isolated enzymatically. The adipose tissue was rinsed in a DPBS solution with antibiotics, ground, and incubated in a collagenase type I solution (100 units/mL) at 37°C until tissue particles were dissolved, but for no more than 1 h. Then, collagenase was inactivated with an albumin solution, centrifuged, and rinsed in the DMEM medium several times, until the lipid layer had been completely removed. The settled cells were suspended and seeded onto Petri dishes that contained the DMEM medium with 10% fetal bovine serum and a solution of antibiotics. On the DMEM medium, adipose tissue cells showed fibroblast-like morphology. After 3–4 passages, the cells were removed from the Petri dishes with a trypsin solution and seeded onto sterile PHA membranes: 10^5 cells per scaffold. Gene expression of collagen type I (Col-1) was determined by real time RT-PCR, to confirm the fibroblast phenotype of the cells and their viability.

In the DMEM medium, adipose tissue cells acquired the fibroblast phenotype and synthesized Col-1, which was confirmed by the real-time RT-PCR in the region of the quantitative evaluation of gene expression to the factors of Col-1 (Figure 5.5). Adipose tissue cells growing on the polystyrene of the culture plates were used as the control. Results of quantification of Col-1 by real-time PCR were confirmed by immunocytochemical staining. Figure 5.6 shows results of immunocytochemical staining with Col-1 antibodies at Day 14 of adipose tissue cell culture. In the cells growing on nonwoven PHA membranes, the rate of collagen synthesis was higher than in the cells growing in the DMEM medium on polystyrene (control).

Thus, nonwoven membranes of PHA terpolymers of both types were found suitable for constructing hybrid tissue engineering grafts carrying epidermal cells.

Experimental wound dressings were prepared from P(3HB/4HB) by using one of the two processes: by pouring polymer solutions on the smooth surface followed by solvent evaporation or by electrospinning (Figure 5.7).

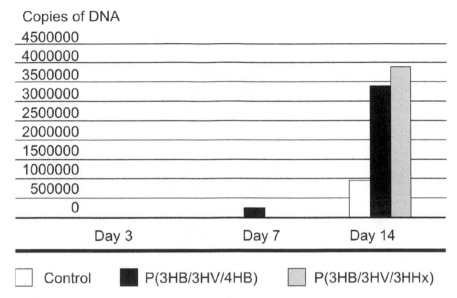

FIGURE 5.5 Results of quantification of mRNA expression to Col-1 factors in cells cultivated on P(3HB/3HV/4HB) and P(3HB/3HV/3HHx) membranes. Results were normalized relative to the housekeeping gene – β-actin (Shishatskaya's data).

SEM images show differences in the surface morphology of wound dressings prepared by different techniques. Solution cast P(3HB/4HB) (9:1) films had round pores, up to 2 μm in diameter. Nonwoven membranes comprised of randomly oriented ultrafine fibers showed distinct fibers of mean diameter of 1.7 μm and the distance between them of 1.3 – 1.67 μm. Results of the comparative study of the surface properties of the wound dressings are given in Table 5.5; and physical/mechanical properties are given in Table 5.6.

The effectiveness of nonwoven polymer membranes prepared by electrospinning from P(3HB/4HB) copolymers was studied on model skin defects created in Wistar rats. The rats were given inhalation anesthetic, and a 2×2-cm section in the interscapular region of each rat was shaved and then excised (both epidermis and dermis, to its full thickness) under aseptic conditions. The average wound area was 4.0–4.5 cm². The experiment was conducted with three groups of animals: two treatment groups, with the grafts of nonwoven membranes and membranes carrying epidermal

Control

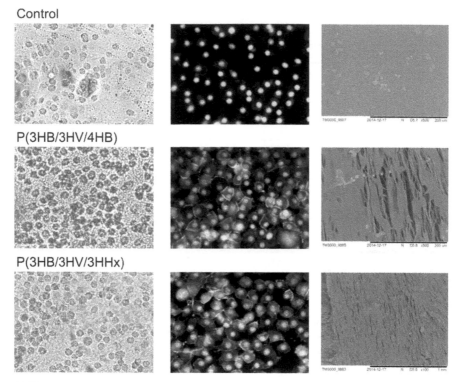

P(3HB/3HV/4HB)

P(3HB/3HV/3HHx)

FIGURE 5.6   Results of immunocytochemical staining of adipose tissue cells with Col-1 antibodies and SEM images at Day 21 of cell culture on elastic nonwoven membranes of PHA terpolymers (Shishatskaya's data).

cells derived from adipose tissue. In the third, control, group, the wound was covered with Voskopran dressings. The dressings were fixed with 4 sutures and covered with sterile gauze. The state of the animals (motion activity, feeding behavior) was assessed daily throughout the experiment. Wounds were photographed and the area of the wound and the rate of healing were determined to monitor wound healing. Conventional histological techniques were used. Investigations included evaluation of the inflammatory reaction and acanthosis (an indicator of increased proliferation of basal and spinous cells); counting of horn cysts (a marker of the lagging cell differentiation in the epidermis under the conditions of accelerated regeneration); quantification of sebaceous glands and hair follicles (a source of undifferentiated epithelial cells that can be involved in regeneration of the epidermis in the case of vast shallow wounds).

FIGURE 5.7 Photographs of experimental wound dressings – solution cast films and nonwoven electrospun membranes (top) P(3HV/4HB) and SEM images of the microstructure (bottom) (Volova's data).

Throughout the postoperative observation period, all animals were healthy and active, displayed normal eating behavior, and moved on their own. After the treatment group animals awakened from anesthesia, they did not show any signs of pain. Complete blood counts showed the following: until Day 4, the leukograms of all animals had demonstrated slight lymphopenia, an increase in band neutrophils to 4%, and an increase in segmented neutrophils to 1.2%. One day after the defect was created, the counts of erythrocytes and hemoglobin decreased, and the animals developed leukocytosis. The counts of erythrocytes and hemoglobin had returned to their initial levels by Day 3 in the treatment groups and by Day 7 in the control. Measurements of the wound area suggested that in the treatment groups, the defects repaired faster (Figure 5.8).

Photographs of the wounds taken during the experiment are shown in Figure 5.9. Inflammation was observed in the wound region in all animals,

TABLE 5.5 Surface Properties and Physical/Mechanical Properties of Experimental Wound Dressings Prepared From P(3HB/4HB) (Shishatskaya's data)

Surface properties:	Film	Nonwoven membrane
Water contact angle	57.4±0.6	72.12±0.55
Surface tension γ, erg/cm^2	43.1	31.19
Interfacial free energy γ_{SL}, erg/cm^2	3.9	8.69
Cohesive forces W_{SL}, erg/cm^2	112.0	95.30
Roughness average Ra, nm	92.909	39.228
Root mean squared roughness Rq, nm	113.062	50.234
Young's modulus, E, MPa	336.69±0.41	1.59±0.30
Elongation to break, ε, %	14.2	72.98±10.33
Tensile strength σ, MPa	4.88±0.43	0.42±0.08

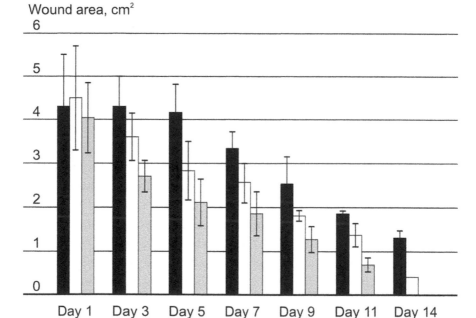

FIGURE 5.8 Dynamics of the decrease in the area of skin defects created in laboratory rats under different types of wound dressings: 1 – control group; 2 – wounds covered with polymer membranes; 3 – wounds covered with polymer membranes carrying cells (Shishatskaya's data).

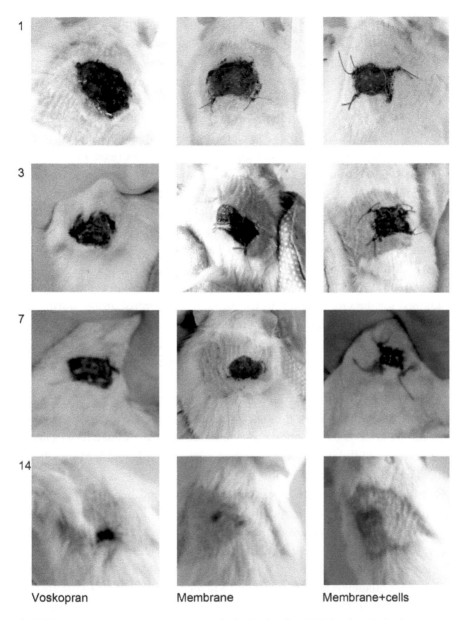

Voskopran   Membrane   Membrane+cells

FIGURE 5.9 Photographs of skin wounds during healing (Shishatskaya's data).

but the hyperemia of the dermis vessels and cell infiltration of the dermis was pronounced to different extents depending on the group. In the control group, large amounts of seropurulent exudate containing leukocytes were localized under the surface necrotic masses. In the group with fibrous grafts, hyperemia of the vessels was less pronounced and there was no purulent exudate. Measurements of the wound surface area showed the most significant changes in the group with electrospun cell-seeded grafts. During the first few days, in this group, the wound decrease rate was the highest – 0.4–0.65 cm^2/d; at Day 3, the wound area was 3.6 cm^2 (81% of the initial wound area). At Day 5, in the group with fibrous grafts, the wound area constituted 62.7% of its initial area, at Day 7 – 57%, and at Day 14 – 9%. In the group with films, the healing rate was somewhat lower, 0.3–0.4 cm^2/d. At Day 3, the wound area decreased by 9% (91% of the initial area), at Day 7, the wound area constituted 78% of the initial area, and at Day 14 – 9.1%, which was comparable with the results obtained for the group with fibrous grafts. In the control group, during the first 24 h post-surgery, the healing rate was 0.16–0.23 cm^2/d.

In the group with nonwoven membranes without cells, the average healing rate was lower, 0.45 cm^2/d. At Day 5, the area of the wounds decreased to 62.7% of the initial wound area, at Day 7 – to 57%, and at Day 14 – to 9.1%. In the control group, with the commercial Voskopran, the wound healing rate was lower than in the treatment groups – 0.16–0.23 cm^2/d; the area of the wounds had decreased by 4% by Day 5. At Day 7, the decrease in the wound area reached 23% and at Day 14, at the end of the experiment, 30% of the initial wound area.

To get insight into the mechanism of the wound healing process, we detected the factors that characterized the degree of inflammatory process, vascularization, and formation of the new connective tissue at the defect site. Real-time RT-PCR was performed in the region of quantification of gene expression to the factors of rat angiogenesis (VEGF), inflammation (TNF-a), collagen type I (Col-1) (an indicator of connective tissue formation), keratin 10 (K10), and keratin 14 (K14) (indicators of the formation of the spinous and granulosa layers of the epidermis during regeneration) in order to quantify the level of the transcription of the genes responsible for inflammation, vascularization, and connective tissue formation. Total RNA was isolated from the granulation tissue of the treatment and control groups by extraction with a guanidinium thiocyanate-phenol-chloroform mixture from the RNA-Extran reagent kit. Then, from the RNA template, using reverse transcriptase, we synthesized cDNA with several types of oligonucleotide primers: with a mixture of random primers – hexa primers – and oligo-dT primer, following the procedure recommended by the manufacturer of Sintol. Negative controls of reverse transcription reaction were prepared for all samples, to confirm the absence of DNA contamination in the initial RNA. Real-time PCR-amplification for quantification of cDNA fragments of rat VEGF, rat TNF-a, rat Col-1, rat K10, and rat K14 was performed by using a CFX-96 thermocycler, according to the manufacturer's protocol, using the color channel of HEX.

During the first 24 h, all animals developed edemas, which were more pronounced in the control group. The histology of the new tissue collected from the defect site at Day 3 showed that in the control animals, that was connective tissue, with very few collagen fibers and blood vessels. Vascularization was more pronounced under the experimental wound dressings. That was confirmed by the PCR of VEGF gene expression, as an indicator of blood vessel formation (Figure 5.10).

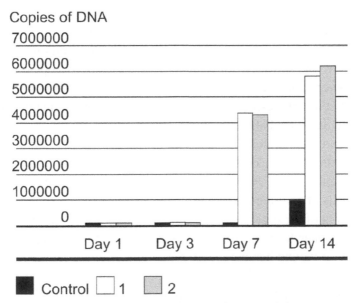

FIGURE 5.10 VEGF gene expression in the control group and the treatment groups (1 – nonwoven polymer membrane; 2 – nonwoven membrane seeded with epidermal cells) during skin wound healing in Wistar rats (Shishatskaya's data).

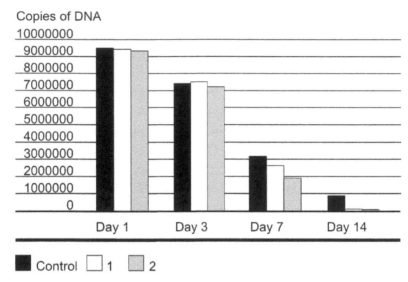

FIGURE 5.11 TNF-α gene expression in the control group and the treatment groups (1 – nonwoven polymer membrane; 2 – nonwoven membrane seeded with epidermal cells) during skin wound healing in Wistar rats (Shishatskaya's data).

In the wounds covered with the polymer membrane dressings of P(3HB/3HV/4HB), the density of the newly formed collagen fibers was higher than in the control. At Day 7, the total cell counts in the dermis under the wound surface in the field of view of the microscope were significantly lower in the treatment groups. There was also a clear shift towards fibroblast cells, suggesting a faster reduction in the inflammatory process, which had developed at the defect site.

The faster inflammation reduction in the treatment groups was confirmed in the PCR-based tests of TNF-α gene expression (inflammation factor) (Figure 5.11). Results of planimetric measurements (Figure 5.8) and analysis of histological sections (Figure 5.12), which suggested faster regeneration of the defects in the treatment groups, were confirmed by the counting of hair follicles, sebaceous glands, and horn cysts, as objective indicators of wound healing (Table 5.7).

Results of PCR-based test of gene expression of collagen type I (Col-1) (an indicator of connective tissue formation), keratin 10 (K10) and keratin 14 (K14) (indicators of the formation of the spinous and granulosa layers

TABLE 5.6 The History of Regeneration of Hair Follicles and Sebaceous Glands in the Wounds Healing Under Different Wound Dressings (Shishatskaya's data)

	The numbers of hair bulbs, sebaceous glands, and horn cysts		
Group	Day 3	Day 7	Day 14
Control	–	5 horn cysts	7 hair bulbs
Voskopran			3 sebaceous glands
			7 horn cysts
Treatment	–	4 hair bulbs	20 hair bulbs
nonwoven membrane		7 sebaceous glands	16 sebaceous glands
			no horn cysts
Treatment	–	6 hair bulbs	24 hair bulbs
nonwoven membrane +cells		8 sebaceous glands	17 sebaceous glands
			no horn cysts

of the epidermis during regeneration) also showed that new tissues developed considerably better and wound healing rate was faster under experimental polymer membranes. The number of Col-1, K10, and K14 cDNA copies was higher than that in the control by factors of 4.0–5.0, 3.5–4.0, and 4.0–4.5, respectively (Figure 5.12).

Sebaceous glands and hair follicles are sources of undifferentiated epithelial cells (structurally similar to basal layer cells) that can be involved in regeneration of the epidermis in the case of vast shallow chemical burns. Regeneration started earlier in the treatment groups, in which hair bulbs and sebaceous glands were detected at Day 7, when only horn cysts, a marker of the lagging cell differentiation in the epidermis, were detected in the control group animals.

The study of nonwoven polymer membranes and membranes carrying epidermal cells (grafts) on laboratory animals showed that they conformed to the wound surface in all areas, protected the wound, and facilitated reparative processes, enabling the wounds to heal faster than the wounds covered with the Voskopran membrane. The polymer membranes prepared and tested in this study can be used to reduce inflammation, enhance angiogenic properties, and facilitate skin regeneration.

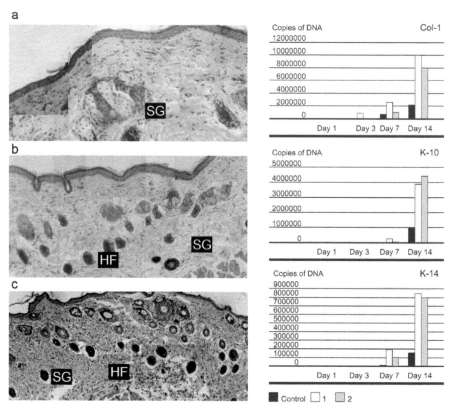

FIGURE 5.12 Histological sections from the skin defect sites. Wound healing (Day 7 post defect creation) under different wound dressings: a – control (Voskopran membrane), b – nonwoven P(3HB/4HB) polymer membrane; c – nonwoven P(3HB/4HB) polymer membrane seeded with epidermal cells. Notations: SG – sebaceous glands, HF – hair follicles. Hematoxylin and eosin staining, 100×. The number of Col-1, K10, and K14 cDNA copies during regeneration of skin defects in Wistar rats (Shishatskaya's data).

5.4   PILOT CLINICAL TRIALS OF NANOMATRICES COMPRISED OF ULTRAFINE PHA FIBERS AS WOUND DRESSINGS IN THE TREATMENT OF SUBCUTANEOUS SEPTIC WOUNDS

To enhance the therapeutic efficacy of wound dressings, they may be loaded with various drugs. In recent years, new processes of physical and chemical modification of wound dressings have been developed. Drugs can be immobilized on the surface of the dressings, enabling their slow release and prolonged therapeutic effect. To treat purulent wounds, the dressings are loaded with antiseptics, and the dressings themselves are fabricated from nonwoven polyvinyl alcohol fibers activated with sodium dichloroisocyanurate or hydrogen peroxide, hydrogels, etc. (Minchenko, 2003). A disadvantage of many wound dressings is that they stick to the wound surface. The adhesion of the dressing to the wound surface may be caused by various reasons, the most frequent of which is "gluing" of the dressing to the wound surface. The "glue" is the exudate, which dries and forms an eschar. As the hydrophilicity of the polymer material constituting the dressing is increased, it adheres to the wound surface more firmly. During the granulation phase, the adhesion of the wound dressing is associated with the growth of the granulation tissue into the pores of the dressing material.

A special group of wound dressings comprises the dressings containing drugs, including antibacterial ones. They prevent the wound from being infected. These are such dressings as Iskom, DDB, and DDBM, which are fabricated from polyethylene film, spray-coated with a mixture of equal proportions of antibiotics (cephalexin, streptomycin, erythromycin, terramycin, tetracycline, vibramycin, synthomycin, neomycin, kanamycin, nystatin, daktarin, canesten, and rivanol) (5%) and talcum (95%). However, such dressings are impermeable to vapor and air, causing exudate accumulation on the wound. The DDBM film, a modification of the DDB film, has small drainage holes. Porous film Foliderm (Folium, St. Petersburg) is fabricated from hydrophobic material, which is permeable to gases and impermeable to bacteria. Films Aseplen, Foliderm, and DDBM can be used as wound dressings in the first phase of wound healing. Most of the films, however, are used as wound dressings in phases 2 and 3, as they protect the wounds without preventing tissue regeneration. Atraumatic absorptive wound dressings can be designed by using hydrophobic non-adhesive polymer materials, which may be loaded with antibacterial drugs.

Researchers of the General Surgery Department at the V.F. Voino-Yasenetsky Krasnoyarsk Federal Medical University carried out pilot clinical trials of the effectiveness of using nonwoven membranes composed of ultrafine PHA fibers, which were prepared by electrospinning.

A preliminary study was conducted to investigate antimicrobial properties of PHA-based wound dressings loaded with ceftriaxone (5, 10, and

15% of the polymer weight). The antimicrobial effect of the dressings was evaluated by the diffusion method, by measuring the areas where microbial culture growth was inhibited (Figure 5.13). Tests were performed with two cultures of Gram-negative bacteria, *Ps.aureginosa* and *E.coli*, potential pathogens to humans, which are recommended for evaluation

Ps.aureginosa   E. coli

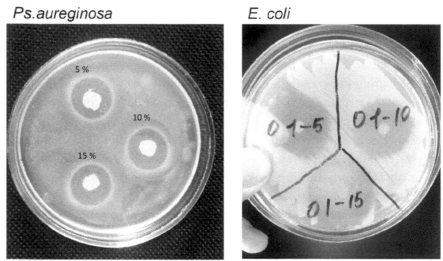

Membrane loaded with ceftriaxone

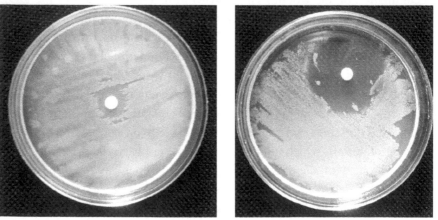

Control: paper disks + ceftriaxone

FIGURE 5.13 Results of testing antimicrobial effects of nonwoven P(3HB/4HB) membranes loaded with ceftriaxone compared with control (Volova's data).

of antibacterial activity of nonwoven and textile materials. Disks 5 mm in diameter were cut out of the electrospun nonwoven membrane and placed on the agar medium with the bacterial cells seeded on it. The cultures were incubated at 37°C for 24 h, and then the inhibition area was measured. Paper disks loaded with ceftriaxone (Bio RAD) to the same concentrations were used as control. Nonwoven disks with no antibiotic were also tested, showing no inhibition regions.

All membranes inhibited the growth of the both microbial cultures. The inhibition regions were very distinct, and their areas were easily determined. Changes in concentrations of the antibiotic in the membranes did not significantly influence the degree of inhibition, and the area of the inhibition regions on the membranes was comparable to that on the paper disks. Thus, dissolution of the drug in chloroform and the subsequent electrospinning did not decrease its efficacy. Of the two species, *E.coli* and *Ps.aureginosa*, *E.coli* was more sensitive to ceftriaxone.

The membranes loaded with ceftriaxone were used in the clinical trial. The purpose of the trial was to evaluate the efficacy of the wound dressing of degradable P(3HB/4HB) loaded with the antibacterial drug, ceftriaxone, in the treatment of purulent skin wounds. The trial was conducted with 22 patients, including 9 patients with acute purulent infections of soft tissues (abscesses, phlegmons, phlegmonous-necrotic erysipelas, trophic ulcers caused by venous disorders and diabetes mellitus), who were treated by using the proposed polymer dressings. The reference group consisted of the patients with the same diseases, who were treated by using wound dressing Voskopran (Russia). The groups were similar with respect to age, gender ratio, the type of purulent diseases of soft tissues, and the area of the wounds. The methods used in the trial were clinical evaluation of the wound healing, wound planimetry, pH measurement, laser Doppler flowmetry, and statistical processing. The patients were observed for between 14 and 45 days. When the patients with purulent diseases of soft tissues were admitted to the clinic, they underwent both the general physical examination and evaluation of the wound processes. The appearance of the wound, the presence or absence of skin hyperemia and infiltration of the wound edges, the time of the beginning of granulation and the type of the granulations, the start and development of epithelialization were evaluated during initial examination. All patients with purulent skin diseases (abscesses, phlegmons, phlegmonous-necrotic erysipelas) received immediate surgical treatment of the wounds in the operating room (lancing of abscesses and removal of as much necrotic tissue as possible, debridement of the wound, and drainage). All wounds were superficial, with the area no more than 1% of the body surface area, and the depth no more than 15 mm. After the wounds were debrided, they were covered with the dressings in accordance with the design of the trial. Subsequently, the medically indicated treatment included infusion, antibacterial, anti-inflammatory, and symptomatic therapies.

Topical medication was scheduled based on the wound-healing phase, according to V.I. Struchkov's classification, accepted in Russia (Struchkov, 1975). In the first two days, the dressing was periodically removed, the wounds were cleansed with a 3% hydrogen peroxide solution, purulent-necrotic masses were removed mechanically, and the dressing was applied again (either Voskopran or an ultrafine PHA film). The treatment in the inflammation phase included the use of a 0.02% PLIVA-sept solution. During the regeneration phase, bandaging was performed once every 2–3 days, and before the wound was covered with the dressing, solcoseryl and methyluracil ointments were applied to it. After the granulation tissue was formed, the defect healed by secondary intention.

The capillary blood flow was examined by laser Doppler flowmetry (LDF) with a BLF 21 monitor (Transonic System Inc., U.S.) equipped with a Type R (right angle) 15-mm-diameter probe to measure perfusion. LDF parameters were recorded by reading the data from the digital display or through the interface on the IBP-compatible computer, with FlowtraceSoftware, WinDaq 100 and WinDaq Playback packages, which are used to see the blood flow data in the display and process them. The capillary blood flow was measured on the skin of a healthy limb or the skin of another healthy body part; these values were taken as the initial data. Then, the blood flow was measured at the boundary between the healthy and hyperemic skin, 10 cm from the wound edge, at the wound edge, and in the wound. Parameters were monitored for 45 days, every 3 days. Measurements were taken at 1 or 2 points, depending on the size of the wound. Perfusion was examined before the treatment and during topical medication in different phases of the wound healing process. A Hi 99181 portable contact pH-meter (Germany) was used to measure pH at 10.0 cm from the wound, at the wound edge, and on the wound surface. Before the examination, excessive exudate was removed by dabbing at the wound with dry gauze. The adjustment of the pH-meter was performed automatically. The results were displayed in the liquid crystal display. Planimetry of the wound was conducted for 45 d, to evaluate the rate of wound healing. Changes in the area of the wound surface were monitored by Popova's method (Popova, 1942), to determine the wound healing rate.

By using the nanomatrix of P(3HB/4HB) as a wound dressing, the researchers managed to optimize the wound healing process, speed up the development of the mature granulation tissue in the wound by 3 and 3–15 days, and increase the frequency of the epithelialization time by 28.7% as compared with the process where Voskopran wound dressing was used. In the group of patients with Voskopran, the body temperature reliably decreased at Day 7 post surgery, reaching normal values at Day 12–17. When the patients were admitted to the clinic, their laboratory data showed high levels of endogenous intoxication. In Group 1, the leukocyte counts reached $18.5\pm2.7\times10^9$/L; by Day 12, this parameter declined to $8.4\pm0.48\times10^9$/L (p<0.05), reaching normal levels, $5.9\pm0.5 \times10^9$/L (p<0.05), by Day 17. During the first 24 h after wound debridement, the overall blood toxicity was reduced and the leukocyte counts decreased, but the differences were non-significant. In the first 24 h of the treatment, the leukocyte index of intoxication reached 5.42 ± 1.2 units, at Day 5, it was 4.17 ± 0.43 units (p<0.05), at Day 10, it was lower by a factor of two, decreasing to normal values only at Day 17 – 1.52 ± 0.34 units (p<0.05). In Group 2, with the experimental nanomatrices, the body temperature of the patient decreased at Day 3 after wound (trophic ulcer) debridement. At Day 6–7, the temperature was no more than 37°C. The leukocyte counts in the blood of the Group 2 patients were $16.1\pm1.2\times10^9$/L (p<0.05) at Day 3, reaching normal levels – $7.8\pm1.2\times10^9$/L (p<0.05) – at Day 10.

Endogenous intoxication in the patients of Group 2 decreased significantly at Day 3 (Figure 5.14). The examination of perfusion showed that in Group 1, tissue perfusion in the center of the wound was higher than at a point 10 cm away from the wound (13.1 ± 1.2 PU). After inflammation reduced, blood perfusion became lower. A significant decrease was recorded at Day 8–9 – 10.2 ± 0.54 PU.

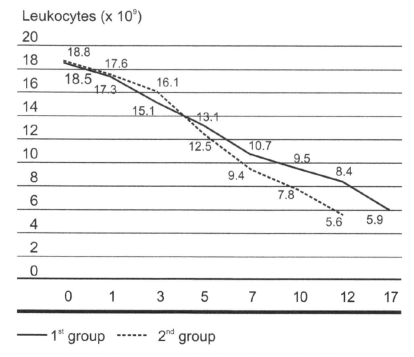

—— 1ˢᵗ group ····· 2ⁿᵈ group

FIGURE 5.14   Leukocyte counts of the Group 1 patients (with Voskopran) and the Group 2 patients (with a P(3HB/4HB) nanomatrix loaded with the antibiotic) (Vinnik's data).

In the Group 2 patients, the values of tissue perfusion of the wounds did not differ significantly; they were higher than the values at the point 10 cm away from the wound; at the wound edge, tissue perfusion reached 13.4±1.4 PU. After inflammation reduced, blood perfusion became lower. A significant decrease was recorded at Day 9–10 – 9.4±0.72 PU (Figure 5.15).

Results of the treatment were evaluated by the clinical indications of the wound healing process, by using wound planimetry followed by calculation of L.N. Popova's index, and cytologically – by measuring pH. The system of evaluation of the wound healing process is given in Table 5.7.

Clinical evaluation of the wound healing process was based on such parameters as the time of disappearance of the edema, formation of the first granulations, filling of the wound by mature granulation tissue, the start of epithelialization, and complete epithelialization. The outcomes of

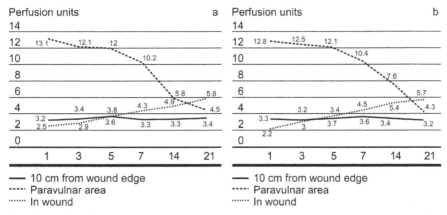

FIGURE 5.15 Characterization of microcirculation in the wound and at the wound edge in the reference group patients (with Voskopran) (a) and the treatment group patients (with a P(3HB/4HB) nanomatrix loaded with the antibiotic) (b) (Vinnik's data).

TABLE 5.7 Evaluation of the Outcome of the Treatment in the Clinical Groups (Vinnik's Data)

Evaluation indicator	Treatment outcomes (days)		
	Good	Adequate	Inadequate
Disappearance of perifocal edema	< 4	5 – 8	> 9
First granulations	< 6	7 – 10	> 10
Mature granulation tissue filling the wounds	< 10	11 – 16	>16
Epithelialization of the wound edges	< 11	12 – 17	> 17
Complete epithelialization	< 30	31 – 35	> 35
L.N. Popova's index in phases 2 and 3 of the wound healing process	> 4%	3 – 4%	< 3%

the treatment showed that in Group 1, the transition from phase 1 to phase 2 of the healing process took a long time: in 6 patients (46.1%), epithelialization was not complete after 45 days (Table 5.8).

The rates of wound healing in the treatment group patients suggested good and adequate outcomes. At Day 45, complete wound epithelialization was observed in 8 patients (88.8%) (Table 5.9).

TABLE 5.8   Evaluation of the Outcome of the Treatment in the Control Group (With Voskopran) (Vinnik's Data)

Evaluation indicator	Treatment outcomes (% patients)		
	Good	Adequate	Inadequate
Disappearance of perifocal edema	–	1 (7.7%)	12 (92.3%)
First granulations	–	1 (7.7%)	12 (92.3%)
Mature granulation tissue filling the wounds	–	4 (30.8%)	9 (69.2%)
Epithelialization of the wound edges	–	2 (15.4%)	11 (84.6%)
Complete epithelialization	3 (23.07%)	4 (30.8%)	6 (46.1%)

TABLE 5.9   Evaluation of the Outcome of the Treatment in the Treatment Group (with PHA Nanomatrix) (Vinnik's Data)

Evaluation indicator	Treatment outcomes (% patients)		
	Good	Adequate	Inadequate
Disappearance of perifocal edema	6 (66.6%)	2 (22.2%)	1 (11.1%)
First granulations	5 (55.5%)	4 (44.4%)	–
Mature granulation tissue filling the wounds	6 (66.6%)	3 (33.3%)	–
Epithelialization of the wound edges		8 (88.8%)	1 (11.1%)
Complete epithelialization	6 (66.6%)	1 (11.1%)	2 (22.2%)

Comparison of changes in pH values with the wound-healing rate showed a definite relationship between these parameters (Figure 5.16). The analysis of correlation between pH and wound healing rate revealed strong correlation between these parameters: as pH of the wound surface increased, the wound-healing rate decreased. Spearman rank correlation was equal to 0.9.

Figures 5.17 and 5.18 show two clinical cases: Patient G., and Patient P., respectively.

Thus, nonwoven membrane composed of ultrafine fibers prepared by electrospinning and loaded with ceftriaxone enables rapid epithelialization of purulent wounds when used as a wound dressing. When the dressing is removed, cell debris and bacteria are removed too. During the regeneration phase, ultrafine PHA film serves as a scaffold for the new tissue on

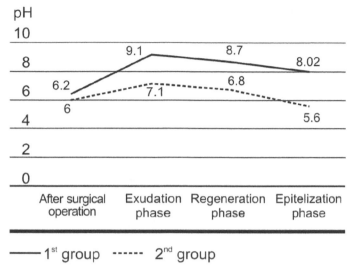

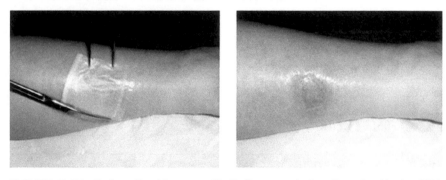

FIGURE 5.16 Comparative characterization of changes in pH in different phases of the wound healing process in different groups of patients (p>0.05) (Vinnik's data).

FIGURE 5.17 Patient G., 48 years old. Bullous erysipelas, Day 6 with the PHA nanomatrix with ceftriaxone – clean wound surface, developed granulation tissue, wound edge epithelialization (Vinnik's data).

the skin and soft tissue defects. The formation of the uniform and sufficiently vascularized tissue is a prerequisite for quicker wound healing and can serve as a basis for the subsequent skin grafting and spontaneous re-epithelialization of superficial wounds. The wound dressing tested in this clinical trial performs important physiological functions of natural skin, provides a barrier against secondary infection, reduces fluid loss, and, at the same time, does not keep the air out.

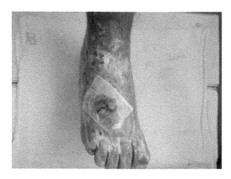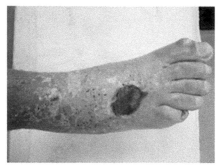

FIGURE 5.18   Patient P., 61 years old. Trophic ulcer of the dorsal surface of the right foot caused by arteriosclerosis obliterans of the vessels of the lower extremities; Day 14 with PHA nanomatrix with ceftriaxone – no pus or fibrin in the ulcer bed, mature pink granulations, wound edge epithelialization (Vinnik's data).

KEYWORDS

- skin defects
- degradable wound dressings
- defect repair
- planimetry
- histology
- clinical trials

REFERENCES

Abaev, Y. K. Rany i ranevaya infektsiya: Spravochnik khirurga (Wounds and wound infection: Surgeon's handbook). Fenix, Moscow, 2006, 428 p. (in Russian).

Arora, A., Prausnitz, M. R., Mitragotri, S. Micro-scale devices for transdermal drug delivery. Int. J. Pharm., 2008, 364, 227–236.

Benfer, M., Kissel, T. Cellular uptake mechanism and knockdown activity of siRNA-loaded biodegradable DEAPA-PVA-g-PLGA nanoparticles. Eur. J. Pharm. Biopharm., 2012, 80, 247–56.

Desai, M. P., Labhasetwar, V., Walter, E., Levy, R. J., Amidon, G. L. The mechanism of uptake of biodegradable microparticles in Caco-2 cells is size dependent. Pharm. Res., 1997, 14, 1568–73.

Eke, G., Kuzmina, A. M., Goreva, A. V., Shishatskaya, E. I., Hasirci, N., Hasirci, V. *In vitro* and Transdermal Penetration of PHBV Micro/Nanoparticles. J. Mater. Sci. Mater. M., 2014, 25, 1471–1481.

Eremenko, R. F., Gaidamaka, A. V., Gladkova, L. V., Gubina, T. N. A study of pharmacological activity of loratadine gel in the model of the contact dermatitis. Provizor (Pharmacist), 1999, 1, 16–19 (in Russian).

Foster, K. A., Yazdanian, M., Audus, K. L. Microparticulate uptake mechanisms of in vitro cell culture models of the respiratory epithelium. J. Pharm. Pharmacol., 2001, 53, 57–66.

Gawkrodger, D. J. Dermatology. Edinburgh Churchill Livingstone Inc. NY, 2002, 1, 109–124.

Goreva, A. V., Shishatskaya, E. I., Volova, T. G., Sinskey, A. J. Characterization of polymeric microparticles based on resorbable polyesters of oxyalkanoic acids as a platform for deposition and delivery of drug. Polym. Sci. A, 2012, 54, 94–105.

Kaem, R. I. Morfologiya rany i ranevaya infektsiya (Wound Morphology and Wound Infection). 2nd revised edition. Meditsina. Moscow, 1990, 451–457 (in Russian).

Minchenko, A. N. Rany. Lecheniye i profilaktika oslozhneniy (Wounds. Treatment and prevention of complications). Spets. Lit. St. Petersburg, 2003, 207 p. (in Russian).

Murueva, A. V., Sheshneva, A. M., Shishatskaya, E. I., Volova, T. G. A study of polymer microcarriers loaded with anti-inflammatory drugs in the treatment of model defects of skin. Byulleten experimentalnoi biologii i meditsiny (Newsletter of Experimental Biology and Medicine), 2014, 5, 614–619 (in Russian).

Nikolaeva, E. D., Shishatskaya, E. I., Mochalov, K. E., Volova, T. G., Sinskey, A. J. A comparative study of cell scaffolds prepared from resorbable polyhydroxyalkanoates with different chemical compositions. Kletochnaya transplantologiya i tkanevaya inzheneriya (Cell Transplantation and Tissue Engineering), 2011, 6(4), 63–67 (in Russian).

Popova, L. N. Kak izmenyayutsya granitsy vnov obrazuyushchegosya epidermisa pri zazhivlenii ran (On changes of the boundaries of the new epidermis during wound healing). Summary of MD thesis, 14.00.27. Moscow, 1942, 16 p. (in Russian).

Prausnitz, M. R., Langer, R. Transdermal drug delivery. Nature Biotechnol., 2008, 26, 1261–1268.

Prow, T. W., Grice, J. E., Lin, L. L., Faye, R., Butler, M., Wurme, E. M. Nanoparticles and microparticles for skin drug delivery. Adv. Drug Deliver. Rev., 2011, 63, 470–491.

Raben, A. S., Alexeeva, O. G., Dueva, L. A. Experimentalnyy allergicheskiy kontaktnyy dermatit (Experimental Contact Dermatitis). Meditsina. Moscow, 1970, 191 p. (in Russian).

Sahoo, S. K., Panyam, J., Prabha, S., Labhasetwar, L. Residual poly-vinyl alcohol associated with poly (D, L-lactide-co-glycolide) nanoparticles affects their physical properties and cellular uptake. J. Control Release, 2002, 82, 105–14.

Sloan, K. B., Wasdo, S. C., Rautio, J. Design for optimized topical delivery: prodrugs and a paradigm change. Pharm. Res., 2006, 26, 2729–2747.

Struchkov, V. I., Grigoryan, A. V., Gostishchev, V. K. Gnoinaya rana (Purulent wound). Meditsina. Moscow, 1975, 311 p. (in Russian).

Tumanov, V. P., German, G. Metodicheskoye rukovodstvo po lecheniyu ran (A guide in wound treatment). Paul Hartmann. Moscow, 2000, 123 p. (in Russian).

Vinnik, Yu. S., Markelova, N. M., Shishatskaya, E. I., Kuznetsov, M. N., Solovyeva, N. S., Zuev, A. P. The use of ultrafine films based on polyhydroxyalkanoates in patients with purulent diseases of soft tissues. Khirurgicheskaya praktika (Surgical practice), 2015, 2, 20–25 (in Russian).

Volova, T. G., Shishatskaya, E. I. Biorazrushayemyye polimery: sintez, svoistva, primeneniye (Biodegradable Polymers: Synthesis, Properties, Applications). Krasnoyarskii Pisatel. Krasnoyarsk, 2011, 49 p. (in Russian).

Volova, T. G., Shishatskaya, E. I. Results of biomedical studies of PHAs produced in the Institute of Biophysics SB RAS and Siberian Federal University. In: *Polyhydroxyalkanoates (PHA): Biosynthesis, Industrial Production and Applications in Medicine.* Chapter 21. Nova Sciences Publ. Inc. NY, USA, 2014, 273–330.

Volova, T. G., Goncharov, D. B., Sukovatyi, A. G., Shabanov, A., Nikolaeva, E. D., Shishatskaya, E. I. Electrospinning of polyhydroxyalkanoate fibrous scaffolds: effect on electrospinning parameters on structure and properties. J. Biomat. Sci. Polym. E., 2014, 25, 370–393.

Wosicka, H., Cal, K. Targeting to the hair follicles: current status and potential. J. Dermatol. Sci., 2010, 57, 83–89.

Xiong, Y. C., Yao, Y. C., Zhan, X. Y., Chen, G. Q. Application of poly-hydroxyalkanoates nanoparticles as intracellular sustained drug-release vectors. J. Biomat. Sci. Polym. E., 2010, 21, 127–140.

Yin, W. K., Feng, S. S. Effects of particle size and surface coating on cellular uptake of polymeric nanoparticles for oral delivery of anticancer drugs. Biomaterials, 2005, 26, 2713–2722.

Yow, H. N., Wu, X., Routh, A. F., Guy, R. H. Dye diffusion from microcapsules with different shell thickness into mammalian skin. Eur. J. Pharm. Biopharm., 2009, 72, 62–68.

CHAPTER 6

POTENTIAL OF POLYHYDROXYALKANOATES FOR BONE DEFECT REPAIR

CONTENTS

6.1 INTRODUCTION

Osteoplasty is an important branch of reconstructive surgery in trauma treatment, orthopedics, maxillofacial surgery, and dentistry. Bone tissue is dense mineralized fibrous tissue, which is very strong mechanically. However, when a force exerted against a bone is stronger than it can structurally withstand, a fracture occurs. Bone tissue is capable of self-healing, but sometimes, tissue loss caused by the injurious factor is too great for the bone to heal. Orthopedic and trauma surgeons use special devices and materials to compensate for this loss and to repair bone defects of various origins.

There is a need for biocompatible materials for closing bone defects caused by surgical intervention or traumas. A special challenge is presented by the large defects of bone tissue due to traumas, cancer, and other diseases. As polyhydroxyalkanoates (PHAs) have high mechanical strength, are slowly degraded *in vivo*, and have piezoelectric effect, they hold great promise for bone regeneration. In the early stages of PHA research, different studies showed the possibility of producing mechanically strong constructions for osteogenesis based on PHAs and PHA composites with hydroxyapatite (HA), which enhances polymer strength (Doyle et al., 1990; Knowles et al., 1992, 1993a, b; Köse et al., 2003a, b, c). Biocompatible and osteoinductive properties of P3HB and P3HB-based composites were proved in experiments on bone defect regeneration in laboratory animals (Mai et al., 2006; Artsis, 2010; Derya et al., 2011). Plates of PHA were effective in closing jaw defects (Yang et al., 2004) and skull defects (Marois et al., 2002; Luklinska et al., 2003). The potential of three-dimensional scaffolds made of PHA was evaluated in rabbit articular cartilage defect model. Engineered polymer constructs inoculated with rabbit chondrocytes were implanted to rabbits; after 16 weeks of implantation, the constructs were filled with white cartilaginous tissue; accumulation of extracellular matrix including type II collagen was observed on the surface of the degrading scaffold (Wang et al., 2008). Important results have been obtained in a study of PHA constructs used to treat experimental osteomyelitis. To prolong the action of antibiotics, they can be encapsulated in Ca-gelatin constructs, bioceramics, and polylactide carriers. Elution of antibiotics in such systems occurs, however, at rather high rates, resulting in ineffective treatment of long-term bone infections. Türesin et al. (2001) investigated PHA delivery systems for prolonged-action antibiotics. Ou et al. (2000) reported that P3HB coating on titanium implants reduced tissue inflammation and facilitated osteogenesis. A similar effect was obtained when P3HB was used to coat ceramic (hydroxyapatite and tricalcium phosphate) implants (Szubert et al., 2014). Successful employment of continuous cell cultures, including precursor cells generating specialized tissues, in experimental biology and medicine has become a prerequisite to the development of novel technologies and approaches for reconstructive orthopedics (Horowitz et al., 1999; Vacanti, 2000). Stem cells hold great promise for bone tissue repair (Terada et al., 2000; Deev et al., 2000; Desyatichenko, Kurdyu-mov, 2008).

Poly-3-hydroxybutyrate has a favorable effect on growth and pro-liferation of mouse osteoblast MC3T3-E-1 cells, improving the quality of bone tissue, which makes it a potentially effective anti-osteoporosis drug. These data were confirmed by the most recent studies, proving that 3-hydroxybutyrate acid and its derivative 3-hydroxybutyrate methyl ester inhibit the development of osteoporosis in mice maintained under simu-lated microgravity, helping preserve bone microstructure and mechanical property (Qian et al., 2014). Ye et al. (2009) investigated the potential of PHA terpolymers P(3HB/3HB/3HHx) as scaffolds for culturing and dif-ferentiation of human adipose-derived stem cells. The authors showed that the scaffolds favored cell attachment and facilitated chondrogenic differentiation of cells. Moreover, in the ectopic test, after the scaffolds were implanted into the subcutaneous layer of mice, they were partially degraded, and cartilage tissue formed both on the surface of the scaffolds and inside them. Composites of P3HB and P(3HB/3HV) with wollastonite were found to be able to support osteoblast cell attachment, also showing good mechanical properties (Reis et al., 2010). Composites of PHA and bioceramics P(3HB/Bioglass) favored osteoblast cell adhesion and *in vivo* formation of calcium-phosphate structures (Misra et al., 2010). Rambo et al. (2012) demonstrated osteogenic potential of the composite mate-rial based on P(3HB/3HV) in polylactide matrix, showing that incorpora-tion of copolymer microspheres into the matrix enhanced the physical/mechanical properties and porosity of the material and, thus, resulted in better MSC proliferation compared with polylactide matrix. Paşcu et al. (2013) reported that scaffolds containing P(3HB/3HV), silk, and nano-hydroxyapatite mimicked the natural bone tissue architecture and facili-tated proliferation of human osteoblasts, forming cell precipitates, which confirmed their biological activity.

Researchers of the Siberian Federal University and the Institute of Bio-physics SB RAS carried out integrated studies of polyhydroxyalkanoates as osteoplastic materials for reconstructive osteogenesis (Shishatskaya et al., 2005, 2006, 2008; Volova et al., 2008, 2011, Shishatskaya et al., 2006). They produced PHA composites with hydroxyapatite and wollastonite and investigated physicochemical and physical/mechanical properties of dense and porous 3D implants prepared by various methods (Shishatskaya et al., 2006; Volova et al., 2008). Experiments with animals proved that the P(3HB/HA) composite had osteoinductive properties, facilitated bone tissue formation during ectopic bone formation (Shishatskaya et al., 2006) and induced osteogenesis (Shishatskaya et al., 2008). As *in vivo* studies showed that PHA had osteoinductive properties and did not cause adverse tissue response when implanted to animals (Shishatskaya et al., 2006), the purpose of the following study was to investigate the possibility of using PHAs to reconstruct defects of cranial flat bones and defects of the tubular bone aggravated by infection.

6.2 A STUDY OF THE EFFECTIVENESS OF USING PHA 3D IMPLANTS TO REGENERATE THE MODEL DEFECT OF SKULL BONES OF LABORATORY ANIMALS

Cranial flat bones have a low self-healing ability because of their biological properties, determined by the embryonal development and histophysiological parameters (Shevtsov et al., 2000). Thus, injuries of skull bones cause long-term disability or death (Gaidar et al., 2001). The fractures of the skull bone tissue are usually ductile fractures: buttonhole fractures with smooth edges. The main mechanisms of flat bone destruction are displacement of layers and tears of the compact bone layers. Human skull bone defects cannot be filled with bone tissue without surgery (Matveeva, 1962; Zotov et al., 1998); therefore, the defects are filled with fibrous connective tissue, followed by scarring. Complex and complicated skull fractures dramatically reduce the self-healing ability of bone tissue, leading to impairment of osteogenesis. In this case, reparative osteogenesis and osteoplastic surgery require specialized materials (Gololobov et al., 2001).

Experiments were conducted on female Wistar rats, 200–250 g each. The rats received 0.2 mL of general anesthetic Zoletil 100 each. The surgery was performed following the ethical guidelines for animal research. The parietal region of the skull was prepared for surgery, and the skin, subcutaneous layer, and epicranial aponeurosis were dissected layer-by-layer. The periosteum was detached with a solution of anesthetic by means of a raspatory. A 5–6-mm-diameter hole was drilled and flushed with physiological saline, without damaging dura mater. Then, implants were placed into the defect. The animals were divided into four groups: Group 1 – negative control, with the defect repaired without an implant; Group 2 – positive control, with the defect filled with commercial drug Colapol; Group 3 – treatment, the defect filled with porous 3D implants of P3HB; Group 4 – treatment, the defect filled with porous 3D implants of P3HB seeded with osteoblastic cells differentiated from the adipose tissue MSCs. The wounds were tightly closed with interrupted sutures and then treated with Vetoquinol Aluspray (Vetoquinol S.A., France). After surgery, the animals were placed into the oxygen cabinet, to restore respiration. During the first 3 days, the animals received antihistamine Claritin-Tavegyl, 1/10 pill each. The reparative osteogenesis process at the implantation site was monitored for 4 months. The animals underwent computed tomography to monitor defect repair; samples of blood were taken to measure hematological, molecular, and biochemical parameters; tissue samples were taken for histology. The animals were sacrificed by using an overdose of an anesthetic in 15 days, 1, 2, 3, and 4 months post-surgery (Volova's data).

No animals died during the postoperative period. None of the animals showed any significant postoperative complications. Feeding behavior and motion activity were restored in 12 h after the surgery. The defect site was moderately edematous, and sutures were being epithelialized. At Day 3, edemas were reduced in all groups of animals. All implants stayed in the defects throughout the observation period. The macroscopic examination did not reveal any tissue suppuration or inflammation at the implant site. Until Day 15 post-surgery, the leukocyte counts in the blood had been increased, with the highest counts recorded in the positive control group. Leukocytosis was mainly caused by the increased counts of band neutrophils: they were higher in the two treatment groups and in the positive control than in the negative control. The monocyte counts were also increased, because of migration of the reserve cells from bone marrow and the marginal blood pool, frequently associated with the musculoskeletal traumas. The lymphocyte counts were decreased, which might be associated with the suppression of the rats' immune system. The lowest lymphocyte counts were recorded in the positive control group ($61.05 \pm 0.31\%$). Later, the lymphocyte counts reached their normal levels. The hematological parameters of the animals were stabilized in 30 days.

Examination of blood phosphatase activity showed a considerable increase in the activity of acid phosphatase (AP) in all animals at Day 15 post-surgery (Figure 6.1). That was caused by local acidosis and the activity of osteoclasts, which are involved in the resorption of bone fragments

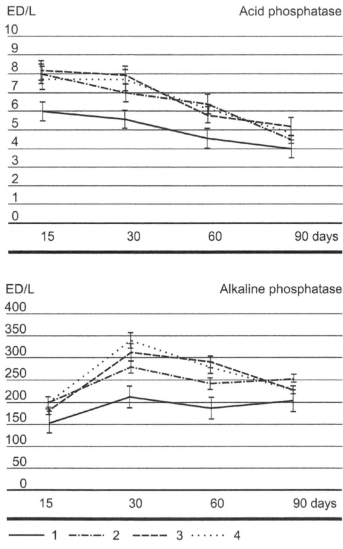

FIGURE 6.1   Levels of acid phosphatase (AP) and alkaline phosphatase (ALP) activities in rats' blood serum: 1 – negative control (with the defect repaired without an implant); 2 – positive control (with the defect filled with Colapol); 3 – treatment (the defect filled with a porous 3D implant of P3HB; 4 – treatment (the defect filled with a porous 3D implant of P3HB seeded with osteoblastic cells) (Volova's data).

and necrotic tissues formed in the course of inflammatory process. High AP levels were recorded in the two treatment groups (7.7 ± 1.2 units/L) and the positive control group (8 ± 0.5 units/L), but the differences between animals in the same group were no higher than 0.50 units/L during the experiment. A slight increase in the AP in the positive control group at Day 60 must have been caused by the complete resorption of Colapol at the defect site. In the negative control group, the AP activity was at the normal level from Day 30 to the end of the experiment. Alkaline phosphatase (ALP) activity in the blood serum of all animals was significantly decreased in the first 15 days, most probably because this enzyme was actively involved in phases 1 and 2 of bone demineralization. After Day 30, however, the ALP activity increased in all animals, which might be associated with the ossification: this process is performed by osteoblasts, releasing ALP. In that phase, alkaline phosphatase activity reached its peak, being involved in the synthesis of extracellular matrix and mucopolysaccharides and in the formation of fibrillary proteins, facilitating mineral salt deposition. The highest ALP activity (340 ± 0.45 units/L) was recorded in Group 4 (hybrid implants: 3D constructs of P3HB + osteoblast cells). In the positive control group, the ALP activity varied during the experiment: at Day 60, it decreased to 239 ± 0.15 units/L, but at the end of the experiment it rose again, reaching 250 ± 0.15 units/L. This might have been caused by complete resorption of collagen constituting Colapol, with no bone defect repair taking place. In the negative control group, a dramatic decrease in the ALP occurred at Day 15, but in later phases, ALP activity in the blood serum was at a normal level.

Bone formation process was analyzed by performing real-time PCR to determine the expression of the genes of BGP (osteocalcin), VEGF (vascular endothelial growth factor), and Col 1 (collagen type I). cDNA expression was determined relative to the expression of the housekeeping gene − β-actin. In 30 days post-surgery, the highest copy number of cDNA (6.410) encoding BGP was recorded in the treatment group with porous implants of P3HB seeded with osteoblasts (Figure 6.2); the lowest copy number was in the negative control group (5.596). The evaluation of the activity after 90 days showed a decrease in the BGP production in all groups, as the major part of BGP was expressed in the phases of bone formation, taking part in osteoid mineralization.

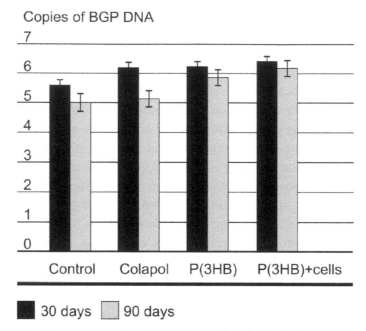

FIGURE 6.2 The copy number of cDNA encoding BGP of the regenerating bone fragments at Days 30 and 90: the negative control group, the positive control group – Colapol; P3HB porous implant; P3HB porous implant with osteoblast cells differentiated from AT-MSCs (Volova's data).

VEGF – vascular endothelial growth factor, stimulating vessel growth – angiogenesis, is also a critical condition for bone regeneration. VEGF stimulates migration and differentiation of human primary osteoblasts (Orlandini, 2007) and serves as a mediator of such osteoinductive factors as TGF-β1, IGF, and FGF-2. The use of tissue-engineering constructs is often limited by poor vascularization, which may be an obstacle to the transport of nutrients. These factors inhibit formation of new bone tissue. In the study described here, at Day 30 post-surgery, the level of expression of cDNA encoding VEGF was low compared to the control, where the cDNA copy number was 1.658. At Day 90, however, the cDNA expression increased in all groups (Figure 6.3).

The major mechanism of bone tissue repair is formation of the bone matrix, which is structurally based on collagen, mainly collagen type I, fibers. At Day 30, the highest copy number (11.719) of cDNA encoding Col 1 was recorded in the group with Colapol; the lowest number was in

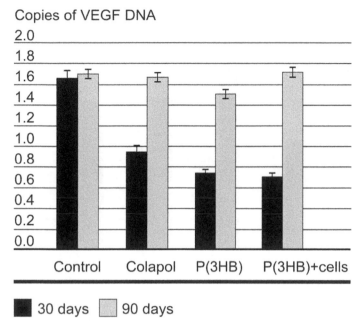

FIGURE 6.3 Copy number of cDNA encoding VEGF of the regenerating bone fragments at Days 30 and 90: the negative control group, the positive control group – Colapol; P3HB porous implant; P3HB porous implant with osteoblast cells differentiated from AT-MSCs (Volova's data).

the negative control group (7.356) (Figure 6.4). In the treatment groups (P3HB implant and P3HB + cells), the cDNA copy number was comparable to that in the group with Colapol. That was the phase of formation of fibrous tissue and resorption of bone fragments. After 90 days, Col 1 production in the negative and positive control groups decreased significantly, remaining almost unchanged in the both treatment groups.

The processes of regeneration of the skull defects differed between groups. At Day 30, in the negative control group, the defect was partly filled by necrotic masses and fibrous tissue; there were foci of leukocyte infiltration (Figure 6.5A).

Regenerative-proliferative changes with a narrow necrotic fringe can be seen on the edges of the defect. At the same time point, in the positive control group, the defect was filled with fibrous tissue containing numerous segmented leukocytes, lymphocytes, and plasma cells, which

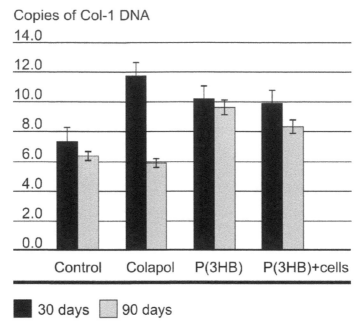

FIGURE 6.4 Copy number of cDNA encoding Col 1 of the regenerating bone fragments at Days 30 and 90: the negative control group, the positive control group – Colapol; P3HB porous implant; P3HB porous implant with osteoblast cells differentiated from AT-MSCs (Volova's data).

produced antibodies in response to the implantation of foreign material. The remaining Colapol was surrounded by a small fibrous capsule with macrophages; necrotic regions were visible at the sites of bone fragment resorption and collagen destruction (Figure 6.5B). The defect was filled by loose coarse fibrous tissue considerably infiltrated by segmented leukocytes, lymphocytes, and plasma cells. Around the remaining Colapol, there were necrotic regions with destroyed collagen and resorbing bone fragments; a thin fibrous capsule infiltrated by mononuclear cells could be seen. At Day 30 after implantation of P3HB porous structure (Figure 6.5C), the defect was also filled by fibrous tissue of varying density. The inflammatory infiltration of the fibrous tissue was less pronounced than in the group with Colapol and was mainly represented by mononuclear cells as well as by macrophages and foreign body giant cells. The fibrous

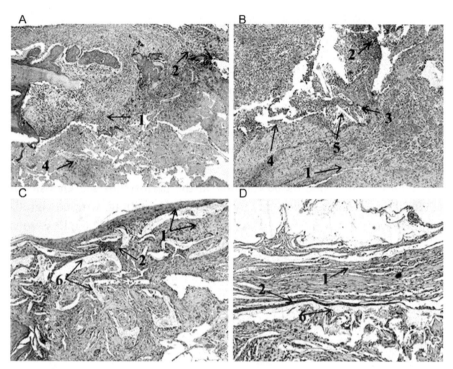

FIGURE 6.5 Histological sections of regenerating bone fragments at Day 30 post surgery: A – negative control (with the defect repaired without an implant); B – positive control (with the defect filled with Colapol); C – treatment (the defect filled with a porous 3D implant of P3HB); D – treatment (the defect filled with a porous 3D implant of P3HB seeded with osteoblastic cells). The arrows point to: 1 – fibrous tissue; 2 – leukocyte infiltration; 3 – plasma cells; 4 – leukocytes; 5 – Colapol; 6 – P3HB porous implant. Magnification 100× (Volova's data).

tissue had grown into the porous structure of the polymer implant and formed reticulofibrous callus. Similar changes were observed at Day 30 in the group with the implant of P3HB with osteoblasts (Figure 6.5D). A developing connective tissue callus with a forming fibrous operculum was observed at the defect site; granulation and fibrous connective tissues were growing into the pores of the implant. Inflammatory response was minimal; a few segmented leukocytes, lymphocytes, and macrophages were observed.

At Day 90, in the negative control group, coarse fibrous tissue with highly ordered collagen was developing (Figure 6.6A). Osteogenesis, mainly due to osteoblast and osteocyte activity, was observed on the edges of the defect. Thin, randomly oriented trabeculae were formed. At the same time point, in the positive control group (with Colapol), the center of the defect was also filled by coarse vascularized fibrous tissue, which contained foci of inflammatory infiltrations, mainly around resorbing old bone fragments (Figure 6.6B). At the edges of the defect, slow formation of lamellar bone tissue was observed. In the treatment group, with P3HB porous implants, the defect region was much smaller than in the controls; it was filled by sufficiently vascularized connective tissue callus, with the new bone tissue forming at the periosteum and endosteum (Figure 6.6C). Residual polymer material was degrading, with no significant inflammatory response observed. In the group with 3D P3HB porous

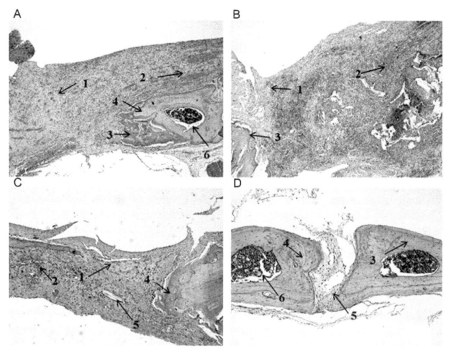

FIGURE 6.6 Histological sections of regenerating bone fragments at Day 90 post surgery: A – negative control; B – positive control (with the defect filled with Colapol); C – treatment (the defect filled with a porous 3D implant of P3HB); D – treatment (the defect filled with a porous 3D implant of P3HB seeded with osteoblastic cells). The arrows point to: 1 – fibrous tissue; 2 – blood vessels; 3 – trabeculae; 4 – new bone tissue; 5 – P3HB implant material; 6 – bone marrow. Magnification 100× (Volova's data).

implants seeded with osteoblasts, bone formation was observed both at the edges of the defect, i.e., centripetally, from the cellular sources of reparative osteogenesis, and in the central regions of the defect (Figure 6.6D). The defect was filled by fibrous connective tissue with numerous vessels and mineralized trabeculae. The edges of the defect consisted of the new bone tissue with osteoblasts synthesizing intercellular substance, osteocytes, and a few Haversian canals. In this group, bone tissue regeneration occurred at the highest rate.

At Day 120, the defects were considerably reduced in all groups (Figure 6.7). In the negative and positive control groups, reparative osteogenesis was incomplete (Figure 6.7A, B). The central part of the regenerating

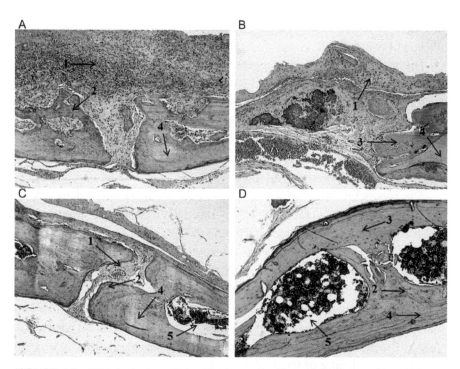

FIGURE 6.7 Histological sections of regenerating bone fragments at Day 120 post surgery: A – negative control; B – positive control (with the defect filled with Colapol); C – treatment (the defect filled with a porous 3D implant of P3HB); D – treatment (the defect filled with a porous 3D implant of P3HB seeded with osteoblastic cells). The arrows point to: 1 – fibrous tissue; 2 – new bone tissue; 3 – osteocytes; 4 – lamellar bone tissue; 5 – bone marrow. Magnification 100× (Volova's data).

tissue was still represented by fibrous connective tissue with the foci of lymphocyte infiltrations around trabeculae. The formation of mature lamellar bone tissue was slow. By contrast, in the treatment group with P3HB 3D porous implants, at Day 120, the fibrous tissue in the central region was completely replaced by the new bone tissue with numerous trabeculae, which merged with each other and with the defect edges (Figure 6.7C). Thus, the new bone tissue definitely prevailed, and the defect was closed. The new bone tissue consisted of mature lamellar bone with osteocytes and the medullary canal. In the group with implants of P3HB and osteoblasts, reparative osteogenesis was complete. The bone defect was completely replaced by well-developed bone tissue with restored organotypic tissue architectonics (Figure 6.7D). There was a slight thickening of periosteum at the defect site; it contained new medullary canals with sufficient amounts of bone marrow, without reactive changes, with high adipocyte content.

The results of histology suggest that porous 3D implants of biocompatible and biodegradable poly-3-hydroxybutyrate facilitate bone regeneration owing to their gradual biodegradation and bone tissue growing into the pores of the implant, with the minimal inflammatory response. Osteoblastic cells seeded onto the implant retain their viability and high proliferation rates for a long time, enabling more effective regeneration of bone tissue and, thus, reducing the time needed for regeneration.

The regeneration of the bone tissue defect was monitored by taking digital images of the rats' skulls by multislice computed tomography (MSCT) with multiplanar reconstruction (MPR). The scanning was done at 120 kV and 70 mA, Rot with low-energy X-ray radiation 0.70s/HE +5.6 mrn/rot, 0.6 mm 0.562:1/0.6sp. Digital 3D projections were obtained with the image resolution of 18 mm; the data were processed with the Radi AntDicom Viewer software. The parameters examined were the size and shape of the bone defect, the uniformity of the new tissue structure, the state of the periosteum, and the presence of fractures and other injuries at the site of the closing of the model cranial defect. The filling of the defect was scanned in the front and axial projections.

Figure 6.8 shows digitized results of determining the defect size by computed scanning of the skull in front projection at different time points of the experiment. At Day 30, 4–5-mm defects are clearly seen in the left

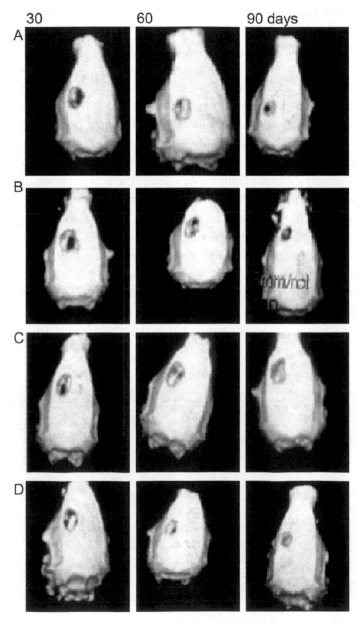

FIGURE 6.8 Computed tomography images (front projection) of the regeneration of bone defects in the rats' skulls at Days 30, 60, and 90: A – negative control; B – positive control (with the defect filled with Colapol); C – treatment (the defect filled with a porous 3D implant of P3HB); D – treatment (the defect filled with a porous 3D implant of P3HB seeded with osteoblastic cells) (Volova's data).

parietal bone in all groups. In the positive control group, the boundaries of the defect are vague, which could be caused by insignificant bone resorption. At Day 60, the CT digital images show the reduction in bone defects in all groups, with some regeneration of the bone tissue at the edges of the cavity. In the treatment group with P3HB implants seeded with osteoblasts, the image shows the greatest amount of the native bone tissue; it can be seen not only at the edges but also on the upper lateral surface of the parietal bone, inside the defect. At Day 90 post-surgery, the bone defect was partly regenerated in the treatment groups. The density of the new bone tissue was different from the density of the cortical bone of the skull; the new bone tissue completely lined the bone defect. At that time point, no significant filling of the bone defect was observed in the negative and positive control groups. The use of Colapol to fill the defect did not result in quicker regeneration compared with the negative control.

Figure 6.9A shows digitized results of scanning the defect size in axial projection, which provides a clearer demonstration, at Day 90 post-surgery. The smallest defects were observed in Groups 3 and 4: 1.9 and 1.4 mm, respectively. At the same time point, the size of the unfilled defect in

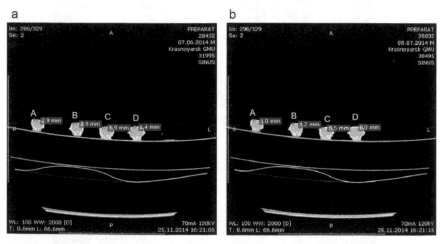

FIGURE 6.9 Computed tomography images (axial projection) of the regeneration of bone defects in the rats' skulls at Days 90 (a) and 120 (b): A – negative control; B – positive control (with the defect filled with Colapol); C – treatment (the defect filled with a porous 3D implant of P3HB); D – treatment (the defect filled with a porous 3D implant of P3HB seeded with osteoblastic cells) (Volova's data).

the control groups was about 2.9 mm. At Day 120, in the treatment group with the P3HB implant, the size of the defect was no more than 10% of its initial size (0.5 mm). In the treatment group with the P3HB implant seeded with osteoblasts, the defect was completely closed. In the negative and positive control groups, the size of the defect was about 1.0–1.2 mm (Figure 6.9B).

The study described above suggests that porous 3D implants of the degradable poly-3-hydroxybutyrate are effective for repairing the slowly regenerating defects of cranial flat bones. Even more effective are tissue-engineering grafts: porous polymer 3D scaffolds seeded with osteoblast cells differentiated from AT-MSCs.

6.3 EVALUATION OF THE EFFECTIVENESS OF USING P3HB IMPLANTS AND FILLING MATERIAL TO REPAIR MODEL DEFECTS OF TUBULAR BONES IN LABORATORY ANIMALS

The diaphysis of the tubular bone is usually subjected to ductile-brittle comminuted fractures. The main mechanism of destruction of tubular bone diaphysis is formation of lacunar-cavernous deformations of the bone matrix. Management of compound fractures (multifragmentary, gun-shot, complicated fractures), which consolidate slowly; treatment of complications caused by fractures (pseudarthrosis, osteomyelitis, etc.); and plastic surgery of bone tissue defects created by resections of tumors and tumor-like growths involve the use of bone-replacing materials, special constructions, and devices. At the present time, surgical reconstruction of the defects in the bones of the limbs frequently causes complications.

Researchers of the Institute of Biophysics SB RAS and the Siberian Federal University studied the effectiveness of using porous 3D implants of P3HB to repair tubular bone defects. A high-purity powdered homopolymer of 3-hydroxybutyric acid P3HB was mixed with crystalline sucrose and subjected to cold pressing, to produce 3D constructs. The constructs were placed in water, sucrose was dissolved, and porous implants were produced.

Osteoplastic properties of P3HB 3D implants were studied in experiments on 4-month-old male chinchilla rabbits (12 animals, 3 per group) with model defects of the diaphysis of the femoral bones. The rabbits received anesthesia (xylazine), the medial anterior surface of the femur was treated with a 70% alcohol solution of chlorohexidine, and a 5-cm long skin incision was made. The muscles were bluntly separated and fixed. A bone defect was created on the anterior surface of the femur with an osteotome. Then, the bone cavity was filled with different materials. In Group 1 (treatment), the defect was filled with P3HB 3D porous implant; in Group 2 – with the hydroxyapatite/collagen composite material (Colapol, Polistom, Russia). In Group 3, the defect was filled with autograft bone chips. The defects were filled completely. The defects were covered with ultrafine film of PHA. The wounds were completely closed layer-by-layer and covered with aseptic dressings. To avoid fractures, the limbs were immobilized with plaster splints. The wounds were dressed once a day, during seven days. During the first three days, the animals received intermuscular injections of Sol. Ketonali 0.1 for relieving pain. During the postoperative period, the animals were observed to check their feeding behavior, motion activity, and the state of the operated limbs. X-ray examination of the operated limbs was performed with a RUM-20 system, at 44 mA, 0.1 kV, and a 1-second exposure. The X-ray examination was performed at Days 30 and 90 post-surgery. Such parameters as the size and shape of the defect, the uniformity of the new bone tissue, the state of the periosteum, cortical plate, and medullar canal, and the presence or absence of fractures at the defect site were monitored. Prior to the morphological examination, the bone tissue samples were decalcified in a Trilon-B solution and then embedded in paraffin. Histological sections, 5–7 μm thick, were stained with hematoxylin and eosin. Microscopic examination was done to evaluate the state of the soft tissues surrounding the bone, periosteum, cortical layer, and medullary canal.

Clinical characterization of the state of the treatment group animals (with P3HB implants), the Group 2 animals (with Colapol), and the Group 3 animals (with autograft bone) is given in Table 6.1. During the early postoperative period, the rabbits were sluggish, moved very little, did not load the affected limb. At Days 3–5, the rabbits resumed normal feeding behavior and locomotor activity; at Day 7–9, the Group 2 animals showed soft tissue edemas at the surgery site, and Group 3 animals had local hyperemia of skin. At Day 15.5±1.2 on average, the edemas and

TABLE 6.1 Characterization of the State of Laboratory Animals in the Postoperative Period (Shishatskaya's data)

Parameter	Days post-surgery (M±SD)		
	Group 1, treatment P3HB 3D implant	Group 2, Colapol	Group 3, autograft bone
Recovery of support ability of the affected limb	5±0.9*	6±0.9*	6±1.2
Healing of the surgical skin wound	7.4±0.9*	7.2±0.9*	8.9±1.7
Reduction in the edema and soft tissue hyperemia	8.75±0.96*	9.28±0.9*	11.7±1.8

Note: * – the data are statistically significant at $p < 0.05$.

hyperemia were reduced. The animals began to load the affected limbs at Day 5.01±1.5 post-surgery. The surgical wounds healed by first intention, no suppuration was recorded, sutures were removed at Day 7. All animals remained alive.

X-ray examination of the rabbits' limbs after surgical intervention showed that in the group with P3HB 3D porous implants, at Day 30, in the middle third of the femoral bone, there was a 2.5-mm round cavity, which was considerably smaller than the initial defect and the defect observed at Day 30 in Group 2 (Colapol) (Figure 6.10). The cavity had well-defined boundaries, without a sclerotic rim. The periosteum was not changed, and there was no periosteal reaction. In Group 2, with Colapol used to fill the defect, the survey radiograph taken at Day 30 showed a 3-mm round cavity in the middle third of the femoral bone. The cavity had well-defined boundaries, without a sclerotic rim. In 2 animals of Group 3, the survey radiograph taken at Day 30 showed a pathologic fracture in the middle third of the femoral bone (where the bone tissue defect had been created), with bone fragments displaced transversely and angularly, with diastasis of the terminal 3–4 mm. The radiograph showed periosteal reaction of bone fragments, local foci of bone tissue destruction, and osteolysis. At Day 90, partial consolidation was observed. The formation of the false joint at the fracture site, chronic periosteal inflammation, and local foci of bone tissue destruction were observed in 2 animals.

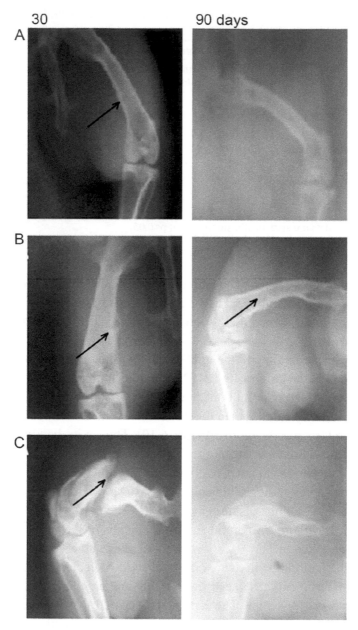

FIGURE 6.10   Radiography data on the regeneration of the model defect of bone tissue: A – Group 1, treatment P3HB 3D implant; B – Group 2 (Colapol); C – Group 3 (autograft bone) (Shishatskaya's data).

Histological examination of the defect site at Day 30 showed that in Group 1, with P3HB porous implants, there was a layer of loose fibrous connective tissue of non-uniform thickness around the implant; the tissue was infiltrated with lymphocytes and a few segmented leukocytes (Figure 6.11). In the fields of view, there was cellular infiltration by a few macrophages, which spread to the porous implant. In Group 2, with Colapol implanted into the defect, at Day 30, the surface of the bone tissue defect was covered by a layer of necrotic masses and fibrin, incorporating bone fragments. Inflammatory infiltration, with polymorphonuclear leukocytes prevailing, was observed around the implant. The bone defect was filled with loose fibrous connective tissue with abundant capillary-type vessels, which was non-uniformly infiltrated by lymphocytes, a few neutrophils, and macrophages. In Group 3 (with autograft bone), at Day 30, the center

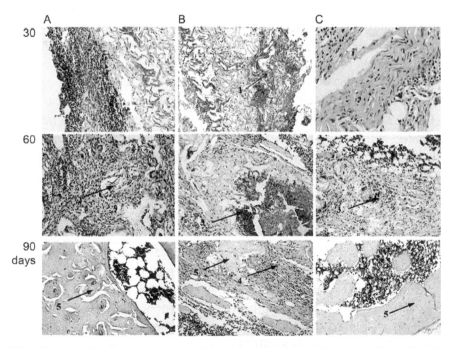

FIGURE 6.11 Histological sections of bone tissue from the defect site at Days 30, 60, and 90 post surgery in A – Group 1, treatment (P3HB 3D implant); B – Group 2 (Colapol); C – Group 3 (autograft bone). Notations on the sections: 1 – leukocyte infiltration; 2 – P3HB fragments; 3 – inflammatory reaction; 4 – coarse fibrous connective tissue; 5 – new bone tissue. Stained with hematoxylin and eosin. Magnification 100× (Shishatskaya's data).

of the defect was filled with coarse connective tissue, with perifocal proliferation of capillary-type vessels and foci of lymphohistocytic infiltrates.

At Day 60, in Group 1, the P3HB 3D implant was surrounded by loose
fibrous connective tissue that contained numerous macrophages, including
giant multinucleate osteoclast-like cells, responsible for lacunar resorption of bone fragments (Figure 6.11A). Small thin-walled capillary-type
vessels were growing from the loose connective tissue into the porous
implant. At the periphery of the implants, there were regions consisting
of dense fibrous connective tissue with osteoblasts. That was indicative
of the starting replacement of the soft connective tissue by osteoid substance. Resorption of the polymer implant, which was accompanied by the
implant break-up, was clearly seen. In Group 2 (Colapol), at Day 60, the
defect cavity was mainly filled with mature connective tissue, which contained Colapol particles surrounded by lymphocytes, polymorphonuclear
leukocytes, and macrophages. Regions of cellular detritus with perifocal
inflammatory reaction were still visible (Figure 6.11B). In Group 3 (autograft bone), the cavity was filled with weakly vascularized coarse fibrous
scar tissue with fragments of immature cartilage (Figure 6.11C).

At Day 90, in Group 1, the defect site was filled with new bone tissue.
The new bone was relatively mature, had lamellar structure, with osteons
(Figure 6.11A). In the center, the new bone was represented by new trabeculae; at the periphery, the bone had a sufficiently ordered structure, and
the boundary with the surrounding bone was indistinct. The medullary
canal was uniformly wide and filled with bone marrow. In that phase, the
reparative processes did not involve histiocytes and multinucleate cells,
and no pronounced inflammatory reaction was observed. In Group 2 (Colapol), at Day 90, rather large regions of loose and dense fibrous connective
tissue were still present in the center of the defect (Figure 6.11B). In Group
3, the defect was filled with the callus of the new bone consisting of chaotically oriented variously shaped trabeculae and cavities with bone marrow (Figure 6.11C). The medullary canal was non-uniformly narrowed.

Thus, X-ray and morphological examination of the artificially created
bone defects showed that P3HB 3D porous implant facilitated bone tissue
regeneration, which occurred at a faster rate than regeneration aided by the
commercial material, without inflammatory process and with the formation of new mature bone.

Reconstructive osteogenesis is often affected by postoperative infection of the defects. Treatment of chronic osteomyelitis has been a challenging task for surgeons for many years. Osteomyelitis comprises between 3 and 10% of surgical infectious bone diseases. It is very difficult to cure chronic osteomyelitis and the so-called implant-associated osteomyelitis, which develop at the implantation site of joint implants, support pins, screws, etc. A disadvantage of surgical reconstruction involving metallic orthopedic implants is that infection may develop at the interface between tissues and the implant, which limits the use of this method of bone tissue repair. The interface region must receive anti-inflammatory and antimicrobial drugs, whose concentrations must be maintained at definite levels for a prolonged time. One of the main challenges in the treatment of chronic osteomyelitis is finding a proper technique for bone defect repair. Such generally accepted methods as myoplasty, fasciocutaneous flap repair and filling of the cavity with synthetic materials do not enable complete anatomical and functional bone regeneration. Therefore, new effective techniques of bone replacement should be found to treat chronic osteomyelitis (Leonova, 2006).

A study was performed to test P3HB-based material as a filling material for reconstructing infected bone tissue defects. The experiments were conducted on 60 4–5-month-old male chinchilla rabbits. The rabbits were divided into 3 groups (20 rabbits in each group): 2 treatment groups and 1 control. In Group 1, P3HB filling material was used; in Group 2 – P3HB filling material + tienam; in Group 3 (control), the defect was filled with autograft bone chips. The rabbits were anesthetized with ketamine–droperidol, and a 3-cm long aperture was created through the anteromedial approach to the tibia. The muscles were bluntly separated and fixed. Osteotomy was performed through a 0.5-cm incision in the anterior surface of the metadiaphyseal region of the tibia; limited separation of periosteum was carried out; the medullary cavity was opened up and bone marrow was removed using Volkmann's curette. A gauze wick with *Staphylococcus aureus* culture (10^9 cells) was placed into the wound. The postoperative wound was closed layer-by-layer. Formation of the model of experimental osteomyelitis took 1 month; the model was 100% reproduced; primary chronic osteomyelitis was established. Then, treatment of bone defects was performed in the rabbits with chronic osteomyelitis. Skin incision was made, and debridement of postoperative scars, fistula, and sinus tracts was performed. Purulent discharges were drained; necrotic tissues and sequestra were removed. During the surgery, samples were collected to identify the causative agent and determine antibiotic sensitivity. Trepanation of the bone was performed to reach the apparently healthy bone; medullary cavities were opened up. The resulting trough-shaped bone cavity and the adjacent medullary cavities were thoroughly curetted using a Volkmann's spoon. After all nonviable tissues were removed, the cavity was washed with an antiseptic solution (plivasept, chlorohexidine solution). Counterincisions were made and irrigators for postoperative drainage were placed at the surgery site. After that, the bone cavity was filled with bone substitute materials: demineralized autologous graft bone taken from the iliac crest (the control group) and experimental materials: P3HB and P3HB/tienam (the treatment groups). The wound was completely closed layer-by-layer. Suction drains were connected to a vacuum apparatus. During the postoperative period, the limbs were immobilized using plaster splints or orthoses. The state of the animals (feeding behavior, locomotor activity, the state of the limbs that had been operated on) was monitored throughout the experiment. After termination at 30, 60, and 90 days postoperatively, bone specimens were retrieved and examined to estimate the state of the surrounding soft tissues, periosteum, cortical layer, and medullary cavity. Bacteriological plating on Chistovich medium (egg-yolk salt agar) was performed at the defect site. Results were evaluated at days 2–4.

During the early postoperative period (up to 3 days), the rabbits remained sluggish, moved very little, did not load the affected limb. At Days 3–5, the rabbits resumed normal feeding behavior and locomotor activity, but they did not load the affected limb. At Days 5–12, 19 rabbits had soft tissue edema at the surgery site; 12 rabbits showed local dermal hyperemia. At Day 16.5 ± 3.2, on average, the edema and hyperemia were reduced. The rabbits began to load the affected limbs at Day 6.22 ± 1.72 post surgery. Surgical wounds healed at Day 11 ± 1.32, on average. All animals remained alive during the postoperative period. In the experiments with P3HB used as a filling material for the experimental osteomyelitis, we observed significantly more rapid healing of postoperative skin wounds and recovery of the support function of the affected limb than in the control group. In the treatment groups, healing of the surgical skin wound took 7.4 ± 0.9 days, on average, after surgery and reduction of the edema and hyperemia – 8.75 ± 0.96 days; while in the control group these processes took 9.9 ± 1.7 and 11.7 ± 1.8 days, respectively. Recovery of the support function of the affected limb was observed at Days 4.28 ± 0.9 and 5.56 ± 1.2 in the treatment and control groups, respectively.

Radiography of the operated limbs revealed the following (Figure 6.12): at Day 30 after P3HB and P3HB/tienam were used to fill the bone defects, the projection image of the bone defect showed clear round areas with distinct boundaries and a cloud-like shadow in the center. At Day 90, radiography showed regeneration of the anatomical structure of the bone in all rabbits. In the control group, at Day 90, the regenerating bone tissue still had non-uniform structure, the periosteum was thickened, and the medullary cavity was undetectable. Results of microbiological investigation of the samples retrieved from the defect sites are given in Table 6.2.

The microbial profile of the control group was investigated by plating microbial associations at Days 30 and 90. These associations mainly contained *S. aureus* (48.2%) and associations of Gram-positive and Gram-negative anaerobic microorganisms and *E. coli*. In the P3HB treatment

TABLE 6.2 Microbiological Analysis of Tissue Samples of the Wounds (Markelova's data)

Groups of animals	Day 30 post-surgery	Day 90 post-surgery
Control group (autograft bone)	6.4×10^7 +/– 110 in 1 g *Staphilococcus aureus*; 4.4×10^7 +/– 110 in 1 g community of Gram (+) and Gram (–) anaerobic microorganisms, *E. coli*	3.2×10^4 +/– 150 in 1 g *Staphilococcus aureus*; 2.8×10^4 +/– 150 in 1 g community of Gram (+) and Gram (–) anaerobic microorganisms, *E. coli*
Treatment group with P3HB filling material	4.2×10^4 +/– 150 in 1 g *Staphilococcus aureus* 4×10^4 +/– 150 in 1 g community of Gram (+) and Gram (–) anaerobic microorganisms, *E. coli*	No growth
Treatment group with P3HB/tienam filling material	No growth	No growth

30   90 days

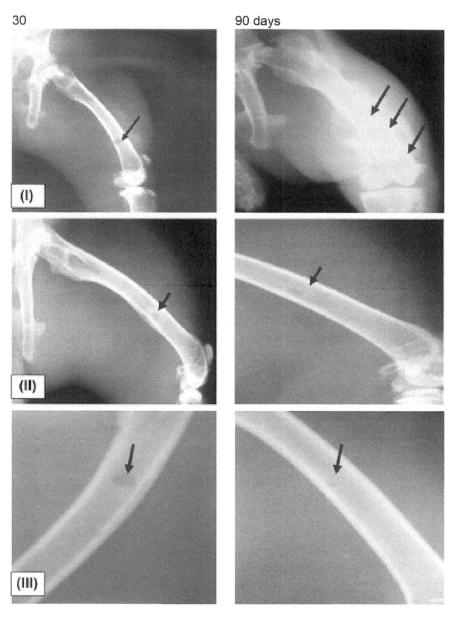

FIGURE 6.12 Radiography data on the regeneration of the model defects of bone tissue infected with Staphilococcus aureus in experiments with different implant materials: I – control group (allograft bone); II – treatment group (P3HB filling material), III – treatment group (P3HB/tienam filling material) (Markelova's data).

group, microbial associations were plated at Day 30; they mainly contained *S. aureus* (44.1%) and associations of Gram-positive and Gram-negative anaerobic microorganisms and *E. coli* (55.9%). At Day 90, bacterioscopic examination of the samples retrieved from this group gave a negative result, that is, the hydrophobic P3HB powder suppressed the infection. Bacterioscopic examination of samples of the material retrieved from the P3HB/tienam group did not show any growth of microorganisms at all time points.

Recovery of the anatomical structure of the bone tissue was only observed in the treatment groups. In the control group, the bone structure was not completely regenerated. Histological examination of tissue sections from the sites of model defects of the treatment groups (with P3HB and P3HB/tienam filled defects) showed that at Day 15, the center of the defect was filled with fibrous connective tissue with pronounced perifocal proliferation of capillary-like vessels and focal lymphohistiocytic infiltration (Figure 6.13III). Primitive osteogenesis, with the formation of oste-

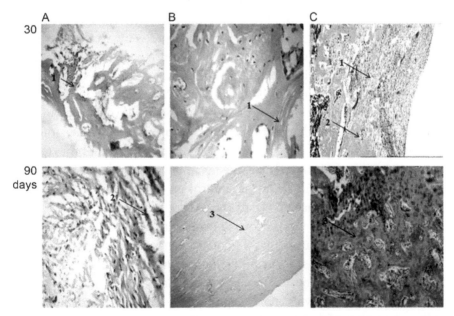

FIGURE 6.13 New bone tissue at the implant sites: A – allograft bone chips; B – P3HB filling material; C – P3HB/tienam filling material. Notations on the sections: 1 – coarse fibrous connective tissue; 2 – immature bone tissue; 3 – lamellar bone tissue. Stained with hematoxylin and eosin. Magnification 100× (Markelova's data).

oid-like masses and bone trabeculae with architectural deformations and chaotic arrangement of osteocytes, was mainly observed around the new vessels. Osteogenesis regions showed a high degree of basophilia of cells and osseomucoid of the developing bone tissue. At that time point, in the control group, the defect was filled by loose connective tissue and, partly, by granulation and fibro-reticular tissue and residual detritus. We observed pronounced leukocyte infiltration of the new tissue, hyperemia of blood vessels with the stasis of blood cells, and perivascular edema; vascular network was weakly developed.

In the treatment groups, at Day 30, the defect became smaller and was filled by fibrous, bone, and cartilage tissues (Figure 6.13). Intercellular substance was arranged in concentric layers around the vessels of the new osteons; bone lamellae were being formed. Towards the periphery of the defect, the bone tissue was more mature and had an osteon-trabecular structure; the osteons were arranged irregularly; the Haversian canals had different widths and were hyperemic; no perivascular edema was observed. In the control group, the defect on the periphery was filled by immature bone tissue, with fibrous scar and cartilage tissue in the center. The new tissue and the surrounding bone tissue were weakly vascularized. The periosteum was thickened and infiltrated with macrophages. At Day 90, all histological sections of the treatment groups showed signs of complete recovery of the bone structure. The periosteum was completely formed and consisted of the outer and inner layers. The bone tissue showed different degrees of maturity and had a lamellar structure; osteons were arranged irregularly; the number of mature fibrocytes and osteocytes increased. Interstitial substance became optically denser and homogenous (Figure 6.13). In the control group, at Day 90, the defect site was seen as the narrowing of the bone at the surgery site. Microscopic examination of the sections showed that the defect was filled by rather mature bone tissue, with some interlayers of cartilage tissue. Dense bone tissue with a few osteons prevailed; the intercellular substance was less homogenous and loose. The cells were mainly represented by mature fibroblasts and osteocytes; no basophilia of cells and intercellular substance was observed. The cortical plate was narrowed; the bone tissue was more mature; the cells were not numerous; sinusoid blood vessels were observed in intertrabecular spaces; the medullary cavity was either dramatically narrowed or covered with a layer of hyaline cartilage.

Thus, the experiments showed that using P3HB to fill the bone defect complicated by chronic osteomyelitis considerably facilitated the curbing of inflammation, bone defect repair, and recovery of the functional properties of the affected limbs and was significantly more effective than using allograft bone. Repair processes in experimental cavities of rabbits' tibias filled with PHA resulted in anatomical and functional bone regeneration by Day 60 post-surgery. The filling of the bone cavity with demineralized autologous bone graft did not result in the regeneration of the bone structure.

6.4 RESULTS OF CLINICAL TRIALS OF USING PHAS TO REPLACE BONE TISSUE

Clinical trials of PHAs were carried out by researchers of the General Surgery Department at the V.F. Voino-Yasenetsky Krasnoyarsk Federal Medical University. Evaluation of the effectiveness of using PHA materials to close bone defects was based on the outcomes of the treatment of 33 patients with the fractures of the upper and lower extremities, including those complicated by chronic osteomyelitis. The patients stayed in the General Surgery and Trauma Surgery Departments of the hospitals between 2008 and 2012.

Reconstruction of the bone tissue defects was done with PHA granulated powder combined with the antibacterial drug Tienam (10:1) to close defects with primary and secondary infections and pressed implants to replace the defects caused by impression fractures and comminuted fractures of the bones of the upper and lower extremities. All patients underwent surgery. The patients were divided into three groups. Group 1 comprised 13 patients with bone tissue defects at the fracture sites, including patients with post-traumatic osteomyelitis who first underwent necrosequestrectomy and then traditional osteoplastic surgery by closing the defect with autograft bone and drainage. Group 2 comprised 11 patients with defects closed with Collapan. Group 3 comprised 9 patients who underwent combined plastic surgery of the bone tissue defect: the defect was filled with

PHA material, and then, periosteal defect was closed with ultrafine films. If indicated, additional fixation at the defect site was performed with titanium plates (Figure 6.14). The effects were monitored by X-ray examination (before the intervention, at Day 30, and at Day 90 post surgery) and based on clinical and laboratory parameters: healing time, the recovery of the weight-bearing function of the limb, the duration of disability and the hospital stay, and the duration of antibacterial treatment.

Based on localization of the defect, the patients were distributed as follows: femur – 14 patients, tibia – 8 (30.4%), humerus – 7 (10.9%), and calcaneus and knee joint – 6 (2.2%). The post-traumatic osteomyelitis was diagnosed in 24 patients, postoperative osteomyelitis – in 7 (28.3%), and gunshot osteomyelitis – in 2 patients.

Although few patients took part in the trial, taking into account the requirements for limited clinical trials, we managed to obtain clinical data confirming the effectiveness of using PHA-based bone replacement material previously proved in the experiment. Clinical outcomes of the treatment are given in Table 6.3.

In Group 1, with the defects filled with autograft bone, 6 patients (50.3%) suffered from the recurrence of the disease during the observation period, and X-ray examination did not show complete filling of the defect. After radical necrosequestrectomy, only in 3 patients (23.1%), the wound healed by first intention.

In Group 2, with defects filled with biocomposite hydroxyapatite-based material containing collagen and antiseptic – Collapan – the surgical

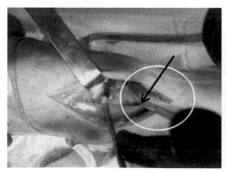

FIGURE 6.14 Implantation of PHA granules into the defect cavity and fixation with a titanium plate (Vinnik's data).

TABLE 6.3   Clinical Outcomes of the Treatment of the Patients (Vinnik's data)

Parameter	Group 1 (n–13)	Group 2 (n–11)	Group 3 (n–9)
Recurrence of disease	6 (46.1%)	4 (36.3%)	1 (11.1%)*
Healing by first intention	3 (23.1%)	6 (54.5%)*	7 (77.8%)*
More than 50% defect closure	4 (30.7%)	6 (54.5%)*	7 (77.7%)*
Complications	6 (46.1%)	5 (45.4%)	2 (22.2%)*
Complete recovery of the function of the limb	5 (38.5%)	5 (45.4%)	6 (66.6%)*

Note: * – differences are significant relative to Group 1.

wound healed by first intention in 54.5% patients, and by secondary intention – in 45.5%. At the end of the observation period (Day 90), radiographs showed that the osteomyelitis cavities were being gradually filled by bone tissue. The incidence of the disease recurrence was considerably lower than in the negative control group – 36.6%. In two cases, the recurrence of disease was caused by the presence of extensive post-osteomyelitis cavities, which were not completely filled with Collapan after necrosequestrectomy; in two other cases, the reason was the development of skin necrosis above the pathological site and exposure of the surgical site.

The best results were obtained in Group 3, in which the defect was filled with the P3HB-based material combined with the antibacterial drug tienam. The incidence of the disease recurrence in Group 3 was the lowest – 11.1%; in 77.7% cases, the wounds healed by first intention. The recurrence of disease was caused by the development of superficial thrombophlebitis of the lower limb. Radiographs showed that the defects were closed in shorter time and that in 77.7% cases, more than 50% defect closure occurred by the end of the observation period. Table 6.4 gives the structure of postoperative complications in the groups of patients.

As can be seen from the data in Table 6.4, the highest incidence of postoperative complications was recorded in Group 1. Filling of the defect with autograft bone did not ensure all necessary conditions for bone tissue repair but provided nutrient medium for infection; the material was quickly eliminated from the cavity. Thus, in Group 1, the total incidence of complications reached 53.8%. Collapan is based on collagen and, thus, it

TABLE 6.4   Characterization of Complications in Patients of Different Groups (Vinnik's data)

Complication	Group 1 (n–13)	Group 2 (n–11)	Group 3 (n–9)
Skin necrosis above the defect	3 (23.1%)	2 (18.2%)	1 (11.1%)*
Abscess, phlegmon of soft tissues	2 (15.4%)	2 (18.2%)	-
Fistula formation	1 (7.7%)	1 (9.1%)	-
Pathological fracture	1 (7.7%)	-	-

facilitates osteogenesis more effectively, but it has no antibacterial properties and, hence, cannot prevent and eradicate intraosseous infection. The total incidence of complications in Group 2 reached 45.4%. P3HB-based bone replacement material combined with antibacterial tienam provided the best conditions for bone tissue regeneration due to its osteoinductive properties, also preventing or reducing inflammatory infection at the bone defect site. In Group 3, only one patient had complications (11.1%). The clinical outcomes influenced the duration of hospital stay of the patients. The average time of hospital stay in Group 1 was 36±14.8 days, in Group 2 – 32.6±11.7 days, and in Group 3 – 28±10.9 days.

Thus, the first results of clinical trials with P3HB used to fill the bone tissue defects suggest that the material has osteoinductive properties and shows slow degradation, enabling gradual replacement of the defect by bone tissue and prolonged effect of the antibacterial drug. That resulted in the reduction in the incidence of disease recurrence and postoperative complications, shorter duration of hospital stay of the patients, and better outcomes of the treatment of the patients with this serious disease. Below we describe the case of the patient of Group 3, which confirms the results of the clinical trial of P3HB implants (Figure 6.15). Patient Y., 50 years of age. Diagnosis: primarily infected multifragmental gunshot fracture of the right femur. Surgical treatment: Step 1: primary surgical treatment, debridement of the wound, wound tamponade by Mikulich's method; Step 2 (in 21 days): total necrosequestrectomy, osteosynthesis, replacement of the osteomyelitis defect with titanium nickelide implant and PHA material with tienam, periosteal reconstruction with ultrafine

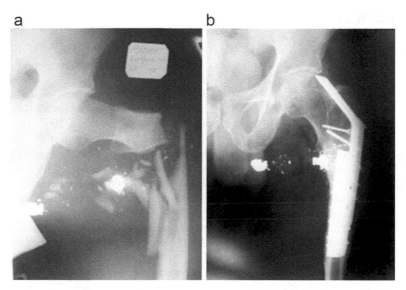

FIGURE 6.15   Patient Y.'s radiographs: a – on the admission to the hospital; b – 4 years after the trauma (Vinnik's data).

PHA film. Outcome: in 90 days, regeneration of the medial cortical layer; axial loading of the leg is allowed. In 4 years, complete consolidation of the fracture, 100% replacement of the osteomyelitis defect, recovery of leg function, no recurrence of osteomyelitis, and no complications.

Thus, the first clinical trials of PHA-based bone replacement material show that it effectively closes bone tissue defects and can be used both under aseptic and septic conditions.

KEYWORDS

- **bone defects**
- **reconstructive osteogenesis**
- **filling material**
- **implants**
- **segmental osteotomy**
- **osteomyelitis**

REFERENCES

Artsis, M. I., Bonartsev, A. P., Iordanskii, A. L., Bonartseva, G. A., Zaikov, G. E. Biodegradation and Medical Application of Microbial Poly(3-Hydroxybutyrate). J. Mol. Cryst. Liquid Cryst., 2012, 555, 232–262.

Deev, R. V., Isaev, A. A., Kochish, A. Yu., Tikhilov, R. M. Cellular technologies in trauma surgery and orthopedics: ways of development. Kletochnaya transplantologiya i tkanevaya inzheneriya (Cell Transplantation and Tissue Engineering), 2007, 18–30 (in Russian).

Derya, B. H., Ebru, K., Hazer, B. Poly(3-hydroxyalkanoate)s: Diversification and biomedical applications A state of the art review. Mat. Sci. Eng. C, 2012, 32, 637–647.

Desyatichenko, K. S., Kurdyumov, S. G. Trends in designing of tissue engineering systems for osteoplasty. Kletochnaya transplantologiya i tkanevaya inzheneriya (Cell Transplantation and Tissue Engineering), 2008, 3, 62–69 (in Russian).

Doyle, C., Tanner, E. T., Bonfield, W. *In vitro* and *in vivo* evaluation of polyhydroxybutyrate and polyhydroxyvalerate reinforced with hidroxyapatite. Biomaterials, 1990, 11, 206–215.

Gaidar, B. V., Parfenov, V. E., Tegza, V.Yu. Neurosurgery in contemporary local armed conflicts. Voenno-meditsinskiy zhurnal (Military Medical Journal), 2002, 12, 28–32 (in Russian).

Gololobov, V. G. Posttraumatic regeneration of bone tissue. Modern approach. Fundamentalnyye i prikladnyye problemy gistologii. Gistogenes i regeneratsiya tkaney (Fundamental and Applied Issues of Histology. Tissue Generation and Regeneration). Military Medical Academy Publishers. St. Petersburg, 2004, 257, 94–109 (in Russian).

Horowitz, D. M., Williams, S. F., Martin, D. P. Peoples, O. P. PHA applications: addressing the price performance issue. I. Tissue engineering. Int. J. Biol. Macromol., 1999a, 25, 11–121.

Horowitz, E. M., Prockop, D. J., Fitzpatrick, L. A. Transplantability and therapeutic effects of bone marrow-derived mesenchymal cells in children with osteogenesis imperfect. Nat. Med, 1999b, 5, 309–313.

Knowles, J. C., Hastings, G. W. *In vitro* and *in vivo* investigation of a range of phosphate glass-reinforced polyhydroxybutyrate based on degradable composites. J. Mater. Sci., 1993a, 4, 102–106.

Knowles, J. C., Hastings, G. W. Physical properties of a degradable composite for orthopedic use which attaches bone. In: T. Takagi Proceeding First Int. Conference on Intelligent Materials. Technomic. Lancaster, 1993b, 495–504.

Knowles, J. C., Hastings, G. W., Ohita, H., Niwa, S., Baeree, N. Development of a degradable composite for orthopedic use: *in vivo* biomechanical and histological evaluation of two bioactive degradable composites based on the polyhydroxybutyrate polymer. Biomaterials, 1992, 13, 491–496.

Köse, G. T., Korkusuz, F., Korkusuz, P., Hasirci, V. In vivo tissue engineering of bone using poly(3-hydroxybutyric acid-co-3-hydroxyvaleric acid) and collagen scaffolds. Tissue Eng., 2004a, 10, 1234–1250.

Köse, G., Kenar, H., Hasirci, N., Hasirci, V. Macroporous poly(3-hydroxybutyrate-co-3-hydroxyvalerate) matrices for bone tissue engineering. Biomaterials, 2003b, 24, 1949–1958.

Köse, G. T., Ber, S., Korkusuz, F., Hasirci, V. Poly(3-hydroxybutyric acid-co-3-hydroxyvaleric acid) based tissue engineering matrices. J. Mater. Sci. Mater. Med., 2003c, 14, 121–126.

Luklinska, Z. B., Schluckwerder, H. *In vivo* response to HA – polyhydroxybutyrate/polyhydroxyvalerate composite. J. Microsc., 2003, 2, 121–129.

Mai, R., Hagedorn, M. G., Gelinsky, M., Werner, C., Turhani, D., Späth, H., Gedrange, T., Lauer, G. Ectopic bone formation in nude rats using human osteoblasts seeded poly(3)hydroxybutyrate embroidery and hydroxyapatite-collagen tapes constructs. J. Craniomaxillofac. Surg., 2006, 34, 101–109.

Markelova, N. M. Obosnovaniye primeneniya vysokotekhnologichnykh izdeliy meditsinskogo naznacheniya iz biodegradiruyemykh polimerov v rekonstruktivnoy khirurgii (experimentalno-klinicheskoye issledovaniye) (Substantiation of using high-tech medical devices of biodegradable polymers in reconstructive surgery (experimental study and clinical trials). MD dissertation in Specialty 14.01.17: Surgery. Krasnoyarsk, Voino-Yasenetsky Krasnoyarsk Federal Medical University, 2013, 272 p. (in Russian).

Marois, Y., Zhang, Z., Vert, M. Synthetic bioabsorbable polymers for implants. Astm Special Technical Publication, 2000, 12–38.

Matveeva, A. I. Zameshcheniye defektov kostey cherepa regeneriruyushchey kostyu (Replacement of skull bone defects with regenerating bone). USSR AS Publishers. Moscow, 1962, 136 p. (in Russian).

Misra, S., Ansari, T., Valappil, S., Mohn, D., Philip, S. E., Stark, W. J., Roy, I., Knowles, J. C., Salih, V., Boccaccini, A. R. Poly(3-hydroxybutyrate) multifunctional composite scaffolds for tissue engineering applications. Biomaterials, 2010, 31, 2806–2815.

Ou, G., Bao, C., Liang, X., Chao, Y., Chen, Z. Histological study on the polyhydroxybutyric ester(PHB) membrane used for guided bone regeneration around titanium dental implants. Hua Xi Kou Qiang Yi Xue Za Zhi, 2000, 18, 215–218.

Paecu, E. I., Stokes, J., McGuinness, G. B. Electrospun composites of PHBV, silk fibroin and nano-hydroxyapatite for bone tissue engineering. Mater. Sci. Eng., 2013, 33, 4905–4916.

Cao, Q., Zhang, J., Liu, H., Wu, Q., Chen, J., Chen, G. Q. The mechanism of anti-osteoporosis effects of 3-hydroxybutyrate and derivatives under simulated microgravity. Biomaterials, 2014, 35, 8273–8283.

Rambo, C. R., Costa, C. M., Carminatti, C., Recouvreuxa, D. O. S., d'Acamporab, A. J., Porto, L. M. Osteointegration of poly-(3-hydroxybutyrate-co-3-hydroxyvalerate) scaffolds incorporated with violacein. Mater. Sci. Eng., 2012, 32, 385–389.

Reis, E., Borges, A., Fonseca, C., Martinez, M., Eleterio, R., Morato, G., Oliveira, P. Biocompatibility, osteointegration, osteoconduction, and biodegradation of a hydroxyapatite-polyhydroxybutyrate composite. Braz. Arch. Biol. Technol., 2010, 53, 817–826.

Shevtsov, V. I., Shreiner, A. A., Smelyshev, K. N. A technique of correcting deformities of long tubular bones. Geniy ortopedii (Genius of orthopedics), 2000, 1, 104–108 (in Russian).

Shishatskaya, E. I. Biocompatible and functional properties of a poly(3-hydroxybutyrate/ hydroxyapatite hybrid composite. Vestnik transplantologii iskusstvennykh organov (Bulletin of Transplantation of Artificial Organs), 2006, 3, 34–38 (in Russian).

Shishatskaya, E. I. Biomedical investigation, application of PHA. Macromol. Symposia, 2008, 269, 65–81.

Shishatskaya, E. I., Belyaev, B. A., Vasiliev, A. D., Mironov, P. V., Volova, T. G. Structure and physicochemical properties of a poly(3-hydroxybutyrate)/hydroxyapatite hybrid composite. Perspektivnyye materialy (Advanced Materials), 2005, 1, 47–51 (in Russian).

Shishatskaya, E. I., Chlusov, I. A., Volova, T. G. A hybrid PHA-hydroxyapatite composite-for biomedical application: production and investigation. J. Biomat. Sci. Polym. E., 2006, 17, 481–498.

Shishatskaya, E. I., Kamendov, I. V., Starosvetsky, S. I., Volova, T. G. A study of osteo-plastic properties of matrices prepared from a resorbable polyester of hydroxybutyric acid. Kletochnaya transplantologiya i tkanevaya inzheneriya (Cell Transplantation and Tissue Engineering), 2008, 5(4), 41–47 (in Russian).

Shishatskaya, E. I., Mironov, P. S., Volova, T. G. Properties of resorbable polyester Bio-plastotan composites with wollastonite and hydroxyapatite. Plasticheskiye massy (Plastics), 2008, 12, 41–43 (in Russian).

Shishatskaya, E. I., Kamendov, I. V., Starosvetsky, S. I., Vinnik Yu.S., Markelova, N. N., Shageev, A. A., Khorzhevsky, V. A., Peryanova, O. V., Shumilova, A. A. An in vivo study of osteoplastic properties of resorbable poly-3-hydroxybutyrate in models of segmental osteotomy and chronic osteomyelitis. Artif. Cells Nanomed. Biotechnol., 2014, 42, 344–355.

Shishatskaya, E. I., Nikolaeva, E. D., Shumilova, A. A., Shabanov, A. V., Volova, T. G. Cultivation and differentiation of bone marrow multipotent mesenchymal cells on scaffolds of the resoebable BIOPLASTOTAN. Kletochnaya transplantologiya i tkanevaya inzheneriya (Cell Transplantation and Tissue Engineering), 2013a, 8(1), 57–65 (in Russian).

Shishatskaya, E. I., Vinnik Yu.S., Markelova, N. M., Shageev, A. A., Kamendov, I. V., Sta-rosvetsky, S. I., Khorzhevsky, V. A., Peryanova, O. V., Shumilova, A. A. An in vivo study of osteoplastic properties of the resorbable poly-3-hydroxybutyrate in mod-els of chronic osteomyelitis. Vrach aspirant (Post-graduate doctor), 2013b, 1.1(56), 127–132 (in Russian).

Shtilman, M. I. Polimery mediko-biologicheskogo naznacheniya (Polymers intended for biomedical applications). Akademkniga Publishers. Moscow, 2006, 399 p. (in Rus-sian).

Shumilova, A. A., Nikolaeva, E. D., Shishatskaya, E. I. Opportunities of polyhydroxyal-kanoates for reconstructive surgery. International Symposium on Biopolymers. Bra-zil. September, 2014.

Shumilova, A. A., Nikolaeva, E. D., Shishatskaya, E. I. Osteoplastic implants based on resorbable polyesters for reconstructive surgery. V International Congress "Bio-materials and Nanobiomaterials: Recent Advances Safety-Toxicology and Ecology Issues" (Bionanotox-2014). May, 2014.

Shumilova, A. A., Shishatskaya, E. I., Markelova, N. M., Vinnik Yu.S., Zuev, A. P., Kirichenko, A. K., Solovyeva, N. S. Osteoplastic properties of macroporous implants based on poly-3-hydroxybutyrate in regeneration of the bone defect of rabbit's tubular bone. Fundamentalnyye issledovaniya (Fundamental research), 2015, 7, 697–706 (in Russian).

Szubert, M., Adamska, K., Szybowicz, M., Jesionowski, T., Buchwald, T., Voelkel, A. The increase of apatite layer formation by the poly(3-hydroxybutyrate) surface modification of hydroxyapatite and β-tricalcium phosphate. Mater. Sci. Eng., 2014, 34, 236–244.

Terada, S., Sato, M., Sevy, A., Vacanti, J. P. Tissue engineering in the twenty-first century. Yonsei Med. J., 2000, 41, 685–691.

Türesin, F., Gürsel, I., Hasirci, V. Biodegradable polyhydroxyalkanoate implants for osteomyelitis therapy: in vitro antibiotic release. J. Biomat. Sci. Polym, E., 2001, 12, 195–207.

Vacanti, C. A, Vacanti, J. P. The science of tissue engineering. Orthop. Clin. North. Am., 2000, 31, 351–356.

Volova, T. G., Shishatskaya, E. I. Biorazrushayemyye polimery: sintez, svoistva, primeneniye (Biodegradable Polymers: Synthesis, Properties, Applications). Krasnoyarskii Pisatel. Krasnoyarsk, 2011, 49 p. (in Russian).

Volova, T. G., Shishatskaya, E. I. Results of biomedical studies of PHAs produced in the Institute of Biophysics SB RAS and Siberian Federal University. In: Polyhydroxyalkanoates (PHA): Biosynthesis, Industrial Production and Applications in Medicine. Chapter 21. Nova Sciences Publ. Inc. NY, USA, 2014, 273–330.

Volova, T. G., Shishatskaya, E. I., Mironov, P. V., Goreva, A. V. structure and physicochemical properties composite polyxydroxybyturate/wollastonite. Perspektivnyye materialy (Advanced Materials), 2009, 1, 43–50 (in Russian).

Volova, T. G., Shishatskaya, E. I., Sinskey, A. J. Degradable Polymers: Production, Properties and Applications. Nova Science Pub., Inc. NY, USA, 2013, 380 p.

Wang, Y., Bian, Y. Z., Wu, Q., Chen, G. Q. Evaluation of three-dimensional scaffolds prepared from poly(3-hydroxybutyrate-co-3-hydroxyhexanoate) for growth of allogeneic chondrocytes for cartilage repair in rabbits. Biomaterials, 2008, 29, 2858–2868.

Yang, M., Zhu, S., Chen, Y. Studies on bone marrow stromal cells affinity of poly (3-hydroxybutyrate-co-3-hydroxyhexanoate). Biomaterials, 2004, 25, 1365–1373.

Ye, Ch., Hu, P., Ma, M. X., Xiang, Y., Liu, R. G., Shang, X. W. PHB/PHBHHx scaffolds and human adipose-derived stem cells for cartilage tissue engineering. Biomaterials, 2009, 30, 4401–4406.

Zotov, Yu. V., Kasumov, R. D., Savelyev, V. I. Khhirurgiya defektov cherepa (Surgery of Skull Defects). IU Publishers, St. Petersburg, 1998, 140 p. (in Russian).

PART IV

PERSPECTIVES FOR USING PHAs
IN ABDOMINAL SURGERY

CHAPTER 7

A STUDY OF MESH IMPLANTS COATED WITH A BIOCOMPATIBLE POLYHYDROXYALKANOATES LAYER

CONTENTS

7.1 INTRODUCTION

Achievement of better outcomes of surgical interventions in abdominal surgery is impossible without using new materials. For instance, surgical treatment of the patients with postoperative ventral hernias of the anterior abdominal wall has been one of the challenges in abdominal surgery. Hernia formation is a complex condition, caused by imbalance between intra-abdominal pressure and resistivity of the abdominal wall. Surgery using local tissues does not ensure stable improvement. The employment of tension-free techniques in surgeries of the postoperative ventral hernias and the use of synthetic materials was a revolution in hernia repair. Meshes can be implanted in super-aponeurotic, sub-aponeurotic, and intramuscular positions. However, plastic surgery of the anterior abdominal wall involving the use of synthetic allografts is a complicated surgical procedure and may cause the development of postoperative complications, both specific and nonspecific ones. Support meshes of a new generation are needed for hernia repair, which is one of the most common surgical operations, amounting to 10–15% of all surgeries. More than 20 million herniotomies are performed in the world, with recurrent hernias appearing in 10–15% (Fedorov and Adamyan, 2000). So-called barrier techniques are being developed to prevent surgical adhesions. Materials used to prepare such meshes should be able to prevent adhesions to internal organs, be resistant to infection, be mechanically strong, and tolerate long-term tension without deep scarring and encapsulation.

Commercial meshes are prepared from synthetic materials, mainly polypropylene. These are such meshes as Cousin, Vipro, and Ultrapro (U.S.). Polypropylene meshes, however, have caused a lot of complications (infection, fistula formation, postsurgical pain, discomfort caused by the foreign body sensation), and research effort was redirected to the development of the so-called "lightweight" polymer meshes used for abdominal wall surgery. A lot of such meshes failed to function adequately. For instance, Timesh (GFE Medzintechnic GmbH, Germany), prepared from polypropylene monofilaments with diameter 0.07 mm, coated by titanium, had low tensile strength (3.7 N/cm) along the row of loops, and its production was phased out. In an attempt to reduce the amount of the material used to construct support meshes, either nonabsorbable monofilaments of smaller diameter were used (to decrease the weight of chemically uniform, mainly polypropylene, meshes) or polypropylene monofilaments were blended with synthetic fibers. Lightweight meshes, produced using a special technology, are prepared from the smallest possible amount of material and are easy to handle and strong. The most commonly used lightweight meshes are Optilene Mesh LP (Braun, Germany), Biomesh light (Cousin, U.S.), Esfil Light (Lintex, Russia). The surface density of the support meshes commonly used in Europe and U.S. varies within a rather narrow range, between 28 and 36 g/m², and thus, there is no obvious leader.

Composite meshes consist of two types of fibers: polypropylene and a material hydrolyzed in host tissues. The implant has a rather dense structure, providing mechanical strength and resistance to deformation in the first 24 h post-surgery. Subsequently, some of the fibers are hydrolyzed and replaced by the ingrowing connective tissue; in this phase, mechanical strength of the mesh is mainly achieved due to the host's scar tissue. Thus, the amount of the foreign body that resides in the host's organism is minimized. Ethicon, Inc. produces vicryl/prolene composite meshes Vipro and Vipro II, which have different braiding patterns and densities. Another type of the mesh produced by Ethicon, Inc. is Ultrapro, a composite monofilament mesh consisting of nonabsorbable prolene monofilament fibers and hydrolyzed polyglecaprone monofilament fibers. The advantages of this mesh are similar to those of Vipro, but, in contrast to vicryl, polyglecaprone fibers have monofilamentous structure, which reduces the "wicking" tendency. Moreover, intact Ultrapro is harder and, thus, more convenient for preperitoneal fixation. As the mesh is partly hydrolyzed, large-size pores are formed, and the resulting connective tissue line is more flexible and physiologically acceptable.

Polytetrafluoroethylene barriers preventing adhesions, such as Preclude, produced by Gore-Tex Surgical Membrane, W. L. Gore and Associates, Flagstaff, AZ, U.S., also have a few drawbacks: polytetrafluoroethylene is hydrophobic, and, thus, its attachment to tissues is not strong enough. Moreover, the material is not biodegradable; the barrier has to be fixed with sutures, and it permanently stays as a foreign body in the abdominal cavity, increasing the chance of adhesion formation and infection as remote postsurgical effects. Hyaluronic acid (HA) and HA-based composites are commonly used as barriers. Blending of HA with phosphate buffer solution resulted in the construction of Sepracoat, the barrier preventing serous inflammation and adhesions in animal models. Interceed (Ethicon Inc., Somerville, NJ, U.S.) is a commercial device based on cellulose; it is a membrane that is completely absorbed in 28 days. Unfortunately, the efficiency of this device is reduced in the presence of blood or a large amount of peritoneal fluid. Carboxymethyl cellulose (CMC) and its derivatives such as Na-CMC, are included in the Public Register of Drugs in Russia and in pharmacopeias in Europe and U.S. CMC-based composite implants are widely used in plastic surgery (NEW-FILL® – a blend of CMC and polylactic acid). Membranes based on CMC blended with hyaluronic acid such as Seprafilm (Genzym Corporation, U.S.) are used as films to cover injured surfaces. The membrane is transformed into gel within 24–48 h, but stays in place until Day 7; it is fully absorbed by Day 28; no suture fixation is needed; the membrane can be effectively used in the presence of blood, significantly reducing the degree of severity of surgical adhesions. Clinical use of this "barrier" has been approved in European countries and North America. This device, however, can induce intra-abdominal abscesses; moreover, in patients with peritonitis, Seprafilm does not prevent adhesions in the abdominal cavity; on the other hand, it does not inhibit healing of interintestinal anastomoses. Oxiplex (FzioMed Inc., San Luis Obispo, U.S.) is a more recently designed membrane, consisting of CMC and polyethylene oxide. Experiments on animals showed that the membrane reduced postoperative adhesion formation in peritonitis patients.

These anti-adhesion barriers are definitely effective, but they have a serious drawback – the high cost: over US$ 1000 per surgery. Materials used to prepare such meshes should be able to prevent adhesions to internal organs, be resistant to infection, be mechanically strong, and tolerate long-term tension without deep scarring and encapsulation. However, the clinical use of synthetic barrier meshes caused different complications (serous exudate, infiltrates, migration and rejection of implants, adhesions to internal organs, etc.) (Sukovatykh et al., 2004; Dubova et al., 2007; Klinge et al., 2002). Lightweight meshes (containing decreased amounts of polypropylene), hyaluronic acid-based implants, and meshes coated with biocompatible materials (titanium, β-glucone, collagen) have been developed to enhance biocompatibility of the implants.

Lintex, a firm in St. Petersburg, Russia, developed a "lightweight" mesh of a new generation – Uniflex Light – consisting of polyvinylidene fluoride monofilaments. This mesh is considerably more biologically inert and flexible than polypropylene meshes. Experiments and clinical trials showed that implantation of this mesh leads to the formation of a thin connective tissue capsule, with very few implant-associated complications (Zhukovsky, 2011). This nonabsorbable mesh prepared from biologically inert materials is intended to be used for plastic surgery of the abdominal wall in repair of inguinal, umbilical, postoperative hernias or for tumor resection. Meshes are also used for plastic surgery of chest wall, diaphragm, and other soft tissues.

Another approach to reducing adverse tissue response to the implant is coating of synthetic polymer meshes with layers of biocompatible resorbable polymers. This approach has received much attention of researchers in Russia. Most Russian PHA researchers suffer polymer shortage, while such meshes are simple to prepare and require very small amounts of the polymer. In his PhD paper, Boskhomzhiev (2010) reported a study of tissue response to the implantation of Lintex-Esfil mesh coated with poly-3-hydroxybutyrate in experiments on rats. Tissue response to the implantation of the mesh was exhibited as mild inflammation, which was less pronounced than inflammation in response to the implantation of the uncoated polypropylene mesh. Thus, P3HB coating can be used to increase biocompatibility of polypropylene meshes. V.D. Dmitriev conducted experiments and clinical trials to investigate Lintex-Esfil 1–010 polypropylene mesh coated with ElastoPOB – P3HB blended with polyethylene glycol) (Dmitriev, 2008; Dmitriev et al., 2007). In experiments on rats, tissue response to the implantation of experimental ElastoPOB-AR meshes was less pronounced than tissue response to the uncoated polypropylene implants. Clinical trials of the modified meshes were performed on patients with inguinal hernias and patients with postsurgical ventral hernias. In both groups of patients, the occurrence of early, medium, and remote postoperative complications was lower than in the patients with uncoated Lintex-Esfil 1–010 mesh implants. A similar improvement was obtained in experiments with an ElastoPOB-coated "Eslan" Dacron mesh intended for extracardial heart modeling (Shumakov et al., 2006; Timerbaev, 2010). Experiments on dogs showed that the occurrence of adhesions in the pericardial cavity was significantly lower in dogs with ElastoPOB®-coated implants.

In this chapter, we present results of experimental research and clinical trials of mesh implants modified with a PHA coating, produced at the Institute of Biophysics SB RAS. Clinical trials were performed in cooperation with researchers of the General Surgery Department at the V.F. Voino-Yasenetsky Krasnoyarsk Federal Medical University.

7.2 POTENTIALS AND FIRST CLINICAL RESULTS OF USING MESH IMPLANTS COATED WITH POLYHYDROXYALKANOATES IN HERNIA REPAIR

Surgical treatment of ventral hernias is of great medical and social significance. The incidence of inguinal hernias is rather high, with male patients prevailing (3–7% of able-bodied male population). Herniotomy is among the most frequently performed surgical procedures in abdominal surgery. Classical hernia repair causes high incidence of recurrent hernias and long-term disability, which makes it very costly (Amid, 2002). Over more than a century of hernia repair experience, many surgical techniques have been proposed, based on reintegration of the abdominal wall in the inguinal region and strengthening of the anterior and posterior walls of the inguinal canal by suturing local tissues. Now, very few surgeons still use the techniques of strengthening the anterior wall of the inguinal canal (Girard's, Martynov's, Kimbarovsky's techniques) as they are pathogenetically unsound and frequently cause recurrent hernias. The techniques of strengthening the posterior wall of the inguinal canal, such as Bassini repair, which was proposed in the late 19th century, proved to be more acceptable, as destruction of the posterior wall was found to be an important factor in the formation of any inguinal hernia (Pryakhin, 2007).

Such well-known approaches as the Bassini, Kukudzhanov, Shouldice, McVay, and other techniques not only strengthen the posterior wall of the inguinal canal, transverse fascia, but also enable partial restoration of the valvular function of the structures constituting the retroinguinal space and create less tissue tension. These approaches are physiologically sound and have no unfavorable effects on the function of the testicle. Therefore, these techniques have been regarded as the "gold standard" of hernia repair for a long time and are still widely used in surgeries on young patients, with the retroinguinal space of small height and insignificant destruction of anatomical structures of the inguinal canal. In the patients with pronounced local tissue deficiency, age-related changes in the tissues, and complicated hernia, these techniques cannot prevent recurrent illness (Strizheletsky, 2006). Thus, new approaches to the surgical treatment of inguinal hernias need to be found. The main reason for the inadequate treatment is the patient's tissue deficiency, especially because many techniques involve the creation of duplications, and in the patients with the wide inguinal space, this causes considerable tissue tension, tearing of the skin, and, hence, recurrent hernia (Shulutko et al., 2003). Implants were proposed as a means to compensate for tissue deficiency. The idea was expressed by Billroth, but no suitable materials had been found until the second half

of the 20th century, when polyester, polypropylene, polytetrafluoroethylene, and other modern surgical materials were synthesized (Yegiev et al.,
2000).

A number of tension-free techniques of inguinal hernia surgery,
involving the use of implants, have been proposed recently, and their
effectiveness has been proved in randomized studies. The use of meshes
prevents tissue tension, somewhat simplifies the surgery, reduces intraoperative trauma incidence, decreases pain in the early postoperative
period, shortens the hospital stay, reduces the rehabilitation period,
decreases the incidence of recurrent hernias, and improves the patients'
quality of life (Seleem, 2003). Thus, surgical procedures using the
patient's tissues, with unavoidable tissue tension, are gradually replaced
by modern tension-free methods with allografts. We would like to
describe in greater detail the most commonly used modern techniques
(Elsebae et al., 2008). All tension-free techniques of hernioplasty can be
divided into several types. According to the mesh position relative to the
transverse fascia, the techniques are divided into anterior (Lichtenstein,
Trabucco, PHS, Rutkov-Robbins techniques) and pre-peritoneal (transabdominal pre-peritoneal (TAPP) and totally extraperitoneal (TEP) techniques); according to the intervention technique, they are divided into
open and laparoscopic techniques.

According to the type of fixation of the mesh, the methods are divided
into suturing (Lichtenstein, PHS, Rutkov-Robbins) and suture-free (Trabucco) techniques (Chistyakov et al., 2008). During the surgery, the
posterior wall of the inguinal canal can be accessed either through the
inguinal canal (Lichtenstein, Trabucco, PHS techniques) or from the retroperitoneal space (laparoscopic techniques, Stoppa procedure). Tension-
free techniques, which are techniques of choice, cause very few recurrent
hernias – in 0.1–0.5% of the patients with primary hernias. Postoperative
ventral hernias (POVHs) constitute up to 20–26% of all external abdominal hernias, and they are the second most common hernias after inguinal
ones. POVHs develop in 10% of all patients after abdominal surgeries.
With such hernias, it is technically difficult to close large muscular apo-
neurotic defects because aponeurosis is weakly pronounced, which may
cause hernia recurrence. Better results can be achieved by additional
strengthening of the weakened tissues of the anterior abdominal wall.

A generally recognized current approach to strengthening the tissues of the anterior abdominal wall is to use full-thickness or split-thickness skin autograft, meshes, and synthetic materials. POVHs typically grow quickly, frequently causing complications. Most of the patients belong to the working-age population, and their treatment is a socioeconomic issue (Shevchenko et al., 2003).

Tension-free hernia repair currently prevails over other approaches. This can be achieved through the replacement of tissue defects and strengthening of the partially destroyed anatomical structures with materials of different origins: autograft skin flaps, transverse fascia flaps, allogenic dura mater, xenograft peritoneum, metal meshes, braided and solid polymer meshes. Surgeons tend to avoid using metal and alloy grafts because they do not have the necessary physical properties, and some of them are quickly destroyed, causing postoperative complications. Autografts and allografts cannot be regarded as ideal implants, either, as in some cases, they were completely resorbed, causing suppuration and rejection, and, thus, annihilating the results of surgery (Yurasov, 2002).

At the present time, the most commonly used materials are synthetic polymers: polypropylene and polytetrafluoroethylene. These materials are strong, elastic, relatively inert biologically and chemically, non-toxic, and easy to sterilize. Their physical and biological properties make them nearly ideal materials for the repair of hernias of the anterior abdominal wall. Surgeons commonly use meshes to close various hernias. Meshes are fabricated from non-resorbable materials (polypropylene, polyvinyl idenfluoride, polytetrafluoroethylene, polyethylene terephthalate, vicryl, etc.), absorbable materials (polyglycolide/polylactide, etc.) and combinations thereof. The use of non-absorbable polymer biomaterials inevitably induces activation of chronic inflammatory cascade (cytokines, proteases) and granulomatous macrophage reaction with multinucleate foreign body giant cells. The outcome of the inflammatory reactions is the development of fibrosis in the mesh site, causing not only fixation of the mesh inside the abdominal wall but also strengthening of the latter. On the other hand, inflammatory reactions cause a number of complications. The most frequently reported complications of the early and late postoperative periods are seromas, mesh disruption and dislocation, adhesions, infections, and pain (Cucci et al., 2002). Previous studies proved that the degree of macrophage reaction and formation of foreign body giant cells are largely determined by the type of the material used to fabricate the mesh. Meshes of some absorbable materials (polyglactin 910) are 50% absorbed at week 3 after implantation and are fully biodegraded in 3 months. Composite meshes give the best results, offering a compromise between the degree of inflammatory reaction and the strengthening of the abdominal wall (Garavello et al., 2001; Avad et al., 2005).

The inflammatory granulomatous macrophage reaction in the region of antigenic stimulation occurs in a diffuse manner. A constant stimulating factor in this case is mesh filaments, which are too large for complete phagocytosis to occur. Giant multinucleate cells, similarly to their precursors – macrophages, are able to synthesize anti-inflammatory cytokines, including the transforming growth factor β (TGF-β), vascular endothelial growth factor (VEDF), and platelet-derived growth factor (PDGF). These factors are major components in fibroblast activation and in activation of tissue fibrosing. A large number of studies based on morphological analysis suggest the important role of metalloproteinases in foreign body reaction and remodeling of the tissue matrix. The most frequently discussed issue is fabrication of implants with the minimal antigenic properties, to reduce the macrophage reaction of the body. One of the most promising lines of research in this field is the use of linear polyesters polyhydroxyalkanoates (PHAs) – biocompatible and biodegradable microbial polymers (Sanjay et al., 2006).

7.3 EXPERIMENTAL BASIS FOR USING PHA-COATED MESHES

The main purpose of implanting meshes is to strengthen the connective tissue and muscles of the anterior abdominal wall. The most important conditions for achieving this is quick growth of the connective tissue into the mesh and the absence of local mesh-associated complications. The major factors are the response of the tissues of the anterior abdominal wall to the implant and its biocompatibility.

Macrophage reaction to different types of commercial meshes, which are commonly used in herniotomy, and to PHA-coated ones was studied in the experiment on 180 chinchilla rabbits (Yakovlev et al., 2010;

Markelova et al., 2012; Markelova, 2013). Characterization of the groups of animals is given in Table 7.1.

Polypropylene meshes Esfil (Lintex, Russia) were implanted to Group 1 and 4 animals; lightweight composite meshes VIPRO II (Ethicon, U.S.) were implanted to Group 2 and 5 animals. P(3HB/3HV)-coated polypropylene meshes Esfil were implanted to Group 3 and 6 animals. The meshes were spray-coated in the laboratory at the Institute of Biophysics SB RAS. The polymer-coated meshes were sterilized with vapors of hydrogen peroxide. The animals were aseptically anesthetized by administration of ether and xylazine (with doses calculated taking into account the animal's body mass and following the manufacturer's recommendations). Meshes (5.0×5.0 cm^2) were implanted in the sublay position in the animals of Groups 1, 2, and 3 and in the intraabdominal position in the animals of Groups 4, 5, and 6. In the first case, the meshes were implanted between the muscles of the abdominal wall and aponeurosis in the middle third, symmetrically relative to the abdominal median line, using the sublay technique. The implant was sutured with four interrupted stitches, using the monofilament made of the material matching the mesh material: Monofil 3–0 suture to fix Esfil meshes, Vicril 3–0 suture to fix VIPRO II meshes, and PHA fibers to fix PHA-coated meshes. The wound was tightly closed layer-by-layer and treated with alcohol (Figure 7.1).

TABLE 7.1   Characterization of the Groups of Experimental Animals (Markelova's data)

Group	Mesh material	Implantation type	Number (n)
1	Polypropylene (Esfil)	Sublay	30
2	Polypropylene and lactide/glycolide copolymer (VIPRO II)	Sublay	30
3	Polypropylene (Esfil) + P(3HB/3HV) coating	Sublay	30
4	Polypropylene (Esfil)	Intraabdominal	30
5	Polypropylene and lactide/glycolide copolymer (VIPRO II)	Intraabdominal	30
6	Polypropylene (Esfil) + P(3HB/3HV) coating	Intraabdominal	30

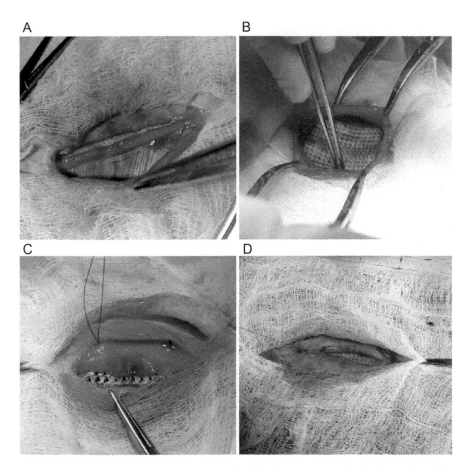

FIGURE 7.1 The sublay technique: A – the bed for the retro-muscular implantation of the mesh has been prepared; B – the mesh of the proper size has been prepared and placed into the bed; C – the mesh is being sutured to the muscular-aponeurotic layer with interrupted stitches; D – the reintegration of the muscular-aponeurotic layer with uninterrupted stitches (Markelova's data).

Intraabdominal implantation of the meshes was performed as follows. The midline laparotomy was done, and the mesh was sutured to the parietal peritoneum with several interrupted stitches, also using the material matching the implant material. The defect of the peritoneum was not sutured, to achieve the conditions similar to the clinical ones, when hernia is repaired by laparotomy. Then, the anterior abdominal wall was

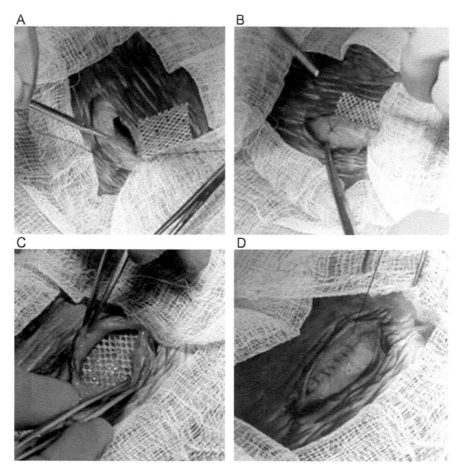

FIGURE 7.2   The intraabdominal technique: A – laparotomy has been performed, and
the mesh of the proper size has been prepared; B – the mesh is being sutured to the
muscular-aponeurotic layer with interrupted stitches; C – the wound with the sutured
mesh; D – reintegration of the muscular-aponeurotic layer with uninterrupted stitches
(Markelova's data).

sutured layer-by-layer, like in the first case (Figure 7.2). The animals were
sacrificed at Days 30, 60, and 90 post-surgery, based on the formation
of the connective tissue framework and tissue response to foreign body
implantation.

Examination of tissue sections did not reveal any macroscopic indications of soft tissue purulence at the implantation site and in the abdominal
cavity in all cases. None of the meshes was displaced; in the animals of all

groups sacrificed at Day 90, the meshes were securely fixed, and there was ingrowth of the surrounding tissues. In the Group 1, 2, and 3 animals, with meshes implanted by the sublay technique, there were very few adhesions. In Group 3, no adhesions were detected in the abdominal cavity; in Groups 1 and 2, adhesions were detected in three (7.5%) and two (5%) animals, respectively. Animals of Group 1 were the first to be operated, and while preparing the retro-muscular bed for the implant, the researcher accidentally made a deeper than necessary cut and slightly injured the lower layer (aponeurosis of the transverse muscle and peritoneum); the cuts 3–5 mm in diameter connected the implant bed with the abdominal cavity. When the abdominal cavities of these rabbits were opened at Day 90 post-surgery, we observed adhesions between the omentum and the mesh (Esfil). As only few adhesions were detected in the animals of Groups 1, 2, and 3, differences between the groups regarding this parameter were not significant.

More significant differences were observed between the animals with the meshes implanted by the intraabdominal technique. Tissue section examination revealed adhesions in the abdominal cavities of almost all animals in Groups 4 and 5. The most common adhesions were connective tissue adhesions between the abdominal wall and the omentum strands at the mesh site. In some of the animals, adhesions were formed with the loops of the small intestine and with the large intestine. In some of the cases, adhesions were found in the regions far away from the implant site and involved the organs that did not directly contact the implant; all kinds of adhesions (flat, membranous, cord-like adhesions, etc.) and conglomerates of the considerably deformed intestine loops were observed. Animals of Group 6, with PHA-coated meshes, did not show significant changes; moderate adhesions were observed only in three rabbits. Quantification of the adhesions is given in Table 7.2.

As can be seen from the data in Table 7.2, the incidence of adhesions in the Group 6 animals was significantly lower than in the Group 4 and 5 animals ($p < 0.01$), and when they did occur, they were statistically significantly less pronounced ($p < 0.001$ and $p < 0.05$). No significant differences in this parameter were found between animals of Groups 4 and 5 ($p > 0.05$). Morphological examination of the implant site after mesh explantation revealed considerable inflammatory infiltration with the infiltrate containing significant counts of macrophage-type cells. These changes were

TABLE 7.2 Adhesion Formation in the Abdominal Cavities of the Group 4, 5, and 6 Animals (Markelova's data)

Group	Mesh type and implantation technique	Number of animals with adhesions in the abdominal cavity	Scoring of the adhesion formation
4 (n=40)	Esfil, Intraabdominal	40 (100%)	2.54±0.78
5 (n=40)	VIPRO II, Intraabdominal	39 (97.5%)	2.19±0.93
6 (n=40)	P3HB-coated Esfil, Intraabdominal	3 (7.5%)	0.87±0.67

observed in all groups of animals, regardless of the material of the implant used and the implantation technique employed. Macrophages aggregated close to the mesh fibers; some of them showed signs of incomplete phago-cytosis and the presence of basophilic intracytoplasmic inclusion bodies (Figure 7.3).

Cellular elements of the inflammatory infiltrate were also represented by cells with morphology of plasmocytes, monocytes, mastocytes, and, more seldom, segmented leukocytes. Extracellular matrix contained fibrous tissue, which was developed to varying degrees and which showed fuchsinophilia when Van Gieson stain was used. In some cases, fibrous

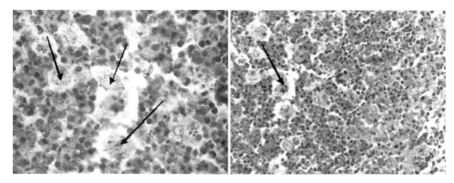

FIGURE 7.3 The histology of the region of composite mesh VIPRO II explantation; the sublay technique; stained with hematoxylin and eosin; the arrows point to macrophages containing phagocytized particles in the region of purulent necrotic detritus; magnification 1000× (Markelova's data).

tissue showed signs of dystrophy, with the fields of hyaline degradation, focal basophilia, and myxomatosis. The degrees of pathomorphological changes differed between and inside the groups, depending on the duration of the postoperative period.

Microscopic examination revealed the following differences between the groups. Implantation of uncoated synthetic materials (Groups 1, 2, and 4, 5) caused the host against implant inflammatory reaction. The immune response was manifested as extensive necrotic areas, pronounced infiltration by CD68[+] macrophages and formation of foreign body giant cells (Figure 7.4).

The most pronounced necrotic and macrophage reactions were observed in the region closest to the explanted mesh site. The same regions showed degradation and defibrization of extracellular matrix and aggregation of mastocytes with signs of degranulation. Many destruction regions with macrophage infiltration were surrounded by a wall of cells

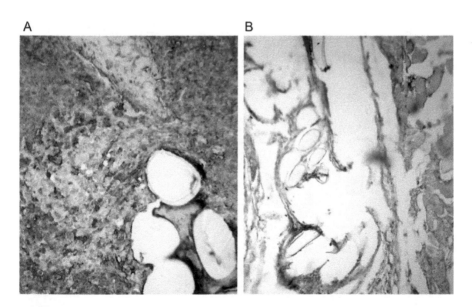

FIGURE 7.4 The histology of the region of explantation of the meshes implanted by the sublay technique: A – infiltration by macrophages exhibiting intense granular expression of CD68+ sites surrounding the mesh, magnification 400×; B – focal infiltration by macrophages exhibiting intense granular expression of CD68+ sites surrounding the mesh, magnification 200×. Immunohistochemical analysis with CD68+ antibodies (Markelova's data).

with monocytoidal differentiation and plasma cells that formed fields and focal aggregates. At some distance from the focus of inflammatory reaction, more or less constant foci of granulomatous productive reaction were observed; they were surrounded by a sclerotic strip and a wall of monocytes and plasma cells (Figure 7.5). The response described above was the most pronounced in the group of animals with the meshes implanted by the intraabdominal technique (Groups 4 and 5).

In the group of animals with PHA-coated meshes (Groups 3 and 6), the inflammatory reaction was less pronounced. The majority of inflammatory cells were morphologically small lymphocytes and plasma cells, which

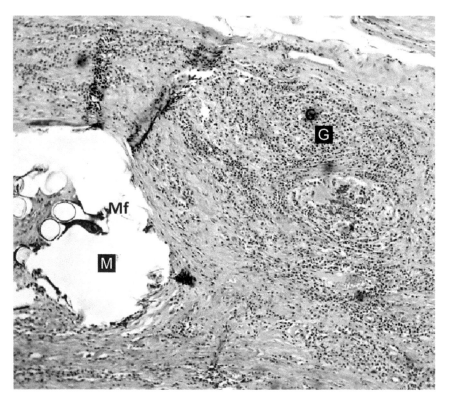

FIGURE 7.5   The histology of the region of explantation of the composite mesh VIPRO II implanted by the sublay technique: productive granulomatous reaction (G) close to the mesh site (M); in the center of the granuloma, there is a multinucleate foreign body giant cell; magnification 40×; stained with hematoxylin and eosin (Markelova's data).

do not express CD68[+]. Tissues surrounding the mesh were destroyed very little, and sclerotic processes were increased, but they were only observed in the implant site, without affecting the surrounding tissues, in contrast to Groups 1 and 2 (Figure 7.6).

In Group 3, with P3HB-coated meshes implanted by the sublay technique, no giant cell reaction was observed (Figure 7.7), while in 9 animals (22.5%) of the group with P3HB-coated meshes implanted by the intraabdominal technique, minimal giant cell reaction was observed, but no granulomatous inflammation foci were detected. The results of morphological analysis were confirmed by the morphometric examination.

The most pronounced CD68[+]-cell infiltration of the tissues surrounding the mesh was observed in Groups 1 and 2, and the least pronounced – in Group 3. At one month, macrophage reaction was evaluated as 1.6 ± 0.14, 2.0 ± 0.33, and 1.4 ± 0.11 scores in Groups 1, 2, and 3, respectively. At two months post-surgery, the scoring was 1.5 ± 0.13, 3.0 ± 0.27, and 1.2 ± 0.04 in

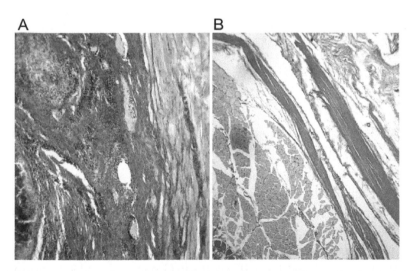

FIGURE 7.6 The histology of the region of explantation of the meshes implanted by the sublay technique: A – Group 2 (VIPRO II, 2 months); pronounced growth of fibrous tissue, spreading to the adjacent muscle layers of the abdominal wall (S), productive granulomatous reaction (G) close to the mesh site (M); B – Group 3 -P3HB-coated Esfil, 2 months; a clearly defined sclerosis region, occupying up to 50% of the abdominal wall thickness, with no signs of sclerotic changes spreading to the muscle layer. Van Gieson staining with picrofuchsin; magnification 100× (Markelova's data).

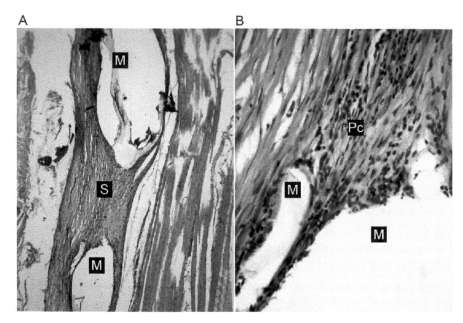

FIGURE 7.7   The histology of the region of explantation of the P3HB-coated meshes implanted by the sublay technique: A – 1 month post-surgery; lymphocyte-plasmocyte infiltrate surrounding the PHA-coated mesh (M) with a distinct sclerotic zone (S), 100×. B – a section of Figure 7.8 A; lymphocyte-plasmocyte infiltration surrounding the fibrous components of the mesh (Pc); magnification 400×. Stained with hematoxylin and eosin (Markelova's data).

Groups 1, 2, and 3, respectively: in Groups 1 and 3, the process became less pronounced, while in Group 2, macrophage counts in the mesh site increased. The differences in the groups were statistically significant ($p < 0.05$), and the least significant difference was observed in Group 3. At three months post-surgery, macrophage infiltration tended to decrease in all groups, with Group 1 having the score comparable to that of Group 3. The macrophage reaction at three months post-surgery was evaluated as 1.1 ± 0.20, 2.8 ± 0.03, and 1.1 ± 0.05 scores in Groups 1, 2, and 3, respectively.

A similar tendency was obtained in the analysis of the extent of the macrophage reaction. At one month, the process was spread least in Group 3, with P3HB-coated meshes, which mostly showed local or moderately spread macrophage reaction (1.3 ± 0.03 scores). The most extensive reaction was observed in Group 2 – 2.9 ± 0.02 scores (in the majority of animals, the reaction extended through all layers of the abdominal wall). The score in Group 1 was close to the score in Group 3 – 1.7±0.12. The differences in the groups were statistically significant (p<0.05). At two months, the extent of the reaction decreased in all groups (1.2±0.04, 2.7±0.52, and 1.2±0.10 scores in Groups 1, 2, and 3, respectively). At three months, the scores were even lower: 1.0±0.02, 1.7±0.04, and 1.3±0.08.

No granulomatous reaction was observed in the animals with P3HB-coated meshes. In Group 2 (with vicryl-polypropylene meshes), however, granulomatous reaction was very strong (p<0.05 relative to Group 3) and did not subside. The granulomatous reaction at the implant site was evaluated as 2.1±0.30, 2.3±0.13, and 2.1±0.02 scores at one, two, and, three months, respectively. In Group 1, the reaction was weaker. At three months, Group 1 animals showed only slight development of the granulo-matous process. The activity of the granulomatous reaction was evaluated at 0.5±0.01, 0.7±0.01, and 1.0±0.02 scores at one, two, and, three months, respectively. No giant cell reaction was detected in Group 3 at any time of the experiment. In Groups 1 and 2, with meshes without PHA coatings, multinucleate foreign body giant cells exhibiting cytoplasmic granular expression of $CD68^+$ were regularly detected in the muscles of the abdomi-nal wall; the reaction was scored as 0.6±0.02 and 1.2±0.48, respectively. The strongest reaction was observed at two months in Group 2 – 3.0±0.43 scores. As the granulomatous process in Group 2 at two months was not clearly pronounced, multinucleate giant cells were discretely located close to the fibers of the mesh. In Group 1, at two months, few giant cells were observed at the mesh site: 0.7±0.01 score. At three months, this parameter was evaluated at 1.0±0.07 and 1.9±0.11 scores in Groups 1 and 2, respec-tively. The giant cell reaction was significantly less pronounced in Group 1 than in Group 2 (p<0.5).

The evaluation of the development of coarse fibrous connective tissue that showed fuchsinophilia when Van Gieson stain was used suggested the following changes. The largest specific volume was occupied by fibrous tissue in Group 3, and it was growing between one and three months, reaching 50% (p<0.05). The average specific volume occupied by fibrous tissue reached 21.5±4.82% at one month post-surgery, 45.5±2.9% at two months, and 52.5±3.91% at three months. In the groups with uncoated meshes, the fibrous tissue growth was no more than 35%; in Group 2, this parameter was stable (32.0±4.12%, 30.0±5.67%, and 32.3±6.09% at one, two, and three months, respectively. In Group 1, it was growing at a rate comparable with the growth rate in Group 3: from 11.0±1.68% at one month to 34.5±4.22% at two months, slightly decreasing at three months – to 29.8±3.34%. The most considerable fibrosing of the abdominal wall tissues surrounding the mesh was observed in Groups 1 and 2, with uncoated meshes, and the process was dynamically progressive. The spread of the sclerotic process in Groups 1 and 2 was evaluated at 1.4±0.44 and 1.3±0.08 scores, respectively, at one month; at two months, it reached 2.1±0.71 and 1.5±0.34 scores. In Group 3, with P3HB-coated meshes, sclerotic processes were only observed at the mesh site; the wide monolayer of fibrous tissue was well-pronounced, and fibrosing of the surrounding tissues was insignificant. The spread of the sclerotic process was evaluated at 1.0±0.09 score at one month and 1.3±0.42 scores at two months post-surgery. These differences between Groups 1, 2, and 3 became especially clear at three months (p<0.05): 2.6±0.52, 2.2±0.47, and 1.5±0.34, respectively. Necrosis spreading to various degrees (usually with the perifocal granulomatous or macrophage reaction) was observed in all groups with uncoated meshes. Significantly more pronounced necrotic processes were observed in the animals of Groups 1 and 2 at all time points (p<0.05); the specific volume occupied by necrotic tissues at the mesh site at one month reached 33.6±4.08 and 32.2±5.82%, respectively. Then, necrotic reactions became weaker, and the decrease was more considerable in Group 1, but the parameters were not comparable with the parameters in Group 3. At two months, the specific volume of necrosis in Groups 1 and 2 was 12.7±2.46 and 17.4±4.63% and at three months – 5.5±1.19 and 12.3±3.07%, respectively. In the animals with PHA-coated meshes, inconsiderable necrosis was observed at one month – 1.0±0.18%; at two months, comparable values were recorded: 1.3±0.32%.

Results of implanting the meshes by the intraabdominal technique showed that the morphometric changes in the tissues were similar to those in the animals to which the meshes were implanted by the sublay technique. In Group 6 with P3HB-coated meshes), the macrophage reaction was less pronounced than in Groups 4 and 5, with uncoated meshes. The differences were statistically significant at all time points of the experiment (p<0.05). Moreover, in Group 6 animals, the macrophage reaction began to decrease at two months, while in Groups 4 and 5, it was increasing even at three months. The average scores of the macrophage reaction in Groups 4, 5, and 6 were 1.5±0.11, 2.0±0.02, and 1.2±0.02 at one month, 2.2±0.23, 2.5±0.01, and 1.4±0.02 at two months, and 2.5±0.03, 2.6±0.14, and 1.1±0.13, respectively, at three months. Similar evaluations were obtained for the spread of the macrophage reaction, which at one month was the highest in Group 5 (2.5±0.05 scores, p<0.05), with comparable levels recorded in Groups 4 and 6 (1.8±0.19 and 2.0±0.54 scores, p>0.05). Later on, in Groups 4 and 5, the macrophage reaction became more extensive (2.6±0.17 and 2.8±0.32 scores at two months and 2.5±0.09 and 2.9±0.14 scores at three months). In Group 6, opposite changes were observed: local and moderately spread macrophage reaction at two months (1.8±0.13 scores) and at three months (1.4±0.08 scores); these values were significantly lower than in Groups 4 and 5 (p<0.01).

Throughout the experiment, in Groups 4 and 5, with uncoated meshes, the frequency of granulomatous reaction in the tissues surrounding the mesh was significantly (by a factor of 1.5–2) higher than in Group 6. In Groups 4 and 5, the average level of granulomatous reaction in the mesh site was 1.1±0.37 and 2.9±0.33 scores at one month, 1.2±0.52 and 2.7±0.54 scores at two months, and 1.1±0.28 and 2.4±0.57 scores, respectively, at three months. In the Group 6 animals, with PHA-coated meshes, no inflammatory reaction with granuloma formation was observed. In Groups 4 and 5, the levels of giant cell reaction were very similar, slightly increasing with time. In these groups, many foreign body giant cells were observed at the explantation site. The levels of the giant cell reaction were 2.4±0.12 and 2.5±0.08 scores at one month, 2.8±0.56 and 2.8±0.34 scores at two months, and 2.9±0.66 and 2.8±0.53 scores, respectively at three months. In Group 6, this parameter was significantly lower at the end of the first month post-surgery (p<0.001) – 1.1±0.08, decreasing to 0.5±0.04 at two months and 0.7±0.02 at three months. Comparison of the specific volumes of coarse fibrous connective tissue suggested fibrosis increase in all groups. At one month, the specific volume of coarse fibrous connective tissue was the highest in Group 4 – 25.5±2.14% – and the lowest in Group 5 – 5.5±1.18%; median levels were recorded in Group 6

– 18.1±2.44%. At two months, this parameter increased in all groups, but to different degrees: in Groups 4 and 6, the volumes were comparable – 36.4±7.22% and 39.2±6.16%, respectively, while in Group 5, it was significantly lower – 20.3±4.11%. At the end of the third month, the values of this parameter remained almost unchanged in Groups 4 and 6 – 38.3±4.12 and 40.1±5.54%, respectively, while in Group 5, the average specific volume occupied by fuchsinophilic tissue increased to 31.1±3.07%. Although in the early phases of the experiment, the degrees of fibrosis were significantly different between groups, and the lowest fibrosing of the surrounding tissues was observed in Group 6 (1.1±0.26 scores at one month and 1.4±0.1 scores at two months) and the highest in Groups 4 and 5 (1.2±0.05 and 2.9±0.47 scores at one month and 2.2±0.12 and 2.8±0.23 scores, respectively, at two months), these differences became insignificant at three months ($p > 0.05$). The spreading of sclerotic processes at three months was evaluated as 2.4±0.78, 2.9±0.56, and 2.5±0.83 scores in Groups 4, 5, and 6, respectively. In Group 6 (with PHA-coated meshes), necrotic reactions were detected only at one month, and they were focal (the specific volume occupied by necrosis was 1.2±0.19%), but at two and three months, no necrosis was detected. Thus, statistically significant differences were obtained between Group 6 and Groups 4 and 5, in which necrosis occupied up to 45% of the specific volume of the tissue at different time points of the experiment. In Groups 4 and 5, the specific volume occupied by necrosis at the mesh site reached 17.2±4.22 and 39.1±5.56% at one month and 23.7±3.12 and 42.4±3.41%, respectively, at two months.

Postmortem examination was done to evaluate the extent of the adhesion development in the abdomens of the animals. In the groups with the meshes implanted by the sublay technique, with the mesh not contacting with the organs in the abdominal cavity, no adhesions were observed regardless of the mesh material. In the groups with the meshes implanted by the intraabdominal technique, statistically significant differences were observed between Group 6, in which adhesions were observed in 7.5% rabbits, and Groups 4 and 5, in which almost all animals (100 and 97.5%) had adhesions.

Thus, intraabdominal implantation of P3HB-coated meshes did not cause formation of adhesions. This may be a considerable advantage in the case of transabdominal hernioplasty, when the mesh cannot be safely covered with the peritoneum, or in the case of retroperitoneal implantation of meshes. The morphological changes detected in this study suggest that implantation of P3HB-coated meshes enables quicker strengthening of the abdominal wall and causes a weaker inflammatory response.

7.4 THE FIRST CLINICAL RESULTS OF IMPLANTING PHA-COATED MESHES

P3HB-coated meshes were used in a clinical trial, in which 112 male patients with unilateral inguinal hernias underwent surgical operations by Lichtenstein technique. The effectiveness of the approach was evaluated by the linear velocity of blood flow in the testicular artery. According to the type of the mesh, the patients were divided into 2 groups: in Group 1, PHA-coated meshes were compared with the polypropylene mesh Esfil (Lintex, Russia); in Group 2, PHA-coated meshes were compared with the lightweight mesh VIPRO II (Ethicon, U.S.), which consisted of polypropylene and resorbable copolymer of lactide and glycolide.

The clinical groups were similar in age, the duration of having hernia, the type of the inguinal hernia, according to the Nyhus classification, and working conditions. On admission to the clinic, all patients underwent the necessary laboratory and instrumental examinations and had no contraindications to surgery. Prior to surgery, all patient underwent ultrasonic examination with an Aloka SSD-4000 ProSound system (ALOKA, Japan), using a linear sensor at a scanning frequency of 7.5 MHz. The B-mode and the

TABLE 7.3 The Distribution of the Patients Between the Groups (Vinnik's data)

Group 1, n–55		Group 2, n–57	
Subgroup 1, reference	Subgroup 1, treatment	Subgroup 2, reference	Subgroup 2, treatment
Polypropylene monofilament mesh Esfil (Lintex, Russia)	P3HB-coated polypropylene monofilament mesh Esfil	Lightweight multifilament mesh VIPRO II (Ethicon, U.S.)	P3HB-coated lightweight multifilament mesh VIPRO II
n–29	n–26	n–29	n–28
Total	**112**		

color Doppler mode were used. In addition to standard parameters, sonography evaluated the state of the inguinal blood vessels on the affected and intact sides. The testicular artery was visualized (Milgalter et al., 2005), and then the linear velocity of the blood flow was determined in the color Doppler mode. The results were recorded in the digital database and the patient's clinical chart. All patients underwent the surgery by Lichtenstein technique I and II. The meshes were sutured with the material matching the implant material. The time of the hospital stay did not exceed 9 days. To evaluate the outcomes of the surgeries, the patients were examined in 30, 180, and 360 days post-surgery. These examinations consisted of sonography and the interview, which included questions on the type and duration of pain, the presence or absence of erectile dysfunction, unpleasant sensation in the groin.

The outcomes of the treatment of Group 1 patients were as follows. In the reference group, pain relief medication lasted 3.1 ± 0.8 days on average; mainly non-opioid analgesics were administered. In the early postoperative period, 2 patients had strong pain (6.9%), 4 patients had scrotum edemas (13.8%), and 2 had hematomas at the surgery site. At Days 5–6 post surgery, 3 patients (10.3%) showed infiltration of the postoperative scar. This complication was cured by an additional administration of antibiotics and application of the compresses soaked in 50% alcohol. Seromas developed in 6 patients (20.7%). This is the most frequent complication after foreign body implantation, as a consequence of aseptic inflammation. In 4 cases, this complication was cured by removing several skin sutures and draining the seroma with a rubber tube drainage. Two patients had to undergo ultrasonically monitored puncture drainage. Thus, early postoperative complications developed in 11 patients of the Group 1 reference subgroup, which constituted 37.9% of the total number of the patients. The hospital stay lasted 9 days, and the sick leave 27.4 ± 2.91 days. The postoperative sonography showed that in all cases, the meshes were visible, and some shrinkage was detected. At Day 180 post-surgery, tissue ingrowth into the mesh was observed. Results of measuring the linear velocity of the blood flow (BFLV) in the testicular artery are given in Table 7.4.

As can be seen from the data in Table 7.4, in the early postoperative period, the patients with polypropylene meshes had an insignificant ($t=1.66$; $p=0.11$) improvement in the testicular artery blood flow on the

TABLE 7.4 BFLV in the Testicular Artery of the Patients of the Group 1 Reference Subgroup (Vinnik's data)

Parameter Time	BFLV in the testicular artery, m/s		Difference from the intact side	
	Intact side	Affected side	Absolute	%
Before surgery	0.216±0.0176	0.190±0.0177	0.11 (0.00; 0.02)	5.1 (0.0; 10.5)
3–4 weeks post surgery	0.216±0.0165	0.192±0.0173	0.013±0.0102	4.59±1.979
6 months post surgery	0.189±0.0144	0.166±0.0208	0.035±0.0165	18.87±8.104
12 months post surgery	0.188±0.0136	0.164±0.0177	0.037±0.0146	19.64±6.998

affected side. In 6 months post-surgery, however, the BFLV in the testicular artery on the affected side was significantly lower than at the previous time point (t=10.60; p<0.001). In some of the patients, the BFLV was only 65% of the normal blood flow velocity (on the intact side). At 12 months post-surgery, no significant changes were observed in the BFLV of the testicular artery (t=0.57; p=0.57). Thus, in the Group 1 reference subgroup patients, the difference between the BFLV of the testicular artery on the intact side and the hernia side increased from 4.59±1.979% to 18.87±8.104% of the normal BFLV, and the impairment was stable, as confirmed by the examination at 12 months post-surgery. The following complications were observed at 12 months post-surgery: 3 patients (12.5%) had chronic pain in the implant site, 2 patients (9.3%) suffered from the pain in the implant site caused by physical activity, 5 patients (22.6%) felt discomfort and complained of foreign body sensation, and 2 patients (9.4%) had erectile dysfunction. Thus, 10 patients, or 39.6% of the total number of patients, had late complications after surgery.

In the Group 1 treatment subgroup, with PHA-coated meshes, none of the 26 patients had intra-surgery complications. In the early postoperative period, pain relief medication was conducted in accordance with the generally accepted standards; the duration of the medication was 3.0 days (3.0 days, 4.0 days). The following early complications were detected: 2 patients (9.5%) had hematomas at the scrotum, and one these patients also had edema (4.8%); 1 patient (4.8%) had strong pain and received longer pain relief medication. Three patients (14.3%) had seromas, which were cured by conservative treatment. Thus, 5 patients (23.8%) of the Group 1 treatment subgroup had early complications. There were no lethal outcomes. The hospital stay lasted 8 days (7.0 days, 9.0 days), and the sick leave 22.8±2.79 days. The postoperative sonography at 3–4 weeks, 6 months, and 12 months showed the following: the meshes were clearly visualized in all patients; no mesh shrinkage was observed; tissue ingrowth into the mesh was observed (at 6 months, mesh spaces were not visualized). The results of measuring the BFLV of the testicular artery before the surgery and at 3–4 weeks, 6 months, and 12 months post-surgery are given in Table 7.5.

The data in Table 7.5 suggest that the patients of the Group 1 treatment subgroup, with P(3HB/3HV)-coated polypropylene meshes fixed with PHA-coated sutures, had an insignificant increase in the BFLV of the testicular artery on the affected side (t=1.92; p=0.07). In 6 months post-surgery, this parameter reached its initial values and then remained unchanged. That was confirmed by sonography at 12 months post-surgery (t=1.00; p=0.32). The following late complications developed in the patients of this group: 1 patient (4.8%) had chronic pain in the implant site, 1 patient (4.8%) sometimes felt pain in the groin caused by physical activity, 2 patients (9.5%) felt discomfort and complained of foreign body sensation, and none of the patients had erectile dysfunction. Thus, 3 patients, or 14.3% of the total number of patients, had late complications after surgery. No recurrent hernias were detected at 12 months post-surgery.

TABLE 7.5   BFLV in the Testicular Artery of the Patients of the Group 1 Treatment Subgroup, M±SD (Vinnik's data)

Parameter Time	BFLV in the testicular artery, m/s		Difference from the intact side	
	Intact side	Affected side	Absolute	%
Before surgery	0.232±0.0023	0.189±0.0080	0.017±0.0063	7.20±2.418
3–4 weeks post surgery	0.212±0.0002	0.192±0.0058	0.015±0.009	5.97±1.901
6 months post surgery	0.212±0.0038	0.189±0.0047	0.018±0.002	7.18±1.913
12 months post surgery	0.214±0.0017	0.188±0.0057	0.017±0.0040	7.68±1.502

In the Group 2 reference subgroup, with lightweight composite meshes VIPRO II, none of the 29 patients had intra-surgery complications. In the early postoperative period, the following complications were observed: 1 patient (4.5%) had strong pain, which was cured by longer pain-relief medication; 7 patients had seromas: in 5 of them, seromas resolved on their own, and 2 patients underwent ultrasonically monitored puncture drainage. One patient (4.5%) had infiltrate in the subcutaneous layer in the suture site, and in another patient (4.5%), the suture suppurated after seroma puncture. Both conditions were cured conservatively, without explanting the meshes. Thus, early complications were observed in 7 patients, or 31.8%, of the Group 2 reference subgroup. All patients of this group received pain relief medication for 3.0 (3.0, 4.0) days. The hospital stay lasted 9 days (8.0 days, 9.0 days), and the sick leave 27.1±1.36 days. There were no lethal outcomes. No recurrent hernias developed after 12 months. The postoperative sonography at 3–4 weeks, 6 months, and 12 months showed the following: the meshes were clearly visualized in all patients; moderate mesh shrinkage was observed in almost all patients; adequate tissue ingrowth into the mesh was observed. The results of measuring the BFLV of the testicular artery are given in Table 7.6.

As can be seen from the data in Table 7.6, in the early postoperative period, the patients with uncoated composite meshes had an insignificant (t=1.31; p=0.20) improvement in the testicular artery blood flow on the

TABLE 7.6 BFLV in the Testicular Artery of the Patients of the Group 2 Reference Subgroup, M±SD (Vinnik's data)

Parameter Time	BFLV in the testicular artery, m/s		Difference from the intact side	
	Intact side	Affected side	Absolute	%
Before surgery	0.209±0.0187	0.199±0.0137	0.012±0.003	4.81±1.371
3–4 weeks post surgery	0.207±0.0193	0.198±0.0152	0.009±0.003	2.46±0.873
6 months post surgery	0.208±0.0179	0.169±0.0144	0.044±0.007	19.29±7.667
12 months post surgery	0.207±0.0174	0.165±0.0145	0.048±0.013	21.61±6.589

affected side. In 6 months post-surgery, however, the BFLV in the testicular artery on the affected side was significantly lower than at the previous time point (t=9.30; p<0.001). In some of the patients, the BFLV was only 30–40% lower than the normal blood flow velocity (on the intact side). At 12 months post-surgery, no significant changes were observed in the BFLV of the testicular artery (t=1.90; p=0.07). Thus, in the Group 2 reference subgroup patients, the difference between the BFLV of the testicular artery on the intact side and the hernia side increased from 2.46±0.873% to 19.29±7.667% of the normal BFLV, and the impairment was stable, as confirmed by the examination at 12 months post-surgery. The following late complications were observed: 2 patients (9.1%) had chronic pain in the implant site, 2 patients (9.1%) suffered from the pain in the implant site caused by physical activity, 4 patients (18.2%) felt discomfort and complained of foreign body sensation, and 3 patients (13.6%) had erectile dysfunction. Thus, 8 patients, or 36.4% of the total number of patients, had late complications after surgery.

In the Group 2 treatment subgroup, with PHA-coated lightweight composite meshes VIPRO II (Ethicon, U.S.), none of the 28 patients had intraoperative complications. The following early complications were detected: 1 patient (4.8%) had strong pain; 2 patients (9.5%) had seromas; 1 patient (4.8%) had infiltrate in the subcutaneous layer in the suture site; 2 patients (9.5%) had hematomas – one patient at the suture site and the other at the scrotum, and that patient also had scrotum edema (4.8%). All complications were cured conservatively. Thus, early complications were observed in 6 patients, or 28.6%, of the Group 2 treatment subgroup. All patients of this group received pain relief medication for 3.0 (2.0, 3.0) days. The hospital stay lasted 8 days (7.0 days, 8.0 days), and the sick leave 23.6±2.41 days. The postoperative sonography at 3–4 weeks, 6 months, and 12 months showed the following: the meshes were clearly visualized in all patients; tissue ingrowth into the mesh was observed; no mesh displacement or considerable shrinkage was observed. The results of measuring the BFLV of the testicular artery are given in Table 7.7.

In the patients of the Group 2 treatment subgroup, with PHA-coated polypropylene meshes fixed with P3HB-coated sutures, no significant changes were recorded in the BFLV in the testicular artery on the hernia side in the early postoperative period (t=0.62; p=0.54). In 6 and 12 months

TABLE 7.7 BFLV in the Testicular Artery of the Patients of the Group 2 Treatment Subgroup, M±SD (Vinnik's data)

Parameter Time	BFLV in the testicular artery, m/s		Difference from the intact side	
	Intact side	Affected side	Absolute	%
Before surgery	0.201±0.0158	0.191±0.0148	0.012±0.0096	5.66±1.849
3–4 weeks post surgery	0.204±0.0164	0.189±0.0156	0.016±0.0012	6.99±1.450
6 months post surgery	0.198±0.0076	0.187±0.0124	0.017±0.009	7.34±2.341
12 months post surgery	0.203±0.0166	0.185±0.0131	0.019±0.006	8.57±1.951

post-surgery, this parameter was not significantly lower than before the surgery (t=1.63; p=0.12 and t=1.64; p=0.12). Thus, during the first 6 months post-surgery, the patients of the Group 2 treatment subgroup had an insignificant decrease in the BFLV of the testicular artery on the surgery side: from 5.66±1.849 to 7.34±2.341% of the normal BFLV (t=1.37; p=0.19). At 12 months, the BFLV remained at the same level and was 8.57±1.951% lower than the normal BFLV. The following late complications were observed: 1 patient (4.8%) had chronic pain in the implant site, 2 patients (9.5%) complained of foreign body sensation, and 1 patient (4.8%) suffered from the pain in the implant site and at the scrotum caused by physical activity. Thus, 3 patients, or 14.3% of the total number of patients, had late complications after surgery. There were no lethal outcomes. No recurrent hernias developed after 12 months.

The average duration of the surgery in the Group 1 reference and treatment subgroups was 56.5±11.75 and 55.2±11.78 min, respectively, i.e., no significant differences were revealed (t=0.35; p=0.73). In the Group 2 reference and treatment subgroups, the average duration of the surgery was 54.1±14.69 and 56.7±11.33 min, respectively, i.e., no significant differences were revealed either (t=0.64; p=0.52). Postoperative pain relief medication was conducted in a standard way in all groups. Then, we counted the number of days during which the patient needed pain relief medication. In the Group 1 reference and treatment subgroups, pain relief medication lasted 3.0 (3.0; 4.0) and 3.0 (2.0; 3.0) days, respectively. In the

Group 2 reference and treatment subgroups, a similar value was obtained: 3.0 (3.0; 4.0) and 3.0 (2.0; 3.0) days, respectively. No significant decrease in the need for pain relief medication of the patients in the treatment subgroups, 1 and 2, relative to the patients in the reference subgroups, 1 and 2, was revealed (Z=1.81; p=0.07 and Z=1.91; p=0.056, respectively). Complications that developed in the patients of the reference and treatment subgroups in the early postoperative period are listed in Table 7.8.

Thus, some of the complications were more numerous in the reference subgroups, and other complications – in the treatment subgroups. However, statistically significant differences were obtained only for the frequency of seroma – a typical complication of abdominal wall repair – and only in the groups with VIPRO II meshes. The frequency of seromas in the patients of the Group 2 treatment subgroup was lower than in the Group 2 reference subgroup (φ*=1.87; p<0.05). The number of patients with postoperative complications did not differ significantly between the reference and treatment subgroups either in Group 1 (φ*=0.71; p>0.05) or in Group 2 (φ*=0.04; p>0.05).

The patients with P3HB-coated meshes and matching sutures had significantly shorter hospital stays and sick leaves than the reference group patients. The durations of the hospital stays and sick leaves of the patients of the Group 1 treatment subgroup were 8.0 (7.0; 9.0) days and 22.8±2.79

TABLE 7.8   The Frequency of Early Complications in the Groups, the Number of Patients with Postoperative Complications (Vinnik's data)

| | Early complications, n (%) | | | |
	Reference group 1	Treatment group 1	Reference group 2	Treatment group 2
Hematoma	2 (6.8%)	3 (11.5%)	-	5 (17.8%)
Seroma	6 (20.6%)	4 (15.3%)	7 (24.1%)	5 (17.8%)
Infiltrate	3 (10.3%)	3 (11.5%)	5 (17.2%)	4 (14.2%)
Suppuration	2 (6.8%)	-	5 (17.2%)	-
Scrotum edema	3 (10.3%)	3 (11.5%)	-	4 (14.2%)
Strong pain	2 (6.8%)	5 (19.2%)	5 (17.2%)	4 (14.2%)
Patients with complications	11 (37.9%)	8 (30.7%)	7 (24.1%)	6 (21.4%)

days, respectively, versus 9.0 (8.5; 9.0) days and 25.3±1.69 days in the Group 1 reference subgroup (Z=3.25; p=0.001 and t=3.73; p<0.001). Similar data were obtained for Group 2: in the treatment group, the durations of the hospital stays and sick leaves were 8.0 (7.0; 8.0) days and 23.2±2.41 days, respectively, while in the reference group, these periods were significantly longer – 9.0 (8.0; 9.0) days and 26.1±1.36 days, respectively (Z=5.61; p<0.001 and t=4.90; p<0.001). In the late postoperative period, the effectiveness of the meshes was evaluated by comparing the following parameters in the groups of patients: the frequency of late complications and the degree of impairment of testicular blood flow. The late complications observed in the patients of the reference and treatment subgroups are listed in Table 7.9.

Such implant-associated complications as foreign body sensation and discomfort at the surgical site were observed in the patients of the Group 1 reference subgroup and the Group 2 reference subgroup in 4 (13.6%) and 3 (10.3%) cases, respectively. In the Group 1 treatment subgroup and the Group 2 treatment subgroup, only three patients in each subgroup (11.5%) had this complication. However, the decrease in the frequency of this complication in the treatment subgroups 1 and 2 was not statistically

TABLE 7.9 The Frequencies of Late Complications in the Groups; the Number of Patients with Complications (Vinnik's data)

	Late complications, n (%)			
	Reference group 1 (N=29)	Treatment group 1 (N=26)	Reference group 2 (N=29)	Treatment group 2 (N=28)
Foreign body sensation and discomfort in the scrotum region	4 (13.6%)	3 (11.5%)	3 (10.3%)	3 (10.7%)
Pain caused by physical activity	2 (6.8%)	2 (7.6%)	3 (10.3%)	2 (7.1%)
Chronic pain	3 (10.3%)	2 (7.6%)	3 (10.3%)	2 (7.1%)
Erectile dysfunction	2 (6.8%)	-	4 (13.6%)	-
Recurrent hernia	-	-	-	-
Patients with complications	9 (31.0%)	4 (15.3%)	7 (24.1%)	4 (14.2%)

significant ($\varphi^*=1.07$; $p>0.05$ and $\varphi^*=0.83$; $p>0.05$, respectively). In the late postoperative period, 2 patients of the Group 1 reference subgroup (6.8%) and 4 patients of the Group 2 reference subgroup (13.6%) complained of different degrees of erectile dysfunction, while none of the treatment subgroup patients had this complication. During one year of follow-up, no recurrent hernias developed in the groups. The frequencies of other late complications (pain in the scrotum region caused by physical activity and chronic pain at the surgery site) were not significantly different between the reference and treatment groups of patients.

The total number of the patients with late complications was significantly lower in the treatment groups, with PHA-coated meshes. During one year, in the Group 1 treatment subgroup, late complications developed in 4 patients (15.3%), while in the Group 1 reference subgroup, late complications developed in 9 patients (31.0%) ($\varphi^*=1.81$; $p<0.05$). In the Group 2 treatment subgroup, the number of the patients with late complications was also significantly lower than in the Group 2 reference subgroup: 4 (14.2%) patients and 7 (24.1%) patients, respectively ($\varphi^*=1.70$; $p<0.05$). The degree of impairment of testicular artery blood flow was evaluated by the changes in the BFLV in the testicular artery at the hernia side, where the spermatic cord emerges from the inguinal ring. Changes in the blood flow linear velocity in the testicular artery on the surgery side that occurred during one year in the patients of all groups are given in Table 7.10.

TABLE 7.10   Changes in the BFLV in the Testicular Artery on the Hernia Side (Vinnik's data)

	BFLV in the testicular artery, M±SD, m/s			
Parameter	Before surgery	3–4 weeks post surgery	6 months post surgery	12 months post surgery
Reference group 1 (N=29)	0.189±0.0177	0.193±0.0173	0.166±0.0208	0.164±0.0177
Treatment group 1 (N=26)	0.187±0.0180	0.191±0.0158	0.189±0.0147	0.187±0.0157
Reference group 2 (N=29)	0.197±0.0137	0.199±0.0152	0.167±0.0144	0.164±0.0145
Treatment group 2 (N=28)	0.189±0.0148	0.191±0.0156	0.186±0.0124	0.187±0.0131

The data in Table 7.10 show that before the surgery, the BFLV values in the Group 1 reference subgroup and the Group 1 treatment subgroup were similar (t=0.41; p=0.68). At 3–4 weeps post-surgery, the BFLV values in the groups were also comparable (t=0.28; p=0.78). At 6 months post-surgery, in the patients of the Group 1 treatment subgroup, with PHA-coated meshes, the BFLV was at the pre-surgery level (t=0.09; p=0.93). In the patients of the Group 1 reference subgroup, the blood flow in the testicular artery on the hernia side was significantly impaired: in 6 months post surgery, the BFLV was lower than the BFLV measured in them before the surgery (t=4.85; p<0.001) and significantly lower than the BFLV of the Group 1 treatment subgroup at 6 months (t=4.66; p<0.001). The measurements of the BFLV in 12 months post-surgery showed the same tendency: the BFLV of the patients of the Group 1 reference subgroup was significantly lower (t=5.03; p<0.001).

Similar changes in the BFLV values were observed in the Group 2 reference subgroup and the Group 2 treatment subgroup. Before the surgery, there were no statistically significant differences between the BFLV values in these two subgroups (t=1.50; p=0.14). In 3–4 weeks, the BFLV measurements showed statistically significant differences in this parameter: in the Group 2 reference subgroup, it increased somewhat (t=2.26; p=0.03). However, in 6 months post-surgery, the BFLV in the Group 2 reference subgroup was significantly worse than the BFLV of this subgroup before the surgery (t=7.10; p<0.001) and the BFLV of the Group 2 treatment subgroup, with PHA-coated meshes, measured at 6 months (t=4.80; p<0.001). In the Group 2 treatment subgroup, in 6 months post-surgery, the BFLV values remained unchanged (t=0.90; p=0.37). In 12 months post-surgery, sonography confirmed the consistency of these changes: the BFLV was significantly lower in the patients of the Group 2 reference subgroup (t=5.42; p<0.001).

These results were confirmed by the data obtained in the study of the patients' quality of life, which was conducted in 12 months after the surgery. The evaluation of the quality of life was based on the scores of the SF-36 health survey. In the reference subgroups, all scores were lower than in the treatment subgroups. In the reference subgroup with the uncoated Esfil meshes, only the evaluation of the physical health showed lower scores than in the corresponding treatment subgroup. However, in the reference subgroup with the uncoated VIPRO II meshes, all scores were lower than in the treatment subgroup, including physical and mental components. This impairment in the quality of life of the reference group patients was caused by the implant-associated complications such as pain caused by physical activity, chronic pain, foreign body sensation and discomfort at the surgery site, and erectile dysfunction.

Thus, the results obtained in this study suggest that implantation of the PHA-coated meshes and the matching sutures can shorten the duration of the hospital stay and sick leave, decrease the frequency of the late implant-associated complications, and avoid the impairment of the patients' quality of life.

KEYWORDS

- herniology
- support meshes
- polymer coating
- implantation
- surgical adhesions
- clinical trials

REFERENCES

Amid, P. K. How to avoid recurrence in Lichtenstein tension-free hernioplasty. Am. J. Surg., 2002, 184, 259–260.

Awad, S. S., Bruckner, B., Fagan, S. P. Transperitoneal view of the PROLENE hernia system open mesh repair. Int. Surg., 2005, 90(3), 63–66.

Chistyakov, A. A., Mitichkin, A. E., Osokin, G. Yu. Trabucco allografts in treatment of front abdominal wall hernias. Almanac of the, A. V. Vishnevsky Institute of Surgery, 2008, 3(2–1), 92–93 (in Russian).

Cucci, M., De Carlo, A., Di Luzio, P., Masella, M., Casciani, E., D'Amico, G. The Trabucco technique in the treatment of inguinal hernias; A Six-Year Experience. Minerva Chir., 2002, 57, 457–459. Processing

Elsebae, M. M., Nasr, M., Said, M. Tension-free repair versus Bassini technique for strangulated inguinal hernia: A controlled randomized study. Int. J. Surg., 2008, 6, 302–305.

Fedorov, V. D., Adamyan, A. A. Evolution of inguinal hernia management. Khirurgiya (Surgery), 2000, 3, 51–53 (in Russian).

Garavello, A., Manfroni, S., Teneriello, G. F., Mero, A., Antonellis, D. Recurrent inguinal hernia after mesh hernioplasty. An emerging problem? Minerva Chir., 2001, 56, 547–552.

Markelova, N. M. Obosnovaniye primeneniya vysokotekhnologichnykh izdeliy meditsinskogo naznacheniya iz biodegradiruyemykh polimerov v rekonstruktivnoy khirurgii (experimentalno-klinicheskoye issledovaniye) (Substantiation of using high-tech medical devices of biodegradable polymers in reconstructive surgery (experimental study and clinical trials). MD dissertation in Specialty 14.01.17: Surgery. Krasnoyarsk, Voino-Yasenetsky Krasnoyarsk Federal Medical University, 2013, 272 p. (in Russian).

Markellova, N. M., Vinnik, Yu. S., Volova, T. G., Markelova, N. M. The effects of different implants on the testicular artery blood flow in the patients after hernia repair surgery. Kreativnaya khirurgiya i onkologiya (Creative surgery and oncology), 2012, 4, 4–7 (in Russian).

Milgalter E., Pearl J.M., Laks H. et al. The inferior epigastric arteries as coronary bypass conduits. Size, preoperative duplex scan assessment of suitability, and early clinical experience // J. Thorac. Cardiovasc. Surg., 1992, 103(3), 463–465.

Pryakhin, A. N. The choice of the technique and technical aspects of using meshes in hernia repair in the treatment of complicated inguinal hernias. Vestnik khirurgii (Surgery Bulletin), 2007, 166(2), 96–99 (in Russian).

Sanjay, P., Harris, D., Jones, P., Woodward, A. Randomized controlled trial comparing prolene hernia system and lichtenstein method for inguinal hernia repair. ANZ J. Surg., 2006, 76, 548–552.

Seleem, M. I. Open mesh-plug technique in inguinal hernia repair-short-term results. S. Afr. J. Surg., 2003, 41, 44–47.

Shevchenko, Y. L., Kharnas, S. S., Egorov, A. V. The choice of the technique for plastic reconstruction of the front abdominal wall in the case of inguinal hernia. Annaly khirurgii (Annals of Surgery), 2003, 1, 20–23 (in Russian).

Shulutko, A. M., El-Saed, A. H. A., Danilov, I. A. Results of the tension-free inguinal hernia repair by Lichtenstein technique. Annaly khirurgii (Annals of Surgery), 2003, 2, 74–77 (in Russian).

Strizheletsky, V. V., Rutenburg, G. M., Guslev, A. B. The position of endo-video-surgical interventions in inguinal hernia repair. Vestnik khirurgii (Surgery Bulletin), 2006, 165, 15–20 (in Russian).

Yakovlev, A. V., Vinnik Yu.S., Markelova, N. M., Shishatskaya, E. I. Treatment of inguinal hernias using polypropylene meshes and meshes coated by polyhydroxyalkanoate-based material. Sibirskoye meditsinskoye obozreniye (Siberian medical review), 2010, 2, 76–80 (in Russian).

Yegiev, V. N., Chizhov, D. V., Rudakova, M. N. Lichtenstein technique in hernia repair. Khirurgiya (Surgery), 2000, 1, 19–21 (in Russian).

Yurasov, A. V., Fedorov, D. A., Shestakov, A. L. Modern tactics of inguinal hernia surgeries. Annaly khirurgii (Annals of Surgery), 2002, 2, 54–59 (in Russian).

CHAPTER 8

A STUDY OF POLYHYDROXYALKANOATES AS MATERIALS FOR DESIGNING FULLY RESORBABLE BILIARY STENTS

CONTENTS

8.1 INTRODUCTION

Polyhydroxyalkanoates (PHAs) can be good candidates for use in surgical reconstruction of bile ducts (Gillams et al., 1990; Glattli et al., 1992). The data reported at the Fourth World Congress of Gastroenterology suggest that the incidence of cholecysto-choledocholithiasis seconds only to atherosclerosis, causing up to 2.5 million annual elective and urgent surgical operations on bile ducts. Among the greatest challenges facing abdominal surgery is treatment of the patients with the obstructive jaundice. Low-invasive biliary operations, bile duct stenting in particular, are becoming more and more common (Ferro et al., 1993; Harewood et al., 2002). Expandable metallic stents are the most commonly used ones, but they can cause adverse effects; mesh stents can be invaded by tumor tissues (Yoshioka et al., 1990), and they cannot be removed in the event of obstruction (Gillams et al., 1990); necrosis of the mucous coat of bile ducts is not uncommon. Biliary stents prepared from polymer materials (silicon, Teflon, polyurethanes, polyethylene, percuflex) (Lammer, Neumayer, 1986) are far from being perfect too. They may cause obstruction for 3–6 months and migrate; moreover, transhepatic surgical intervention is too traumatic. Using PHA stents for biliary surgery has not been reported in the literature until recently.

Experimental research and clinical trials of tubular biliary stents fabricated from PHA were performed by researchers of the General Surgery Department at the V.F. Voino-Yasenetsky Krasnoyarsk Federal Medical University.

8.2 BILIARY STENTING AND DRAINAGE TO TREAT THE OBSTRUCTIVE JAUNDICE DUE TO TUMORS AND OTHER CAUSES

Among the greatest challenges facing medicine is treatment of the patients with the obstructive jaundice. Achievements of the contemporary surgery improved results of the treatment, decreased complications and death rates (Kubyshkin et al., 2008; Vinnik et al., 2001; Nikolsky et al., 2013). Decompression of bile ducts gives time to diagnose the cause of the obstructive jaundice by instrumental examination of the patients (Kurzawinski et al., 1992). Contemporary abdominal surgery may involve not only biliary-enteric bypass procedures and pancreaticoduodonal resections, but also complex liver resections (even subtotal ones), liver transplantation, and even simultaneous liver and pancreas transplantations. Moreover, thanks to the achievements of reconstructive biliary surgery, bi-, tri-, tetra-, and even penta-hepaticojejunostomies can be performed (Korolev et al., 2012; Kabanov et al., 2013; Harewood et al., 2002). Yet, there are many unresolved issues that complicate the treatment of the obstructive jaundice

(Ferro et al., 1993). Only 25–30% of the patients with the cancer-caused obstructive jaundice can be subjected to radical surgery. In 38–46% of all patients, the tumor blockades the common bile duct in the gate of the liver, and extends to the adjacent organs and large vessels, which makes the radical surgery very complicated or even impossible. In this case, only palliative therapy is possible. Low-invasive biliary operations, bile duct stenting in particular, may be a solution (Kubyshkin et al., 2003; Yoshioka et al., 1990). This intervention can have a palliative effect, similar to that achieved by the major surgery, but it is devoid of limitations mentioned above. Moreover, in contrast to major surgery, biliary stenting can be performed again if the obstructive jaundice recurs. Another argument in favor of biliary stenting is that the total cost of the treatment is several-fold lower than the cost of the palliative surgical interventions in the case of the blockage of the bile ducts by tumors (Izrailov et al., 2007; Gillams et al., 1990). In the case of benign biliary strictures, biliary interventions are more limited than in the case of cancer blockages. However, in recent years, the number of laparoscopic surgeries of the bile ducts has increased dramatically, but the number of complications has increased in direct proportion to them. Among the most serious complications is the injury of the common bile duct, whose reconstruction is done during abdominal surgery. In the late postoperative period, this causes formation of the biliary stricture in 15–40% of the patients (Nikolsky et al., 2013; Neuhaus et al., 1998). As the famous French surgeon Mondor said, "a patient with the dissected common bile duct is a dying patient." As repeated surgical reconstructions are obviously technically complicated and traumatic and their outcomes are unpredictable, this is a serious concern in the treatment of patients with bile duct blockages. Therefore, biliary interventions, especially common bile duct stenting, may be a way out of this situation (Murai et al., 1991).

The choice of the technique of biliary stenting is considerably determined by the facilities available in the hospital. In the hospitals with well-developed radiosurgery, the technique of choice is percutaneous transhepatic intervention (Tulin et al., 2007; Yamamoto, 1992). The hospitals specializing in laparoscopic surgery prefer to perform retrograde biliary surgery (Kurzawinski et al., 1992). Early biliary stents were catheters, which were available when the idea of implantation originated and which were used for transhepatic drainage. They were named synthetic stents. The most commonly used materials for fabricating stents were the smooth-surface Teflon, polyurethane, polyethylene, and percuflex (Aripova, 2007; Glattli et al., 1992). Such stents are still widely used. Clinical use of synthetic stents also revealed their drawbacks: they are obstructed in 3–6 months; they often migrate and become contaminated; the transhepatic intervention is too traumatic, as the stent and the guide tube are rather large.

The first report of the clinical use of biliary self-expandable nitinol stents was published in 1989. That stent had the following properties: it was compatible with body tissues and corrosion resistant; it was inserted by image-guided percutaneous transhepatic placement through the angiographic catheter, shaped as a straight line, which then, under the effect of body temperature, expanded, adapting itself to a necessary shape due to the shape-memory property of nitinol. Its pressure on the wall of the bile duct could be regulated; it was capable of self-locking where necessary; it had good collapse resistance to external pressure; there were no bile salts on its surface (Vinnik et al., 2008; Markelova et al., 2012). Later, however, the drawbacks became evident too. First, in the case of the bile duct cancer, implantation of self-expandable metallic stents, which were mesh constructs, did not prevent the tumor tissue from growing through the stent, causing recurrent obstructive jaundice. Second, with the tissue growing through the stent and blockading it, it is impossible to remove it endoscopically and even surgically. The only way out in these cases was to insert another stent, using the stent-in-stent technique. That was the main reason why such stents were not used in the case of benign strictures of the common bile duct and biliodigestive anastomoses (Zakharash et al., 2009; Glattli et al., 1991). Third, almost all models of self-expandable stents have rather low lateral rigidity, which is often insufficient to keep the bile duct open when it is squeezed by the dense external tumor. In these cases, some authors suggested using balloon-expandable metallic stents. Unfortunately, they also had some limitations. With time, the mesh constructs sunk into the bile duct wall infiltrated by the tumor and, in some cases, could cause necrosis of the mucous coat. Therefore, in recent years, synthetic biliary stents have attracted the attention of researchers (Zakharash, 2009; Zagainov, 2011; Kotenko, 2007; Glattl, 1991). Contemporary biliary surgery uses all available stent types, including synthetic, self-expandable, and balloon-expandable ones. The choice of the stent type depends on the clinical case and the type of the bile duct condition. The Carey-Coons stent is the synthetic biliary stent that is easy to remove if it is obstructed. It has synthetic sutures that fasten it to a subcutaneously implanted anchor button. If it is necessary to remove the stent, the surgeon slits the skin above the button and uses the recanalization catheter to pull the fixing suture and the stent through the transhepatic canal into the bile ducts; then, the space previously occupied by the stent is washed and re-stented.

The literature review suggests that biliary stenting is a very effective way to manage bile duct obstruction. Therefore, a common task for surgeons and biotechnologists is to find new materials for designing biliary stents.

8.3 AN EXPERIMENTAL STUDY OF USING PHA TUBES AS BILIARY STENTS

We present results of experimental research and clinical trials of biliary stents, produced at the Institute of Biophysics SB RAS. Clinical trials were performed in cooperation with researchers of the General Surgery Department at the V.F. Voino-Yasenetsky Krasnoyarsk Federal Medical University. Experiments with P3HB biliary stents were performed on 20 adult mongrel dogs. The dogs were divided into groups as shown in Table 8.1. After choledochotomy, PHA stents were implanted to the supraduodenal part of the common bile duct and fixed to its wall with PHA suture material.

TABLE 8.1 Groups of Animals (Markelova's data)

Group	Number of animals, n	Characterization of the group
Negative control	N=7	Intact animals
Positive control	N=5	Silicone biliary stent, anastomosis between the gallbladder and the duodenum
Treatment	N=8	Bioplastotan biliary stent, anastomosis between the gallbladder and the duodenum
Total	N=20	

The animals were anesthetized with calypsol–droperidol, and the surgery was performed at room temperature, under aseptic conditions. After supramedian laparotomy, the common bile duct was identified, and choledochotomy was performed. In the treatment group, a P3HB stent was implanted to the supraduodenal part of the common bile duct and fixed to its wall with PHA suture material (Figure 8.1).

A silicon stent was implanted to the animals of the positive control group; the stent was sutured using Vicryl, a resorbable suture material. The final stage of the operation included control of hemostasis and foreign bodies, layer-by-layer tight closure of the wound, and application of aseptic dressing. Complete blood counts (for hemoglobin, color index, ESR, blood cells) were performed before the surgery, and at Days 7, 30, 60, and 100 post-surgery. The biochemical blood profile was examined by using conventional techniques: whole blood protein, glucose, urea, total bilirubin, amylase activity, and transaminases were determined. Indicators of the state of the nonspecific immunity of the animals were phagocytic activity of lymphocytes, determined with HCT test, and spontaneous and induced luminol-dependent chemiluminescence of lymphocytes [the time of luminescence peak (i-max) and the area under the chemiluminescence

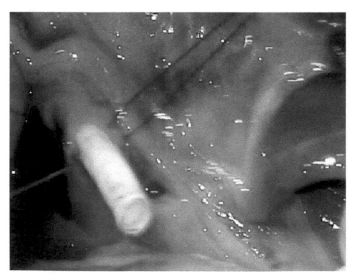

FIGURE 8.1   Insertion of the P3HB stent (Markelova's data).

curve (s-max)]. The animals were observed for 100 days. Then, the positive control and treatment animals were sacrificed by the IV administration of sodium thiopental. Morphological methods of tissue examination included macroscopic description (postmortem results) and morphometric characterization of the tissue specimens (the tissues of the common bile duct wall at the stent implantation site, liver, and duodenum). All animals survived the operation; no intra-surgery complications were detected. During the postmortem, we checked for the presence of exudate and commissural process in the free abdominal cavity and subhepatic spatium and examined the appearance of the common bile duct at the site of implantation and the appearance of liver and duodenum. After the opening of the common bile duct lumen, the following parameters were observed: the common bile duct wall thickness, the presence of visible inflammatory changes, the state of the common bile duct mucous membrane, the reliability of the attachment of the stent to the common bile duct wall, the presence of defects in stent construction as a result of polymer biodegradation, the size of stent lumen, the presence of lumen diminution as a result of sludge and concrement sedimentation, and the thickness of the stent wall. No postmortem was performed on the negative control animals.

Three animals of the positive control group had an insignificant amount of serous exudate in the abdominal cavity (up to 30–40 mL) and a moderate commissural process in the subhepatic spatium. Infiltration, lumen expansion, and cicatricial changes were observed in the common bile duct at the site of stent implantation. In two animals of this group, the stents had migrated towards the ampulla of Vater. The macroscopic examination of the liver showed that it had normal appearance. Two animals had moderate hepatomegaly. The opening of the common bile duct lumens of all animals showed that the common bile duct wall was thickened; excrescence of connective tissue and infiltration were observed. The silicone stent was easily extracted from the lumen. The common bile duct mucous membrane at the site of contact with the stent was of pale pink color; there were atrophied zones detected; and 3 animals had the stent lumens narrowed by 40–50% as a result of sludge and bile component sedimentation. The stents were fragile and had sediments of bile salts and pigments.

After the treatment group animals were sacrificed, the postmortem did not reveal any exudates or cicatricial changes in the free abdominal cavity and subhepatic spatium. The common bile duct had normal appearance at the site of stent implantation; no expansion, inflammatory reaction or cicatricial processes were visualized. All P3HB stents stayed at their initial sites of implantation; no cases of migration were detected (Figure 8.2).

No macroscopic changes were revealed during the examination of liver and duodenum. The common bile duct lumen was retained in the place of stent implantation in all animals and had a normal size (0.4–0.5 mm); no deformations, strictures, cicatricial or inflammatory changes in the stent implantation region were observed (Figure 8.3).

While extracting the stents, we registered a leaky adhesion to the common bile duct mucous membrane; a small effort was enough to extract the stents from the common bile duct lumen, which contained clear bile. The stents retained their initial physical properties; they were not subject to calcification processes; there was no narrowing of stent lumen, its diameter being 3.5±0.1 mm. Biodegradation did not cause any defects, except that the stent wall became thinner: the average wall thickness was 0.05–0.08 mm. No stent lumen narrowed areas were detected (Figure 8.4).

Parameters of the peripheral blood of the negative control, positive control, and treatment group dogs were generally within the range of the

FIGURE 8.2   The P3HB stent in the implantation site, Day 100 (Markelova's data).

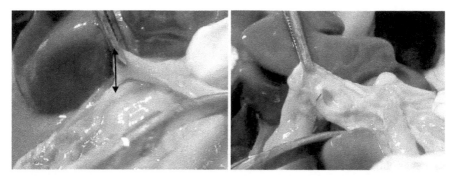

FIGURE 8.3   The common bile duct lumen after P3HB stent removal, Day 100 (Markelova's data).

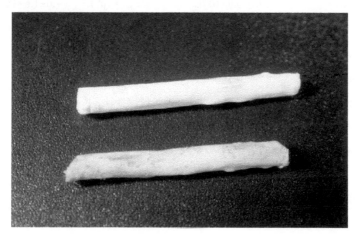

FIGURE 8.4   P3HB stent removed from the common bile duct (Markelova's data).

physiological norm. An insignificant increase in the white blood count (from 10 to 12×10^9/L) and ESR (up to 10–15 mm/h) was recorded on Day 7 post-surgery in the positive control and treatment animals compared to the negative control. By the end of the period of observation, the animals of the positive control group had a moderately increased ESR (up to 10.1 ± 3.3 mm/h). No shifts in the leucograms of the positive control and treatment animals were recorded during the whole period of observation (Table 8.2). Determination of urea and whole protein in blood serum suggested that P3HB-based implants did not produce any adverse effects on nitrogen metabolism or kidney function. There were no significant differences in the parameters characterizing the function of pancreas (blood amylase) between the negative control group and the positive control and treatment groups (Tables 8.2–8.3).

TABLE 8.2 Complete Blood Counts of the Positive Control and Treatment Animals (Markelova's data)

Parameters	Negative control	Positive control	Treatment
Hemoglobin	142±3.5 g/L	137±1.9 g/L	149±2.8 g/L
Erythrocytes	5.12±1.2×10^{12}/L	4.89±1.5×10^{12}/L	5.68±0.9×10^{12}/L
Color index	1.0±0.1	0.9±0.4	0.8±0.2
Leukocytes	7.2±0.9×10^9/L	9.8[1]±1.3×10^9/L	7.6±1.6×10^9/L
ESR	3.2±1.6 mm/h	10.1[1]±3.3 mm/h	4.7±2.1 mm/h
Leukogram			
Band cells, %	2.3±0.1	5.2±0.2	3.1±0.3
Segmented cells, %	68.4±3.6	64.2±4.1	67.9±3.9
Monocytes, %	2.5±0.3	1.1±0.2	3.2±0.1
Lymphocytes, %	26.8±3.2	29.9±4.3[1]	24.8±4.8
Eosinophils, %	2.1±0.1	1.3±0.3	2.2±0.2
Plasmocytes, %	1.1±0.1	1.4±0.1	1.2±0.1

Note: [1] – the differences are significant relative to the negative control.

TABLE 8.3 The Biochemical Blood Profile of the Animals at Day 100 (Markelova's Data)

Parameters	Negative control	Positive control	Treatment
Amylase	64.6±4.39 g×h/L	89.9±6.7 g×h/L	73.3±5.9 g×h/L
Total protein	71.9±3.2 g/L	59±4.6 g/L	67.6±4.8 g/L
Total bilirubin	8.8±1.8 mmol/L	39.91±4.3 mmol/L	9.1±2.0 mmol/L
Glucose	4.2±0.9 mmol/L	4.9±1.9 mmol/L	4.5±1.2 mmol/L
Urea	4.9±1.7 mmol/L	7.1±2.3 mmol/L	5.6±1.8 mmol/L
AST	0.88±0.4 IU	1.01±1.9 IU	0.96±0.2 IU
ALT	0.84±0.7 IU	2.15±3.7 IU	0.86±0.6 IU

Note: [1] – the differences are significant relative to the negative control.

From Day 30 onward, the animals of the positive control group had moderately increased total bilirubin levels caused by bile flow disorder. At Day 100, the animals of this group demonstrated an increased ALT level, which indicated development of cholestasis and compromised liver function (Table 8.3). Registration of liver function parameters of the animals in the course of the experiment did not reveal any pathological deviations in the treatment group. The total bilirubin (the main cholestasis marker) of the animals was within the limits of physiological norm. The activity of hepatic enzymes (ALT and AST) in the course of the experiment did not reveal any AST deviations. The increased ALT activity recorded in the first period of observation in the positive control and treatment animals could have been caused either by operational trauma or by the toxic effect of anesthetics. No statistical differences were recorded between these two groups. Results of the influence of the P3HB stents on the nonspecific immunity of the animals are given in Table 8.4. The nonspecific immunity of the positive control and treatment animals was somewhat increased, as suggested by the higher phagocytosis and induced chemiluminescence levels. No inhibition of phagocytic activity was detected, suggesting the absence of chronic antigenic load and immunological paralysis.

Morphological examination showed that all specimens of the positive control animals had signs of inflammatory cellular reaction and fibrosis. The common bile duct mucous membrane was atrophied, with necrotic patches. The liver had signs of cholestasis, beam and hepatocyte fractures. In the zone of anastomoses, there were signs of inflammatory cellular reaction, a large number of leucocytes and macrophages, and coarse scar tissue. Morphological examination of the common bile duct area of the treatment animals, with P3HB stents, did not reveal any pathological changes (Figure 8.5).

Thus, experiments proved the safety and effectiveness of using P3HB tubes as biliary stents.

8.4 THE FIRST RESULTS OF USING PHA BILIARY STENTS IN THE PATIENTS WITH THE EXTRAHEPATIC BILIARY OBSTRUCTION

The clinical trial included 32 patients with the obstructive jaundice caused by cancer and 21 patients with the cicatricial strictures of the common bile duct. The description of the groups of patients is given in Table 8.5.

TABLE 8.4 Parameters of Phagocytic Activity and Luminol-Dependent Chemiluminescence of Blood Lymphocytes of the Animals at Day 100 Post-Surgery (Markelova's data)

Parameters	Negative control		Positive control		Treatment	
Phagocytosis	32±3.6		58±5.4[1]		53±4.3[1]	
Complete 30 min	21±2.1		47±3.2[1]		44±5.4[1]	
Complete 90 min	19±2.3		45±3.4[1]		42±3.1[1]	
Index of phagocytosis completeness	1.02±0.25		1.08±0.13[1]		1.05±0.25	
HCT test						
Spontaneous	413±13.2		434±16.8		426±14.2	
Stimulated	445±11.2		471±12.7		465±12.4	
Stimulation index (S_{cn}/S_{cm})	1.06±0.32		1.08±0.23		1.09±0.31	
Chemilumines-cence	Spont.	Induced	Spont.	Induced	Spont.	Induced
t-max – time of reaching the peak (s)	255±10.2	298±9.6	341±12.4[1]	311±14.3[1]	240±10,5	285±11.7
i-max – number of pulses at the peak	1511±32.8	1487±38.9	3248±56.9[1]	4248±45.7[1]	1457±27.9	1329±34.1
s-max – peak area	$1.71\pm0.4\times10^5$	$1.92\pm0.6\times10^5$	$3.88\pm0.9\times10^5$	$5.08[1]\pm0.6\times10^5$	$1.67\pm0.6\times10^5$	$1.8\pm0.5\times10^5$
Stimulation index (S_{cn}/S_{cm})	1.08±0.24		1.31±0.21		1.07±0.14	

Note: [1] – the differences are significant relative to the negative control.

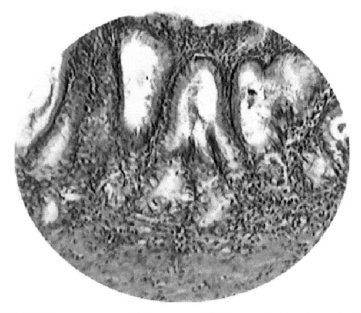

FIGURE 8.5 The mucous coat of the common bile duct at the site of stent implantation in the treatment group animals (Markelova's data).

TABLE 8.5 Groups of Patients with the Obstructive Jaundice (Vinnik's data)

Group	Group characterization	Number of patients
Reference group	Obstructive jaundice, conventional techniques of decompression of extrahepatic bile ducts (EPST, choledochotomy, common bile duct drainage, biliodigestive anastomosis)	17
Treatment group	Obstructive jaundice, biliary stenting with P3HB stents	15
	Total	**32**

The groups were similar in the age composition, sex ratio, and severity, duration, and cause of the obstructive jaundice (Tables 8.6–8.8).

The number of women in the groups was larger than the number of men, and all patients were elderly, which increased the surgical risk.

In 8 patients, the obstructive jaundice was caused by the complicated cholelithiasis and in 3 – by the cancer in the head of the pancreas with the distal blockage of the common bile duct. Surgeries performed on the patients in the reference group and the treatment group are listed in Tables 8.9 and 8.10.

TABLE 8.6   Gender Distribution of Patients with the Obstructive Jaundice (Vinnik's data)

Gender	Reference group	Treatment group
Male	8 (47.0%)	10 (64.0%)
Female	9 (53.0%)	5 (36.0%)

TABLE 8.7   The Age Distribution of the Patients with the Obstructive Jaundice (Vinnik's data)

Group	Average age of the patients
Reference group	66.3±13.4
Treatment group	68.2±15.8

TABLE 8.8   The Distribution of the Patients Based on the Causation of the Obstructive Jaundice (Vinnik's data)

Causes of the obstructive jaundice	Reference group	Treatment group
Cancer in the head of the pancreas, distal blockage of the common bile duct	2 (11.0%)	4 (26.7%)
Cancer of hepatic common bile duct, Klatskin tumor	3 (14.0%)	2 (13.3%)
Gallbladder cancer	2 (11.0%)	-
Cholelithiasis, cholecysto-chole-dochal lithiasis, cholangitis, cicatricial stricture of the common bile duct	10 (64%)	9 (60%)

TABLE 8.9   The Type of Surgery in the Reference Group (Vinnik's data)

Surgery	Number
EPST, nasobiliary drainage	10 (64.0%)
Laparotomy. Cholecysto-entero-anastomosis. Anterior Braun astroenteroanastomosis.	4 (16.7%)
Laparotomy. Long looped Roux Braun holedojejunal anastomosis.	3 (16.7%)

TABLE 8.10   The Type of Surgery in the Treatment Group (Vinnik's data)

Surgery	Number
Laparotomy. Cholecystectomy. Choledochotomy. Choledochoduodenal anastomosis by the Yurash-Vinogradov method on the PHA stent.	9 (60%)
Laparotomy. Cholecystectomy. Choledochotomy, stenting of the common bile duct with the PHA stent. Vishnevsky"s drainage of the common bile duct.	4 (26.7%)
Laparotomy. Choledochotomy, bougienage of the hepatic common bile duct, stenting of the hepatic common bile duct with the PHA stent. Vishnevsky's drainage of the common bile duct.	2 (13.3%)

As can be seen from the data in Tables 8.9 and 8.10, the extent of surgical intervention differed depending on the cause of the obstructive jaundice and the degree of hepatic common bile duct obstruction. In 2 patients with the cancer of the distal common bile duct, stenting was impossible because the stricture was over 3 cm long. In these cases, biliodigestive anastomosis was performed, and stent frames were constructed to prevent restenosis and anastomosis failure.

In one case, when the patient had tumor-caused obstruction, the bougienage of the stricture was performed, and the stent was implanted without creating biliodigestive anastomosis (Figure 8.6). Such minimal intervention is possible is the stricture is short and high.

In this case, in addition to the bougienage of the stricture and stent implantation, the Klatskin tumor was removed. For the short low cicatricial stricture, we developed the following approach to stenting of the common bile duct.

The surgical site was prepared for the minimally invasive access using a Mini-assistant kit (Russia). Choledochotomy (cholecystectomy if indicated) was performed following the retrograde pancreato-cholangiography, which was done to determine exactly the extent of the common bile duct obstruction. Then, a Fogarty biliary probe with a reinforced cuff was passed through the stricture into the duodenum, the stricture was balloon-dilated, the Fogarty biliary probe was again passed into the duodenum and inflated; then, the stent was inserted through the probe as far as it would go and pushed forward slightly to make sure that it was in proper position (Figures 8.7 and 8.8).

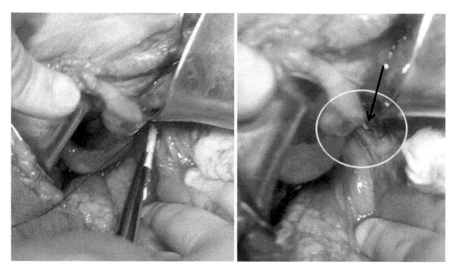

FIGURE 8.6   Implantation of the biliary stent of P3HB (Vinnik's data).

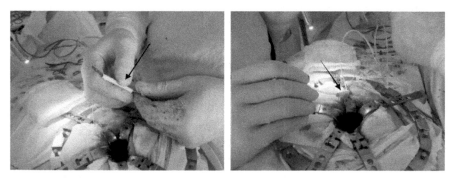

FIGURE 8.7   Inserting the P3HB stent through the modified Fogarty biliary probe (Vinnik's data).

Then, cholangiography was performed to check the patency of the common bile duct; the common bile duct was drained by one of the conventional techniques; the subhepatic space was drained with polyvinylchloride irrigators; the wound was tightly sutured layer-by-layer and covered with an aseptic dressing. During the 38-day follow-up, the state of all patients was monitored by determining general clinical parameters (complete blood count), biochemical blood profile, blood coagulation (coagulogram). In 2–3 weeks post-surgery, fistulous cholangiography was

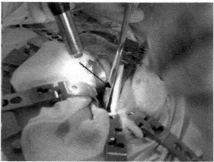

FIGURE 8.8   Implanting the P3HB stent into the common bile duct (Vinnik's data).

performed, and the drainage was removed. After that, the patency of the common bile duct was monitored endoscopically (with an Olympus side-viewing duodenoscope). Total bilirubin was measured at Days 3, 7, 10, 14, 28, and 38. The effectiveness of the treatment of the patients with the obstructive jaundice was evaluated by clinical laboratory data, instrumental examination, and analysis of the outcome. As the average hospital stay of the patients with the obstructive jaundice lasted 32±12.4 days, the duration of the follow-up was 28 days. However, the main clinical and laboratory parameters of all patients had been monitored until they were discharged (died) (a patient that has undergone the surgery on extrahepatic bile ducts must stay in the hospital for 38 days, according to the medical-economic standards).

Analysis of the laboratory, clinical, and instrumental examination data suggested the following. The main criterion of the effectiveness of inter-vention in the case of the obstructive jaundice is blood bilirubin level. The recovery of the normal total bilirubin level suggests successful manage-ment of autointoxication, hepatorenal syndrome, and multiple organ fail-ure. Results of bilirubin measurements at different time points are given in Table 8.11.

As can be seen from the data in Table 8.11, in the reference group, the average bilirubin level had not reached normal values even by Day 28, while in the treatment group, it had decreased to the upper normal limit by Day 14. Thus, in the patients with the obstruction of the distal com-mon bile duct, passage of the bile into the duodenum is more effectively

TABLE 8.11   Total Bilirubin Levels of the Patients at Different Time points (Vinnik's data)

Group	Follow-up days					
	On admission	Day 3	Day 7	Day 10	Day 14	Day 28
Reference group (mmol/L)	189.4±11.4	166.2±18.3	149.8±.19.2	112.8±9.5	45.6±21.6	40.6±18.7
Treatment group (mmol/L)	204.2±10.9	149.9±8.7[1]	98.9±11.2[1]	62.7±7.9[1]	35.0±8.6[1]	27.8±6.15[1]

Note: [1] – the differences are significant at p<0.05 relative to the reference group.

achieved by stenting of the common bile duct than by conventional surgical procedures. In the patients that underwent implantation of stents P3HB by the original technique, the level of bilirubin decreased at a significantly higher rate.

Another important indicator in diagnosing diseases of the hepato-pancreato-duodenal region is the level of blood amylase. Amylase increase indicates pancreatic dysfunction, and this is one of the major diagnostic criteria of acute pancreatitis. The normal amylase level must not exceed 8 g·h/L. Blood amylase levels of the patients are given in Table 8.12. The significantly lower levels of blood amylase in the treatment group patients may suggest that the rapid decrease in bilirubin levels enabled recovery of the function of the pancreas. This organ inevitably suffers in the event of biliary hypertension, which may lead to a very serious complication – pancreonecrosis. This complication, which develops under autointoxication and dysfunction of the liver and kidneys of the patients with the obstructive jaundice, may result in decompensation, multiple organ failure, and death.

Urea is an important indicator of the excretory function of the kidneys (the ability of kidneys to remove unnecessary materials with urine). Urea is a product of protein metabolism, and urea increase is indicative of catabolic processes in the body and an impairment of the excretory function of kidneys, i.e., renal failure. Thus, the level of urea in blood is an indicator of the state of kidneys and liver and the extent of catabolic processes.

TABLE 8.12   Blood Amylase (g·h/L) in the Patients (Vinnik's data)

Group	Days On admission	Day 3	Day 7	Day 10	Day 14	Day 28
Reference group	15.5±2.5	14.7±3.4	12.9±2.8	14.4±2.9	8.4±1.8	10.5±1.9
Treatment group	7.7±1.3	5.0±10.9	5.9±10.7	4.2±10.8	4.4±10.6	3.7±10.4

Note: [1] – the differences are significant at p<0.05 relative to the reference group.

The normal urea level in the blood of adults ranges between 2.5 and 6.4 mmol/L. We measured urea levels in the blood of the patients throughout the follow-up. Results are listed in Table 8.13. The levels of urea in the reference group patients had decreased to the normal values by Day 28, while in the treatment group patients, the urea levels had decreased to the normal values by Day 14. It is important that changes in this parameter correlated with changes in the bilirubin level, i.e., the more rapid bilirubin decrease was accompanied by the improvement in the kidney function.

ALT is an intracellular aminotransferase enzyme catalyzing the amino acid – keto acid transamination by transferring an amino group. ALT catalyzes the transfer of an amino group from alanine to α-ketoglutarate, the products of this reversible transamination reaction being pyruvate and glutamate. Transamination occurs in the presence of coenzyme – pyridoxal phosphate, the active form of vitamin B_6. The highest levels of ALT are measured in the liver and kidneys. Lower levels are detected in the heart, skeletal muscles, pancreas, spleen, lungs, and erythrocytes. ALT is

TABLE 8.13   Urea Levels (mmol/L) in the Patients (Vinnik's data)

Group	Days On admission	Day 3	Day 7	Day 10	Day 14	Day 28
Reference group	7.4±1.2	8.9±1.1	7.5±1.3	9.2±0.9	6.6±0.8	5.6±0.9
Treatment group	5.6±1.1	4.9±10.9	7.0±10.8	7.0±10.7	5.8±10.8	5.4±10.6

Note: [1] – the differences are significant at p<0.05 relative to the reference group

an intracellular enzyme, and its levels in the blood serum of healthy people are low. When, however, ALT-rich cells (liver, kidneys, myocardium, skeletal muscles) are damaged or destroyed, the enzyme appears in the blood, reaching elevated levels. As ALT is not organ-specific, the level of its serum activity is not always correlated with the severity of the disease affecting the organ (the extent of necrosis). Although both ALT and AST levels are increased in the case of the liver damage, ALT is a more specific marker of liver diseases than AST. ALT levels in the blood of the patients with the obstructive jaundice are given in Table 8.14. The ALT levels at all time points, between the admission and the end of the follow-up, were considerably higher than the normal levels in both groups, suggesting liver dysfunction under hyperbilirubinemia. However, ALT levels of the treatment group patients became significantly lower than ALT levels of the reference group patients from Day 10 onward. The most considerable increase in the ALT levels was recorded between Days 7 and 14 in both groups.

A similar trend was observed in changes of the levels of AST, which is also a marker of cell destruction (Table 8.15).

TABLE 8.14   ALT Levels in the Patients (Vinnik's data)

Group	Days					
	On admission	Day 3	Day 7	Day 10	Day 14	Day 28
Reference group	1.5±0.4	2.3±0.6	2.0±0.5	2.2±0.3	2.1±0.7	2.0±0.4
Treatment group	1.8±0.2	1.9±10.3	2.1±0.4	1.7±10.2	1.5±10.2	1.3±10.1

Note: [1] – the differences are significant at $p<0.05$ relative to the reference group.

TABLE 8.15   AST Levels in the Patients (Vinnik's data)

Group	Days					
	On admission	Day 3	Day 7	Day 10	Day 14	Day 28
Reference group	1.2±0.2	1.5±0.1	1.4±0.3	1.3±0.1	1.3±0.3	1.1±0.2
Treatment group	0.9±0.2	1.1±0.3	1.4±0.3	1.3±0.2	0.9±10.1	0.8±10.2

Note: [1] – the differences are significant at $p<0.05$ relative to the reference group

Development of anemia in the patients with the obstructive jaundice, especially with cancer as the causative factor, is an unfavorable diagnostic criterion, suggesting decompensation of the patient's state. The extent of anemia is directly related to how long the hyperbilirubinemia has persisted and to its level. Total hemoglobin concentrations were not significantly different between groups. However, during the postoperative period, one of the treatment group patients bled from the varicose veins in the esophagus, which did not happen in the reference group, and this may be the reason for the similar average hemoglobin levels in the groups. Total hemoglobin levels are given in Table 8.16.

Leukocytosis, as the major indicator of inflammatory activity and the autointoxication syndrome, was lower in the treatment group than in the reference group from Day 3 onward, but decreased to its normal values only by Days 14–28. In the reference group, at Day 28, leucocyte counts were insignificantly higher than their normal levels (Table 8.17).

TABLE 8.16 Hemoglobin Concentrations (g/L) in the Patients (Vinnik's data)

Group	Days					
	On admission	Day 3	Day 7	Day 10	Day 14	Day 28
Reference group	141.5±9.4	120.5± 10.4	114.3± 10.2	115.8±8.6	117.8±9.4	123.4±8.2
Treatment group	142.3± 10.1	130± 16.4	122.8± 15.9	114.3±5.6	118.3±4.6	119.3±16.2

Note: [1] – the differences are significant at $p<0.05$ relative to the reference group.

TABLE 8.17 Leukocyte Counts (*109/L) of the Patients (Vinnik's data)

Group	Days					
	On admission	Day 3	Day 7	Day 10	Day 14	Day 28
Reference group	11.0±2.4	10.6±3.1	10.5±2.8	10.9±3.2	9.6±3.3	8.2±2.1
Treatment group	5.1±0.8	8.9±11.2	9.5±11.1	9.9±10.9	7.5±10.7	5.2±10.8

Note: [1] – the differences are significant at $p<0.05$ relative to the reference group.

Analysis of the postoperative complications and the death rate shows that the most unfavorable outcomes were observed in the patients with tumors of the common bile duct. The initially higher levels of bilirubinemia and the longer persistence of the obstructive jaundice syndrome caused such complications as progressive hepatorenal and multiple organ failures, encephalopathy, and hemorrhagic complications, which considerably aggravated the disease and impaired the effects of the surgical intervention. In the treatment group, the total number of complications was significantly lower, as can be seen from the data in Table 8.18.

The death rate in the reference group was 33.3% (2 out of 6 patients died). The main causes of death were complications of the underlying illnesses: the hepatorenal syndrome decompensated in one patient, and the other patient died from progressive multiple organ failure. None of the patients died in the treatment group. In the patients with the predictably unfavorable outcomes (with the cancer), the proposed approach reduced bilirubinemia without palliative surgery (biliodigestive anastomosis), minimized the surgical trauma, and did not cause rapid decompensation, which favorably influenced the treatment outcomes. The average time of the hospital stay in the reference group was 33.2±2.6 days, and in the treatment group – 30.7±1.8 days (Table 8.19).

TABLE 8.18 Postoperative Complications in the Patients (Vinnik's data)

Complication	Reference group	Treatment group
Acute pancreatitis	1 (16.7%)	-
Acute cholangitis	1 (16.7%)	-
Progressive hepatorenal failure	1 (16.7%)	-
Abscess of the postoperative scar	-	1 (20%)
Multiple organ failure and pulmonary edema	1 (16.7%)	-
Bleeding esophageal varices	-	1 (20%)

TABLE 8.19 The Average Time of the Hospital Stay (Vinnik's data)

Group	Average hospital stay, days
Reference group	33.2±2.6
Treatment group	30.7±1.8

Thus, implantation of experimental fully resorbable P3HB stents to treat patients with the obstructive jaundice reduced the number of postoperative complications and the death rate and improved treatment outcomes for this group of patients. The results obtained in this study can be used as the basis for further research in this field.

KEYWORDS

- abdominal surgery
- bile ducts
- biliary obstruction
- stenting
- biliary stent
- reducing postoperative complications

REFERENCES

Aripova, N. U., Ismailov, U. S., Magzumov, I. Kh. Minimally invasive surgery to treat the pathologies of the bile excretion system. Annaly khirurgicheskoy gepatologii (Annals of Surgical Hepatology), 2007, 12, 39 (in Russian).

Ferro, C., Ambrogi, C., Perona, F., Barile, A., Cianni, R. Malignantprosthesis: Wallstent vs Strecker's stent. Radiol. Med. (Torino), 1993, 85, 644–647.

Gillams, A., Dick, R., Dooley, J. S., Wallsten, H., el-Din, A. Self-expandable stainless steel braided endoprosthesis for biliary strictures. Radiology, 1990, 174, 137–140.

Glattli, A., Mouton, W., Schweizer, W., Baer, H. U., Gilg, M., Triller, J., Blumgart, L. H. Percutaneous transhepatic inserted self-expanding metal endoprosthesis in the palliative treatment of malignant obstructive jaundice. Schweiz. Med. Wochenschr., 1992, 122, 663–666.

Harewood, G. C., Baron T.H., LeRoy, A. J., Petersen, B. T. Cost-effectiveness analysis of alternative strategies for palliation of distal biliary obstruction after a failed cannulation attempt. Am. J. Gastroenterol., 2002, 97, 1701–1707.

Izrailov, R. E., Gurchenkova, P. E., Konina, T. N. Using different modifications of nitinol stents in patients with the tumor-caused obstructive jaundice of bile ducts: mistakes, dangers, and complications. Annaly khirurgicheskoy gepatologii (Annals of Surgical Hepatology), 2007, 12, 72 (in Russian).

Kabanov, M. Y., Solovyev, I. A., Yakovleva, D. M. The choice of the extent of minimally invasive drainage interventions in the patients with the cancer in the head of the pancreas. Vestnik Rossiyskoy Voenno-Meditsinskoy Akademii (Bulletin of the Russian Military-Medical Academy), 2013, 41, 97–101 (in Russian).

Korolev, M. P., Fedotov, L. E., Avanesyan, R. G. The effects of integrated techniques of minimally invasive intervention in the treatment of injuries and strictures of bile ducts. Vestnik khirurgii im. I. I. Grekova (I. I. Grekov bulletin of surgery), 2012, 171, 20–27 (in Russian).

Kubyshkin, V. A., Vishnevsky, V. A. Rak podzheludochnoy zhelezy (Cancer of pancreas). Medpraktika. Moscow, 2003, 375 p. (in Russian).

Kurzawinski T., Deery, A., Dooley, J., Dick, R., Hobbs, K., Davidson, B. A prospective controlled study comparing brush and bile exfoliative cytology for diagnosing bile duct strictures. Gut, 1992, 33, 1675–1677.

Lammer, J., Neumayer, K. Biliary drainage endoprostheses: experience with 201 placements. Radiology, 1986, 159, 625–629.

Markelova, N. M. Obosnovaniye primeneniya vysokotekhnologichnykh izdeliy meditsinskogo naznacheniya iz biodegradiruyemykh polimerov v rekonstruktivnoy khirurgii (experimentalno-klinicheskoye issledovaniye) (Substantiation of using high-tech medical devices of biodegradable polymers in reconstructive surgery (experimental study and clinical trials). MD dissertation in Specialty 14.01.17: Surgery. Krasnoyarsk, Voino-Yasenetsky Krasnoyarsk Federal Medical University, 2013, 272 p. (in Russian).

Markellova, N. M., Volova, T. G., Shishatskaya, E. I., Markelova, N. M., Beletsky, I. I. A new approach to biliary stenting. Annaly khirurgicheskoy gepatologii (Annals of Surgical Hepatology), 2009, 204–205 (in Russian).

Murai, R. Hashigchi, F., Kusujama, A. Percutaneous stenting for malignant biliary stenosis. Surg. Endosc., 1991, 5, 140–142.

Neuhaus, H., Hagenmuller, F., Classen, M. Self-expanding and expandable bile duct prostheses. Gastroenterol., 1991, 29, 306–310.

Tulin, A. I., Zeravs, N., Kupchs, K. Endoscopic and percutaneous transhepatic stenting of bile ducts. Annaly khirurgicheskoy gepatologii (Annals of Surgical Hepatology), 2007, 12, 53–61 (in Russian).

Vinnik, Yu. S., Markelova, N. M., Volova, T. G. On using a novel biocompatible polymer as a biliary stent. Proceedings of the Conference on Important Issues of Modern Surgery. Moscow-Krasnoyarsk, 2008, 109–110 (in Russian).

Vinnik, Yu. S., Volova, T. G., Markelova, N. M. Experimental substantiation of using biliary stents of bioresorbable polyhydroxyalkanoates. Proceedings of the 11th Congress of Surgeons of the Russian Federation, Volgograd, 2011, 72–73 (in Russian).

Yoshioka, T., Sakaguchi, H., Yoshimura, H., Tamada, T., Ohishi, H., Uchida, H., Wallace, S. Expandable metallic biliary endoprostheses: preliminary clinical evaluation. Radiology, 1990, 117, 253–257.

Zakharash, M. P., Nichitailo, M. E., Zakharash, Y. M. Techniques of biliary decompression and stenting in the integrated treatment of patients with the obstructive jaundice caused by the cancer in the head of the pancreas. Ukrainskiy zhurnal khirurgii (Ukrainian Journal of Surgery), 2009, 3, 70–72 (in Ukrainian).

CONCLUSION

The range of materials used in medical applications is very wide, encompassing both natural and artificial materials such as metals, ceramics, synthetic, and naturally occurring polymers, various composites, etc.

Materials intended to interact with living systems, which are used to fabricate medical products and devices, are named biomaterials. In spite of the considerable progress of biomaterials science, biomaterials are not produced in sufficient quantities, and no fully biocompatible material has been created yet.

Synthetic and natural high-molecular-weight compounds are increasingly used in medicine. The diversity of polymers, their widely varying stereo-configuration and molecular weight, and the feasibility of producing various composites with different materials provide the basis for designing a wide range of novel materials that would have new valuable properties.

As estimated by WHO, no more than 1% of the commercially produced polymers are used in medical applications. Early medical applications of polymers dealt primarily with their use as materials enhancing the properties of medical items: glassware was replaced by elastic and unbreakable polyolefin plasticware, and nonwoven materials came into wide use. Present-day medical applications of polymers are very wide. They are used to produce portable devices, clinical equipment and instruments, items for personal hygiene, equipment for medical analysis, artificial organs (kidneys, blood vessels, valves, pacemakers, heart-lung machines), materials for dentistry, etc.

Development of new technologies in reconstructive medicine has given an impetus to the broadening of the range of polymers to be used for constructing various implants and prostheses.

This book is devoted to natural biodegradable materials – polyhydroxyalkanoates (PHAs), which have many attractive properties for biomedical applications. This is a new class of biodegradable and biocompatible polyesters, whose physicochemical properties can vary greatly depending on the PHA composition. PHAs have considerable advantages over poly-lactides. In addition to poly-3-hydroxybutyrate, there are many promising PHA copolymers, which, depending on their monomeric composition, have different basic properties (degree of crystallinity, melting point, ductility, mechanical strength, biodegradation rate, etc.).

Based on their extensive fundamental research and favorable results of biomedical tests, the authors carried our experiments and limited clinical trials of PHA-based materials and devices for medical applications (modified meshes for hernia repair, absorbable monofilament sutures, biliary stents, bone replacement implants, etc.). Each of these products may be a product of choice in general surgery, trauma surgery, pharmacology, oncology, etc. There is an acute need for them on the market and in clinical practice. Although these products were designed for different surgical applications – hernia repair, hepatic surgery, purulent surgery, and trauma surgery, they may be also used to replace/regenerate tissue defects and restore the functions of the organs, i.e., in reconstructive surgery.

INDEX